Ravi P. Agarwal

Donal O'Regan

An Introduction to Ordinary Differential Equations

 Springer

Ravi P. Agarwal
Florida Institute of Technology
Department of Mathematical Sciences
150 West University Blvd.
Melbourne, FL 32901
agarwal@fit.edu

Donal O'Regan
National University of Ireland, Galway
Mathematics Department
University Road
Galway, Ireland
donal.oregan@nuigalway.ie

ISBN: 978-0-387-71275-8   e-ISBN: 978-0-387-71276-5
DOI: 10.1007/978-0-387-71276-5

Library of Congress Control Number: 2007935982

Mathematics Subject Classification (2000): 34-01, 34-XX

Universitext

Dedicated to

Ramesh C. Gupta

on his 65th birthday

Preface

Ordinary differential equations serve as mathematical models for many exciting "real-world" problems, not only in science and technology, but also in such diverse fields as economics, psychology, defense, and demography. Rapid growth in the theory of differential equations and in its applications to almost every branch of knowledge has resulted in a continued interest in its study by students in many disciplines. This has given ordinary differential equations a distinct place in mathematics curricula all over the world and it is now being taught at various levels in almost every institution of higher learning.

Hundreds of books on ordinary differential equations are available. However, the majority of these are elementary texts which provide a battery of techniques for finding explicit solutions. The size of some of these books has grown dramatically—to the extent that students are often lost in deciding where to start. This is all due to the addition of repetitive examples and exercises, and colorful pictures. The advanced books are either on specialized topics or are encyclopedic in character. In fact, there are hardly any rigorous and perspicuous introductory texts available which can be used directly in class for students of applied sciences. Thus, in an effort to bring the subject to a wide audience we provide a compact, but thorough, introduction to the subject in *An Introduction to Ordinary Differential Equations*. This book is intended for readers who have had a course in calculus, and hence it can be used for a senior undergraduate course. It should also be suitable for a beginning graduate course, because in undergraduate courses, students do not have any exposure to various intricate concepts, perhaps due to an inadequate level of mathematical sophistication.

The subject matter has been organized in the form of theorems and their proofs, and the presentation is rather unconventional. It comprises 42 class-tested lectures which the first author has given to mostly math-major students at various institutions all over the globe over a period of almost 35 years. These lectures provide flexibility in the choice of material for a one-semester course. It is our belief that the content in each lecture, together with the problems therein, provides fairly adequate coverage of the topic under study.

A brief description of the topics covered in this book is as follows: Introductory Lecture 1 explains basic terms and notations that are used throughout the book. Lecture 2 contains a concise account of the historical development of the subject. Lecture 3 deals with exact differential equations, while first-order equations are studied in Lectures 4 and 5. Lecture 6 discusses second-order linear differential equations; variation of parameters is used here to find solutions to nonhomogeneous equations. Lectures 3–6 form the core of any course on differential equations.

Many very simple differential equations cannot be solved as finite combinations of elementary functions. It is, therefore, of prime importance to know whether a given differential equation has a solution. This aspect of the existence of solutions for first-order initial value problems is dealt with in Lectures 8 and 9. Once the existence of solutions has been established, it is natural to provide conditions under which a given problem has precisely one solution; this is the content of Lecture 10. In an attempt to make the presentation self-contained, the required mathematical preliminaries have been included in Lecture 7. Differential inequalities, which are important in the study of qualitative as well as quantitative properties of solutions, are discussed in Lecture 11. Continuity and differentiability of solutions with respect to initial conditions are examined in Lecture 12.

Preliminary results from algebra and analysis required for the study of differential systems are contained in Lectures 13 and 14. Lectures 15 and 16 extend existence–uniqueness results and examine continuous dependence on initial data for the systems of first-order initial value problems. Basic properties of solutions of linear differential systems are given in Lecture 17. Lecture 18 deals with the fundamental matrix solution, and some methods for its computation in the constant-coefficient case are discussed in Lecture 19. These computational algorithms do not use the Jordan form and can easily be mastered by students. In Lecture 20 necessary and sufficient conditions are provided so that a linear system has only periodic solutions. Lectures 21 and 22 contain restrictions on the known quantities so that solutions of a linear system remain bounded or ultimately approach zero. Lectures 23–29 are devoted to a self-contained introductory stability theory for autonomous and nonautonomous systems. Here two-dimensional linear systems form the basis for the phase plane analysis. In addition to the study of periodic solutions and limit cycles, the direct method of Lyapunov is developed and illustrated.

Higher-order exact and adjoint equations are introduced in Lecture 30, and the oscillatory behavior of solutions of second-order equations is featured in Lecture 31.

The last major topic covered in this book is that of boundary value problems involving second-order differential equations. After linear boundary value problems are introduced in Lecture 32, Green's function and its construction is discussed in Lecture 33. Lecture 34 describes conditions that guarantee the existence of solutions of degenerate boundary value problems. The concept of the generalized Green's function is also featured here. Lecture 35 presents some maximum principles for second-order linear differential inequalities and illustrates their importance in initial and boundary value problems. Lectures 36 and 37 are devoted to the study of Sturm–Liouville problems, while eigenfunction expansion is the subject of Lectures 38 and 39. A detailed discussion of nonlinear boundary value problems is contained in Lectures 40 and 41.

Finally, Lecture 42 addresses some topics for further study which extend the material of this text and are of current research interest.

Two types of problems are included in the book—those which illustrate the general theory, and others designed to fill out text material. The problems form an integral part of the book, and every reader is urged to attempt most, if not all of them. For the convenience of the reader we have provided answers or hints for all the problems, except those few marked with an asterisk.

In writing a book of this nature no originality can be claimed. Our goal has been made to present the subject as simply, clearly, and accurately as possible. Illustrative examples are usually very simple and are aimed at the average student.

It is earnestly hoped that *An Introduction to Ordinary Differential Equations* will provide an inquisitive reader with a starting point in this rich, vast, and ever-expanding field of knowledge.

We would like to express our appreciation to Professors M. Bohner, A. Cabada, M. Cecchi, J. Diblik, L. Erbe, J. Henderson, Wan-Tong Li, Xianyi Li, M. Migda, Ch. G. Philos, S. Stanek, C. C. Tisdell, and P. J. Y. Wong for their suggestions and criticisms. We also want to thank Ms. Vaishali Damle at Springer New York for her support and cooperation.

Ravi P. Agarwal
Donal O'Regan

Contents

Lecture 1
Introduction

An *ordinary differential equation* (ordinary DE hereafter) is a relation containing one real independent variable $x \in \mathbb{R} = (-\infty, \infty)$, the real dependent variable y, and some of its derivatives $y', y'', \ldots, y^{(n)}$ ($' = d/dx$). For example,

$$xy' + 3y = 6x^3 \tag{1.1}$$

$$y'^2 - 4y = 0 \tag{1.2}$$

$$x^2 y'' - 3xy' + 3y = 0 \tag{1.3}$$

$$2x^2 y'' - y'^2 = 0. \tag{1.4}$$

The *order* of an ordinary DE is defined to be the order of the highest derivative in the equation. Thus, equations (1.1) and (1.2) are first order, whereas (1.3) and (1.4) are second order.

Besides ordinary DEs, if the relation has more than one independent variable, then it is called a partial DE. In these lectures we shall discuss only ordinary DEs, and so the word *ordinary* will be dropped.

In general, an nth-order DE can be written as

$$F(x, y, y', y'', \ldots, y^{(n)}) = 0, \tag{1.5}$$

where F is a known function.

A functional relation between the dependent variable y and the independent variable x, that, in some interval J, satisfies the given DE is said to be a *solution* of the equation. A solution may be defined in either of the following intervals: (α, β), $[\alpha, \beta)$, $(\alpha, \beta]$, $[\alpha, \beta]$, (α, ∞), $[\alpha, \infty)$, $(-\infty, \beta)$, $(-\infty, \beta]$, $(-\infty, \infty)$, where $\alpha, \beta \in \mathbb{R}$ and $\alpha < \beta$. For example, the function $y(x) = 7e^x + x^2 + 2x + 2$ is a solution of the DE $y' = y - x^2$ in $J = \mathbb{R}$. Similarly, the function $y(x) = x \tan(x + 3)$ is a solution of the DE $xy' = x^2 + y^2 + y$ in $J = (-\pi/2 - 3, \pi/2 - 3)$. The *general solution* of an nth-order DE depends on n arbitrary constants; i.e., the solution y depends on x and the real constants $c_1, c_2, \ldots, c_n$. For example, the function $y(x) = x^2 + ce^x$ is the general solution of the DE $y' = y - x^2 + 2x$ in $J = \mathbb{R}$. Similarly, $y(x) = x^3 + c/x^3$, $y(x) = x^2 + cx + c^2/4$, $y(x) = c_1 x + c_2 x^3$ and $y(x) = (2x/c_1) - (2/c_1^2) \ln(1 + c_1 x) + c_2$ are the general solutions of the DEs (1.1)–(1.4), respectively. Obviously, the general solution of (1.1) is defined in any interval which does not include the point 0, whereas the general

R.P. Agarwal and D. O'Regan, *An Introduction to Ordinary Differential Equations*, doi: 10.1007/978-0-387-71276-5_1, © Springer Science + Business Media, LLC 2008

solutions of (1.2) and (1.3) are defined in $J = \mathbb{R}$. The general solution of (1.4) imposes a restriction on the constant c_1 as well as on the variable x. In fact, it is defined if $c_1 \neq 0$ and $1 + c_1 x > 0$.

The function $y(x) = x^3$ is a *particular solution* of the equation (1.1), and it can be obtained easily by taking the particular value of c as 0 in the general solution. Similarly, if $c = 0$ and $c_1 = 0$, $c_2 = 1$ in the general solutions of (1.2) and (1.3), then x^2 and x^3, respectively, are the particular solutions of the equations (1.2) and (1.3). It is interesting to note that $y(x) = x^2$ is a solution of the DE (1.4); however, it is not included in the general solution of (1.4). This "extra" solution, which cannot be obtained by assigning particular values of the constants, is called a *singular solution* of (1.4). As an another example, for the DE $y'^2 - xy' + y = 0$ the general solution is $y(x) = cx - c^2$, which represents a family of straight lines, and $y(x) = x^2/4$ is a singular solution which represents a parabola. Thus, in the "general solution," the word *general* must not be taken in the sense of *complete*. A totality of all solutions of a DE is called a *complete solution*.

A DE of the first order may be written as $F(x, y, y') = 0$. The function $y = \phi(x)$ is called an *explicit solution* provided $F(x, \phi(x), \phi'(x)) = 0$ in J.

A relation of the form $\psi(x, y) = 0$ is said to be an *implicit solution* of $F(x, y, y') = 0$ provided it determines one or more functions $y = \phi(x)$ which satisfy $F(x, \phi(x), \phi'(x)) \equiv 0$. It is frequently difficult, if not impossible, to solve $\psi(x, y) = 0$ for y. Nevertheless, we can test the solution by obtaining y' by implicit differentiation: $\psi_x + \psi_y y' = 0$, or $y' = -\psi_x/\psi_y$, and check if $F(x, y, -\psi_x/\psi_y) \equiv 0$.

The pair of equations $x = x(t)$, $y = y(t)$ is said to be a *parametric solution* of $F(x, y, y') = 0$ when $F(x(t), y(t), (dy/dt)/(dx/dt)) \equiv 0$.

Consider the equation $y'''^2 - 2y'y''' + 3y''^3 = 0$. This is a third-order DE and we say that this is of degree 2, whereas the second-order DE $xy'' + 2y' + 3y - 6e^x = 0$ is of degree 1. In general, if a DE has the form of an algebraic equation of degree k in the highest derivative, then we say that the given DE is of *degree* k.

We shall always assume that the DE (1.5) can be solved explicitly for $y^{(n)}$ in terms of the remaining $(n + 1)$ quantities as

$$y^{(n)} = f(x, y, y', \ldots, y^{(n-1)}), \tag{1.6}$$

where f is a known function. This will at least avoid having equation (1.5) represent more than one equation of the form (1.6); e.g., $y'^2 = 4y$ represents two DEs, $y' = \pm 2\sqrt{y}$.

Differential equations are classified into two groups: linear and nonlinear. A DE is said to be *linear* if it is linear in y and all its derivatives.

Thus, an nth-order linear DE has the form

$$\mathcal{P}_n[y] \; = \; p_0(x)y^{(n)} + p_1(x)y^{(n-1)} + \cdots + p_n(x)y \; = \; r(x). \qquad (1.7)$$

Obviously, DEs (1.1) and (1.3) are linear, whereas (1.2) and (1.4) are non-linear. It may be remarked that every linear equation is of degree 1, but every equation of degree 1 is not necessarily linear. In (1.7) if the function $r(x) \equiv 0$, then it is called a *homogeneous* DE, otherwise it is said to be a *nonhomogeneous* DE.

In applications we are usually interested in a solution of the DE (1.6) satisfying some additional requirements called *initial* or *boundary conditions*. By initial conditions for (1.6) we mean n conditions of the form

$$y(x_0) \; = \; y_0, \quad y'(x_0) \; = \; y_1, \ldots, y^{(n-1)}(x_0) \; = \; y_{n-1}, \qquad (1.8)$$

where $y_0, \ldots, y_{n-1}$ and x_0 are given constants. A problem consisting of the DE (1.6) together with the initial conditions (1.8) is called an *initial value problem*. It is common to seek a solution $y(x)$ of the initial value problem (1.6), (1.8) in an interval J which contains the point x_0.

Consider the first-order DE $xy' - 3y + 3 = 0$: it is disconcerting to notice that it has no solution satisfying the initial condition $y(0) = 0$; just one solution $y(x) \equiv 1$ satisfying $y(1) = 1$; and an infinite number of solutions $y(x) = 1 + cx^3$ satisfying $y(0) = 1$. Such diverse behavior leads to the essential question about the existence of solutions. If we deal with a DE which can be solved in a closed form (in terms of permutation and combination of known functions x^n, e^x, $\sin x$), then the answer to the question of existence of solutions is immediate. However, unfortunately the class of solvable DEs is very small, and today we often come across DEs so complicated that they can only be solved, if at all, with the aid of a computer. Any attempt to solve a DE with no solution is surely a futile exercise, and the data so produced will not only be meaningless, but actually chaotic. Therefore, in the theory of DEs, the first basic problem is to provide sufficient conditions so that a given initial value problem has at least one solution. For this, we shall give several easily verifiable sets of sufficient conditions so that the first-order DE

$$y' \; = \; f(x, y) \qquad (1.9)$$

together with the initial condition

$$y(x_0) \; = \; y_0 \qquad (1.10)$$

has at least one solution. Fortunately, these results can be extended to the systems of such initial value problems which in particular include the problem (1.6), (1.8).

Once the existence of a solution of the initial value problem (1.9), (1.10) has been established it is natural to ask, "Does the problem have just one solution?" From an applications point of view the uniqueness of a solution is as important as the existence of the solution itself. Thus, we shall provide several results which guarantee the uniqueness of solutions of the problem (1.9), (1.10), and extend them to systems of such initial value problems.

Of course, the existence of a unique solution of (1.9), (1.10) is a qualitative property and does not suggest the procedure for the construction of the solution. Unfortunately, we shall not be dealing with this aspect of DEs in these lectures. However, we will discuss elementary theory of differential inequalities, which provides upper and lower bounds for the unknown solutions and guarantees the existence of *maximal* and *minimal solutions*.

There is another important property that experience suggests as a requirement for mathematical formulation of a physical situation to be met. As a matter of fact, any experiment cannot be repeated exactly in the same way. But if the initial conditions in an experiment are almost exactly the same, the outcome is expected to be almost the same. It is, therefore, desirable that the solution of a given mathematical model should have this property. In technical terms it amounts to saying that the solution of a DE ought to *depend continuously on the initial data*. We shall provide sufficient conditions so that the solutions of (1.9) depend continuously on the initial conditions. The generalization of these results to systems of DEs is also straightforward.

For the linear first-order differential systems which include the DE (1.7) as a special case, linear algebra allows one to describe the structure of the family of all solutions. We shall devote several lectures to examining this important qualitative aspect of solutions. Our discussion especially includes the *periodicity of solutions*, i.e., the graph of a solution repeats itself in successive intervals of a fixed length. Various results in these lectures provide a background for treating nonlinear first-order differential systems in subsequent lectures.

In many problems and applications we are interested in the behavior of the solutions of the differential systems as x approaches infinity. This ultimate behavior is termed the *asymptotic behavior* of the solutions. Specifically, we shall provide sufficient conditions for the known quantities in a given differential system so that all its solutions remain bounded or tend to zero as $x \to \infty$. The asymptotic behavior of the solutions of perturbed differential systems is also featured in detail.

The property of continuous dependence on the initial conditions of solutions of differential systems implies that a small change in the initial conditions brings only small changes in the solutions in a finite interval $[x_0, x_0 + \alpha]$. Satisfaction of such a property for all $x \geq x_0$ leads us to the notion of *stability* in the sense of Lyapunov. This aspect of the qualitative theory of DEs is introduced and examined in several lectures. In particular, here we give an elementary treatment of two-dimensional differential systems which includes the *phase portrait analysis* and discuss in detail Lyapunov's direct method for autonomous as well as nonautonomous systems.

A solution $y(x)$ of a given DE is said to be *oscillatory* if it has no last zero; i.e., if $y(x_1) = 0$, then there exists an $x_2 > x_1$ such that $y(x_2) = 0$, and the equation itself is called oscillatory if every solution is oscillatory. The oscillatory property of solutions of differential equations is an important qualitative aspect which has wide applications. We shall provide sufficiency criteria for oscillation of all solutions of second-order linear DEs and show how easily these results can be applied in practice.

We observe that the initial conditions (1.8) are prescribed at the same point x_0, but in many problems of practical interest, these n conditions are prescribed at two (or more) distinct points of the interval J. These conditions are called *boundary conditions*, and a problem consisting of DE (1.6) together with n boundary conditions is called a *boundary value problem*. For example, the problem of determining a solution $y(x)$ of the second-order DE $y'' + \pi^2 y = 0$ in the interval $[0, 1/2]$ which has preassigned values at 0 and $1/2$: $y(0) = 0$, $y(1/2) = 1$ constitutes a boundary value problem. The existence and uniqueness theory of solutions of boundary value problems is more complex than that for the initial value problems; thus, we shall restrict ourselves only to the second-order linear and nonlinear DEs. In our treatment for nonhomogeneous, and specially for nonlinear boundary value problems, we will need their integral representations, and for this *Green's functions* play a very important role. Therefore, we shall present the construction of Green's functions systematically. We shall also discuss degenerate linear boundary value problems, which appear frequently in applications.

In calculus the following result is well known: If $y \in C^{(2)}[\alpha, \beta]$, $y''(x) > 0$ in (α, β), and $y(x)$ attains its maximum at an interior point of $[\alpha, \beta]$, then $y(x)$ is identically constant in $[\alpha, \beta]$. Extensions of this result to differential equations and inequalities are known as *maximum principles*. We shall prove some maximum principles for second-order differential inequalities and show how these results can be applied to obtain lower and upper bounds for the solutions of second-order initial and boundary value problems which cannot be solved explicitly.

If the coefficients of the DE and/or of the boundary conditions depend on a parameter, then one of the fundamental problems of mathematical physics is to determine the value(s) of the parameter for which nontrivial solutions exist. These special values of the parameter are called *eigenvalues* and the corresponding nontrivial solutions are said to be *eigenfunctions*. One of the most studied problems of this type is the *Sturm–Liouville prob-*

lem, specially because it has an infinite number of real eigenvalues which can be arranged as a monotonically increasing sequence, and the corresponding eigenfunctions generate a complete set of *orthogonal functions*. We shall use this fundamental property of eigenfunctions to represent functions in terms of *Fourier series*. We shall also find the solutions of nonhomogeneous boundary value problems in terms of eigenfunctions of the corresponding Sturm–Liouville problem. This leads to a very important result in the literature known as Fredholm's alternative.

Lecture 2
Historical Notes

One of the major problems in which scientists of antiquity were involved was the study of planetary motions. In particular, predicting the precise time at which a lunar eclipse occurs was a matter of considerable prestige and a great opportunity for an astronomer to demonstrate his skills. This event had great religious significance, and rites and sacrifices were performed. To make an accurate prediction, it was necessary to find the true instantaneous motion of the moon at a particular point of time. In this connection we can trace back as far as, Bhaskara II (486AD), who conceived the differentiation of the function $\sin t$. He was also aware that a variable attains its maximum value at the point where the differential vanishes. The roots of the mean value theorem were also known to him. The idea of using integral calculus to find the value of π and the areas of curved surfaces and the volumes was also known to Bhaskara II. Later Madhava (1340–1429AD) developed the limit passage to infinity, which is the kernel of modern classical analysis. Thus, the beginning of calculus goes back at least 12 centuries before the phenomenal development of modern mathematics that occurred in Europe around the time of Newton and Leibniz. This raises doubts about prevailing theories that, in spite of so much information being known hundreds of years before Newton and Leibniz, scientists never came across differential equations. The information which historians have recorded is as follows:

The founder of the differential calculus, Newton, also laid the foundation stone of DEs, then known as fluxional equations. Some of the first-order DEs treated by him in the year 1671 were

$$y' = 1 - 3x + y + x^2 + xy \tag{2.1}$$

$$3x^2 - 2ax + ay - 3y^2 y' + axy' = 0 \tag{2.2}$$

$$y' = 1 + \frac{y}{a} + \frac{xy}{a^2} + \frac{x^2 y}{a^3} + \frac{x^3 y}{a^4}, \quad \text{etc.} \tag{2.3}$$

$$\begin{aligned} y' = &-3x + 3xy + y^2 - xy^2 + y^3 - xy^3 + y^4 - xy^4 \\ &+6x^2 y - 6x^2 + 8x^3 y - 8x^3 + 10x^4 y - 10x^4, \quad \text{etc.} \end{aligned} \tag{2.4}$$

He also classified first-order DEs into three classes: the first class was composed of those equations in which y' is a function of only one variable, x alone or y alone, e.g.,

$$y' = f(x), \quad y' = f(y); \tag{2.5}$$

R.P. Agarwal and D. O'Regan, *An Introduction to Ordinary Differential Equations*,
doi: 10.1007/978-0-387-71276-5_2, © Springer Science + Business Media, LLC 2008

the second class embraced those equations in which y' is a function of both x and y, i.e., (1.9); and the third is made up of partial DEs of the first order.

About five years later, in 1676, another independent inventor of calculus, Leibniz, coined the term *differential equation* to denote a relationship between the differentials dx and dy of two variables x and y. This was in connection with the study of geometrical problems such as the inverse tangent problem, i.e., finding a curve whose tangent satisfies certain conditions. For instance, if the distance between any point $P(x, y)$ on the curve $y(x)$ and the point where the tangent at P crosses the axis of x (length of the tangent) is a constant a, then y should satisfy first-order nonlinear DE

$$y' = -\frac{y}{\sqrt{a^2 - y^2}}. \tag{2.6}$$

In 1691, he implicitly used the method of separation of variables to show that the DEs of the form

$$y\frac{dx}{dy} = X(x)Y(y) \tag{2.7}$$

can be reduced to quadratures. One year later he integrated linear homogeneous first-order DEs, and soon afterward nonhomogeneous linear first-order DEs.

Among the devoted followers of Leibniz were the brothers James and John Bernoulli, who played a significant part in the development of the theory of DEs and the use of such equations in the solution of physical problems. In 1690, James Bernoulli showed that the problem of determining the isochrone, i.e., the curve in a vertical plane such that a particle will slide from any point on the curve to its lowest point in exactly the same time, is equivalent to that of solving a first-order nonlinear DE

$$dy(b^2y - a^3)^{1/2} = dx\, a^{3/2}. \tag{2.8}$$

Equation (2.8) expresses the equality of two differentials from which Bernoulli concluded the equality of the integrals of the two members of the equation and used the word *integral* for the first time on record.

In 1696 John Bernoulli invited the brightest mathematicians of the world (Europe) to solve the brachistochrone (quickest descent) problem: to find the curve connecting two points A and B that do not lie on a vertical line and possessing the property that a moving particle slides down the curve from A to B in the shortest time, ignoring friction and resistance of the medium. In order to solve this problem, one year later John Bernoulli imagined thin layers of homogeneous media, he knew from optics (Fermat's principle) that a light ray with speed ν obeying the law of Snellius,

$$\sin\alpha = K\nu,$$

passes through in the shortest time. Since the speed is known to be proportional to the square root of the fallen height, he obtained by passing through thinner and thinner layers

$$\sin \alpha \;=\; \frac{1}{\sqrt{1+y'^2}} \;=\; K\sqrt{2g(y-h)}, \tag{2.9}$$

a differential equation of the first order. Among others who also solved the brachistochrone problem are James Bernoulli, Leibniz, Newton, and L'Hospital.

The term "separation of variables" is essentially due to John Bernoulli; he also circumvented dx/x, not well understood at that time, by first applying an integrating factor. However, the discovery of integrating factors proved almost as troublesome as solving a DE.

The problem of finding the general solution of what is now called *Bernoulli's equation*,

$$a\,dy \;=\; yp\,dx + bqy^n\,dx, \tag{2.10}$$

in which a and b are constants, and p and q are functions of x alone, was proposed by James Bernoulli in 1695 and solved by Leibniz and John Bernoulli by using different substitutions for the dependent variable y. Thus, the roots of the general tactic "change of the dependent variable" had already appeared in 1696–1697. The problem of determining the orthogonal trajectories of a one-parameter family of curves was also solved by John Bernoulli in 1698. And by the end of the 17th century most of the known methods of solving first-order DEs had been developed.

Numerous applications of the use of DEs in finding the solutions of geometric problems were made before 1720. Some of the DEs formulated in this way were of second or higher order; e.g., the ancient Greek's isoperimetric problem of finding the closed plane curve of given length that encloses the largest area led to a DE of third order. This third-order DE of James Bernoulli (1696) was reduced to one of the second order by John Bernoulli. In 1761 John Bernoulli reported the second-order DE

$$y'' \;=\; \frac{2y}{x^2} \tag{2.11}$$

to Leibniz, which gave rise to three types of curves—parabolas, hyperbolas, and a class of curves of the third order.

As early as 1712, Riccati considered the second-order DE

$$f(y, y', y'') \;=\; 0 \tag{2.12}$$

and treated y as an independent variable, p $(= y')$ as the dependent variable, and making use of the relationship $y'' = p\,dp/dy$, he converted the

DE (2.12) into the form

$$f\left(y, p, p\left(\frac{dp}{dy}\right)\right) = 0, \tag{2.13}$$

which is a first-order DE in p.

The particular DE

$$y' = p(x)y^2 + q(x)y + r(x) \tag{2.14}$$

christened by d'Alembert as the *Riccati equation* has been studied by a number of mathematicians, including several of the Bernoullis, Riccati himself, as well as his son Vincenzo. By 1723 at the latest, it was recognized that (2.14) cannot be solved in terms of elementary functions. However, later it was Euler who called attention to the fact that if a particular solution $y_1 = y_1(x)$ of (2.14) is known, then the substitution $y = y_1 + z^{-1}$ converts the Riccati equation into a first-order linear DE in z, which leads to its general solution. He also pointed out that if two particular solutions of (2.14) are known, then the general solution is expressible in terms of simple quadrature.

For the first time in 1715, Taylor unexpectedly noted the singular solutions of DEs. Later in 1734, a class of equations with interesting properties was found by the precocious mathematician Clairaut. He was motivated by the movement of a rectangular wedge, which led him to DEs of the form

$$y = xy' + f(y'). \tag{2.15}$$

In (2.15) the substitution $p = y'$, followed by differentiation of the terms of the equation with respect to x, will lead to a first-order DE in x, p and dp/dx. Its general solution $y = cx + f(c)$ is a collection of straight lines. The Clairaut DE has also a singular solution which in parametric form can be written as $x = -f'(t)$, $y = f(t) - tf'(t)$. D'Alembert found the singular solution of the somewhat more general type of DE

$$y = xg(y') + f(y'), \tag{2.16}$$

which is known as *D'Alembert's equation*.

Starting from 1728, Euler contributed many important ideas to DEs: various methods of reduction of order, notion of an integrating factor, theory of linear equations of arbitrary order, power series solutions, and the discovery that a first-order nonlinear DE with square roots of quartics as coefficients, e.g.,

$$(1 - x^4)^{1/2}y' + (1 - y^4)^{1/2} = 0, \tag{2.17}$$

has an algebraic solution. Euler also invented the method of variation of parameters, which was elevated to a general procedure by Lagrange in

1774. Most of the modern theory of linear differential systems appears in D'Alembert's work of 1748, while the concept of adjoint equations was introduced by Lagrange in 1762.

The *Jacobi equation*

$$(a_1 + b_1 x + c_1 y)(x\,dy - y\,dx) - (a_2 + b_2 x + c_2 y)dy + (a_3 + b_3 x + c_3 y)dx = 0$$

in which the coefficients a_i, b_i, c_i, $i = 1, 2, 3$ are constants was studied in 1842, and is closely connected with the Bernoulli equation. Another important DE which was studied by Darboux in 1878 is

$$-L\,dy + M\,dx + N(x\,dy - y\,dx) = 0,$$

where L, M, N are polynomials in x and y of maximum degree m.

Thus, in early stages mathematicians were engaged in formulating DEs and solving them, tacitly assuming that a solution always existed. The rigorous proof of the existence and uniqueness of solutions of the first-order initial value problem (1.9), (1.10) was first presented by Cauchy in his lectures of 1820–1830. The proof exhibits a theoretical means for constructing the solution to any desired degree of accuracy. He also extended his process to the systems of such initial value problems. In 1876, Lipschitz improved Cauchy's technique with a view toward making it more practical. In 1893, Picard developed an existence theory based on a different method of successive approximations, which is considered more constructive than that of Cauchy–Lipschitz. Other significant contributors to the method of successive approximations are Liouville (1838), Caqué (1864), Fuchs (1870), Peano (1888), and Bôcher (1902).

The pioneering work of Cauchy, Lipschitz, and Picard is of a qualitative nature. Instead of finding a solution explicitly, it provides sufficient conditions on the known quantities which ensure the existence of a solution. In the last hundred years this work has resulted in an extensive growth in the qualitative study of DEs. Besides existence and uniqueness results, additional sufficient conditions (rarely necessary) to analyze the properties of solutions, e.g., asymptotic behavior, oscillatory behavior, stability, etc., have also been carefully examined. Among other mathematicians who have contributed significantly in the development of the qualitative theory of DEs we would like to mention the names of R. Bellman, I. Bendixson, G. D. Birkhoff, L. Cesari, R. Conti, T. H. Gronwall, J. Hale, P. Hartman, E. Kamke, V. Lakshmikantham, J. LaSalle, S. Lefschetz, N. Levinson, A. Lyapunov, G. Peano, H. Poincáre, G. Sansone, B. Van der Pol, A. Wintner, and W. Walter.

Finally the last three significant stages of development in the theory of DEs, opened with the application of Lie's (1870–1880s) theory of continuous groups to DEs, particularly those of Hamilton–Jacobi dynamics;

Picard's attempt (1880) to construct for linear DEs an analog of the Galois theory of algebraic equations; and the theory, started in 1930s, that paralleled the modern development of abstract algebra. Thus, the theory of DEs has emerged as a major discipline of modern pure mathematics. Nevertheless, the study of DEs continues to contribute to the solutions of practical problems in almost every branch of science and technology, arts and social science, and medicine. In the last fifty years, some of these problems have led to the creation of various types of new DEs, some which are of current research interest.

Lecture 3
Exact Equations

Let, in the DE of first order and first degree (1.9), the function $f(x,y) = -M(x,y)/N(x,y)$, so that it can be written as

$$M(x,y) + N(x,y)y' = 0, \qquad (3.1)$$

where M and N are continuous functions having continuous partial derivatives M_y and N_x in the rectangle $S : |x - x_0| < a$, $|y - y_0| < b$ $(0 < a,\ b < \infty)$.

Equation (3.1) is said to be *exact* if there exists a function $u(x,y)$ such that

$$u_x(x,y) = M(x,y) \quad \text{and} \quad u_y(x,y) = N(x,y). \qquad (3.2)$$

The nomenclature comes from the fact that

$$M + Ny' = u_x + u_y y'$$

is exactly the derivative du/dx.

Once the DE (3.1) is exact its implicit solution is

$$u(x,y) = c. \qquad (3.3)$$

If (3.1) is exact, then from (3.2) we have $u_{xy} = M_y$ and $u_{yx} = N_x$. Since M_y and N_x are continuous, we must have $u_{xy} = u_{yx}$; i.e., for (3.1) to be exact it is necessary that

$$M_y = N_x. \qquad (3.4)$$

Conversely, if M and N satisfy (3.4) then the equation (3.1) is exact. To establish this we shall exhibit a function u satisfying (3.2). We integrate both sides of $u_x = M$ with respect to x, to obtain

$$u(x,y) = \int_{x_0}^{x} M(s,y)ds + g(y). \qquad (3.5)$$

Here $g(y)$ is an arbitrary function of y and plays the role of the constant of integration. We shall obtain g by using the equation $u_y = N$. Indeed, we have

$$\frac{\partial}{\partial y} \int_{x_0}^{x} M(s,y)ds + g'(y) = N(x,y) \qquad (3.6)$$

R.P. Agarwal and D. O'Regan, *An Introduction to Ordinary Differential Equations*,
doi: 10.1007/978-0-387-71276-5_3, © Springer Science + Business Media, LLC 2008

and since

$$N_x - \frac{\partial^2}{\partial x \partial y} \int_{x_0}^{x} M(s,y)ds \;=\; N_x - M_y \;=\; 0$$

the function

$$N(x,y) - \frac{\partial}{\partial y} \int_{x_0}^{x} M(s,y)ds$$

must depend on y alone. Therefore, g can be obtained from (3.6), and finally the function u satisfying (3.2) is given by (3.5).

We summarize this important result in the following theorem.

Theorem 3.1.   Let the functions $M(x,y)$ and $N(x,y)$ together with their partial derivatives $M_y(x,y)$ and $N_x(x,y)$ be continuous in the rectangle $S : |x - x_0| < a, \; |y - y_0| < b \; (0 < a, \; b < \infty)$. Then the DE (3.1) is exact if and only if condition (3.4) is satisfied.

Obviously, in this result S may be replaced by any region which does not include any "hole."

The above proof of this theorem is, in fact, constructive, i.e., we can explicitly find a solution of (3.1). For this, we compute $g(y)$ from (3.6),

$$g(y) \;=\; \int_{y_0}^{y} N(x,t)dt - \int_{x_0}^{x} M(s,y)ds + \int_{x_0}^{x} M(s,y_0)ds + g(y_0).$$

Therefore, from (3.5) it follows that

$$u(x,y) \;=\; \int_{y_0}^{y} N(x,t)dt + \int_{x_0}^{x} M(s,y_0)ds + g(y_0)$$

and hence a solution of the exact equation (3.1) is given by

$$\int_{y_0}^{y} N(x,t)dt + \int_{x_0}^{x} M(s,y_0)ds \;=\; c, \qquad (3.7)$$

where c is an arbitrary constant.

In (3.7) the choice of x_0 and y_0 is at our disposal, except that these must be chosen so that the integrals remain proper.

Example 3.1.   In the DE

$$(y + 2xe^y) + x(1 + xe^y)y' \;=\; 0,$$

$M = y + 2xe^y$ and $N = x(1 + xe^y)$, so that $M_y = N_x = 1 + 2xe^y$ for all $(x,y) \in S = \mathbb{R}^2$. Thus, the given DE is exact in $\mathbb{R}^2$. Taking $(x_0, y_0) = (0,0)$ in (3.7), we obtain

$$\int_{0}^{y} (x + x^2 e^t)dt + \int_{0}^{x} 2sds \;=\; xy + x^2 e^y \;=\; c$$

a solution of the given DE.

The equation $2y + xy' = 0$ is not exact, but if we multiply it by x, then $2xy + x^2 y' = d(x^2 y)/dx = 0$ is an exact DE. The multiplier x here is called an integrating factor. For the DE (3.1) a nonzero function $\mu(x, y)$ is called an *integrating factor* if the equivalent DE

$$\mu M + \mu N y' = 0 \tag{3.8}$$

is exact.

If $u(x, y) = c$ is a solution of (3.1), then y' computed from (3.1) and $u_x + u_y y' = 0$ must be the same, i.e.,

$$\frac{1}{M} u_x = \frac{1}{N} u_y =: \mu, \tag{3.9}$$

where μ is some function of x and y. Thus, we have

$$\mu(M + N y') = u_x + u_y y' = \frac{du}{dx}$$

and hence the equation (3.8) is exact, and an integrating factor μ of (3.1) is given by (3.9).

Further, let $\phi(u)$ be any continuous function of u, then

$$\mu\phi(u)(M + N y') = \phi(u)\frac{du}{dx} = \frac{d}{dx}\int^u \phi(s)\,ds.$$

Hence, $\mu\phi(u)$ is an integrating factor of (3.1). Since ϕ is an arbitrary function, we have established the following result.

Theorem 3.2. If the DE (3.1) has $u(x, y) = c$ as its solution, then it admits an infinite number of integrating factors.

The function $\mu(x, y)$ is an integrating factor of (3.1) provided (3.8) is exact, i.e., if and only if

$$(\mu M)_y = (\mu N)_x. \tag{3.10}$$

This implies that an integrating factor must satisfy the equation

$$N\mu_x - M\mu_y = \mu(M_y - N_x). \tag{3.11}$$

A solution of (3.11) gives an integrating factor of (3.1), but finding a solution of the partial DE (3.11) is by no means an easier task. However, a particular nonzero solution of (3.11) is all we need for the solution of (3.1).

If we assume $\mu = X(x)Y(y)$, then from (3.11) we have

$$\frac{N}{X}\frac{dX}{dx} - \frac{M}{Y}\frac{dY}{dy} = M_y - N_x. \tag{3.12}$$

Hence, if

$$M_y - N_x = Ng(x) - Mh(y) \tag{3.13}$$

then (3.12) is satisfied provided

$$\frac{1}{X}\frac{dX}{dx} = g(x) \quad \text{and} \quad \frac{1}{Y}\frac{dY}{dy} = h(y),$$

i.e.,

$$X = e^{\int g(x)dx} \quad \text{and} \quad Y = e^{\int h(y)dy}. \tag{3.14}$$

We illustrate this in the following example.

Example 3.2. Let the DE

$$(y - y^2) + xy' = 0$$

admit an integrating factor of the form $\mu = x^m y^n$. In such a case (3.12) becomes $m - (1 - y)n = -2y$, and hence $m = n = -2$, and $\mu = x^{-2}y^{-2}$. This gives an exact DE

$$x^{-2}(y^{-1} - 1) + x^{-1}y^{-2}y' = 0,$$

whose solution using (3.7) is given by $\int_1^y x^{-1}t^{-2}dt = c$, which is the same as $y = 1/(1 - cx)$.

One may also look for an integrating factor of the form $\mu = \mu(v)$, where v is a known function of x and y. Then (3.11) leads to

$$\frac{1}{\mu}\frac{d\mu}{dv} = \frac{M_y - N_x}{Nv_x - Mv_y}. \tag{3.15}$$

Thus, if the expression in the right side of (3.15) is a function of v alone, say, $\phi(v)$ then the integrating factor is given by

$$\mu = e^{\int \phi(v)dv}. \tag{3.16}$$

Some special cases of v and the corresponding $\phi(v)$ are given in the following table:

$v:$	x	y	$x - y$	xy	$\dfrac{x}{y}$	$x^2 + y^2$
$\phi(v):$	$\dfrac{M_y - N_x}{N}$	$\dfrac{M_y - N_x}{-M}$	$\dfrac{M_y - N_x}{N + M}$	$\dfrac{M_y - N_x}{yN - xM}$	$\dfrac{(M_y - N_x)y^2}{yN + xM}$	$\dfrac{M_y - N_x}{2(xN - yM)}$.

If the expression in the second row is a function of the corresponding v in the first row, then (3.1) has the integrating factor μ given in (3.16).

Example 3.3. For the DE

$$(x^2y + y + 1) + x(1 + x^2)y' = 0,$$

we have

$$\frac{M_y - N_x}{N} = \frac{1}{x(1 + x^2)}(x^2 + 1 - 1 - 3x^2) = -\frac{2x}{1 + x^2},$$

which is a function of x. Hence, from the above table and (3.16) we find $\mu = (1 + x^2)^{-1}$. Thus, the DE

$$\left(y + \frac{1}{1 + x^2}\right) + xy' = 0$$

is exact whose solution is given by $xy + \tan^{-1}x = c$.

Example 3.4. For the DE in Example 3.2, we have

$$\frac{M_y - N_x}{-M} = \frac{2}{1 - y},$$

which is a function of y. Hence, from the above table and (3.16) we find $\mu = (1 - y)^{-2}$. Thus, the DE

$$\frac{y}{1 - y} + \frac{x}{(1 - y)^2}y' = 0$$

is exact whose solution is once again given by $y = 1/(1 - cx)$.

Example 3.5. For the DE

$$(xy^3 + 2x^2y^2 - y^2) + (x^2y^2 + 2x^3y - 2x^2)y' = 0,$$

we have

$$\frac{M_y - N_x}{yN - xM} = 1 - \frac{2}{xy},$$

which is a function of xy. Hence, from the above table and (3.16) we find $\mu = x^{-2}y^{-2}e^{xy}$. The resulting exact DE

$$(yx^{-1} + 2 - x^{-2})e^{xy} + (1 + 2xy^{-1} - 2y^{-2})e^{xy}y' = 0$$

has a solution $e^{xy}(x^{-1} + 2y^{-1}) = c$.

Next we shall prove an interesting result which enables us to find the solution of (3.1) provided it admits two linearly independent integrating factors. For this, we need the following lemma.

Lemma 3.3.   Suppose (3.1) is exact and has an integrating factor $\mu(x, y)$ ($\neq$ constant), then $\mu(x, y) = c$ is a solution of (3.1).

Proof.   In view of the hypothesis, condition (3.11) implies that $N\mu_x = M\mu_y$. Multiplying (3.1) by μ_y, we find

$$M\mu_y + N\mu_y y' = N(\mu_x + \mu_y y') = N\frac{d\mu}{dx} = 0$$

and this implies the lemma.   ∎

Theorem 3.4.   If $\mu_1(x, y)$ and $\mu_2(x, y)$ are two integrating factors of (3.1) such that their ratio is not a constant, then $\mu_1(x, y) = c\mu_2(x, y)$ is a solution of (3.1).

Proof.   Clearly, the DEs (i) $\mu_1 M + \mu_1 N y' = 0$, and (ii) $\mu_2 M + \mu_2 N y' = 0$ are exact. Multiplication of (ii) by μ_1/μ_2 converts it to the exact equation (i). Thus, the exact DE (ii) admits an integrating factor μ_1/μ_2 and Lemma 3.3 implies that $\mu_1/\mu_2 = c$ is a solution of (ii), i.e., of (3.1).   ∎

To illustrate the importance of Theorem 3.4, we consider the DE in Example 3.2. It has two integrating factors, $\mu_1 = x^{-2}y^{-2}$ and $\mu_2 = (1 - y)^{-2}$, which are obtained in Examples 3.2 and 3.4, respectively. Hence, its solution is given by $(1-y)^2 = c^2 x^2 y^2$, which is the same as the one obtained in Example 3.2, as it should be.

We finish this lecture with the remark that, generally, integrating factors of (3.1) are obtained by "trial-and-error" methods.

Problems

3.1.   Let the hypothesis of Theorem 3.1 be satisfied and the equation (3.1) be exact. Show that the solution of the DE (3.1) is given by

$$\int_{x_0}^{x} M(s, y)ds + \int_{y_0}^{y} N(x_0, t)dt = c.$$

3.2.   Solve the following initial value problems:

(i) $(3x^2 y + 8xy^2) + (x^3 + 8x^2 y + 12y^2)y' = 0,$ $y(2) = 1.$
(ii) $(4x^3 e^{x+y} + x^4 e^{x+y} + 2x) + (x^4 e^{x+y} + 2y)y' = 0,$ $y(0) = 1.$
(iii) $(x - y\cos x) - \sin xy' = 0,$ $y(\pi/2) = 1.$
(iv) $(ye^{xy} + 4y^3) + (xe^{xy} + 12xy^2 - 2y)y' = 0,$ $y(0) = 2.$

3.3.   In the following DEs determine the constant a so that each equation is exact, and solve the resulting DEs:

(i) $(x^3 + 3xy) + (ax^2 + 4y)y' = 0.$

(ii) $(x^{-2} + y^{-2}) + (ax + 1)y^{-3}y' = 0.$

3.4. The DE $(e^x \sec y - \tan y) + y' = 0$ has an integrating factor of the form $e^{-ax} \cos y$ for some constant a. Find a, and then solve the DE.

3.5. The DE $(4x + 3y^2) + 2xyy' = 0$ has an integrating factor of the form x^n, where n is a positive integer. Find n, and then solve the DE.

3.6. Verify that $(x^2 + y^2)^{-1}$ is an integrating factor of a DE of the form $(y + xf(x^2 + y^2)) + (yf(x^2 + y^2) - x)y' = 0$. Hence, solve the DE

$$(y + x(x^2 + y^2)^2) + (y(x^2 + y^2)^2 - x)y' = 0.$$

3.7. If p and q are functions of x, then the DE $(py - qy^n) + y' = 0$ admits an integrating factor of the form $X(x)Y(y)$. Find the functions X and Y.

3.8. Solve the following DEs for which the type of integrating factor has been indicated:

(i) $(x - y^2) + 2xyy' = 0 \quad [\mu(x)].$

(ii) $y + (y^2 - x)y' = 0 \quad [\mu(y)].$

(iii) $y + x(1 - 3x^2y^2)y' = 0 \quad [\mu(xy)].$

(iv) $(3xy + y^2) + (3xy + x^2)y' = 0 \quad [\mu(x + y)].$

(v) $(x + x^4 + 2x^2y^2 + y^4) + yy' = 0 \quad [\mu(x^2 + y^2)].$

(vi) $(4xy + 3y^4) + (2x^2 + 5xy^3)y' = 0 \quad [\mu(x^m y^n)].$

3.9. By differentiating the equation

$$\int \frac{g(xy) + h(xy)}{g(xy) - h(xy)} \frac{d(xy)}{(xy)} + \ln\left(\frac{x}{y}\right) = c$$

with respect to x, verify that $2/(xy(g(xy) - h(xy)))$ is an integrating factor of the DE $yg(xy) + xh(xy)y' = 0$. Hence, solve the DE

$$(x^2y^2 + xy + 1)y + (x^2y^2 - xy + 1)xy' = 0.$$

3.10. Show the following:

(i) $u(x, y) = c$ is the general solution of the DE (3.1) if and only if $Mu_y = Nu_x$.

(ii) The DE (3.1) has an integrating factor $(M^2 + N^2)^{-1}$ if $M_x = N_y$ and $M_y = -N_x$.

Answers or Hints

3.1. Begin with $u_y = N$ and obtain an analog of (3.7).

3.2. (i) $x^3y+4x^2y^2+4y^3 = 28$. (ii) $x^4e^{x+y}+x^2+y^2 = 1$. (iii) $x^2-2y\sin x = (\pi^2/4) - 2$. (iv) $e^{xy} + 4xy^3 - y^2 + 3 = 0$.

3.3. (i) $3/2$, $x^4 + 6x^2y + 8y^2 = c$. (ii) -2, $y^2 = x(2x - 1)/[2(cx + 1)]$.

3.4. 1, $\sin^{-1}((c - x)e^x)$.

3.5. 2, $x^4 + x^3y^2 = c$.

3.6. $4\tan^{-1}(x/y) + (x^2 + y^2)^2 = c$.

3.7. $\exp\left((1 - n)\int p\,dx\right)$, y^{-n}.

3.8. (i) x^{-2}, $y^2 + x\ln x = cx$. (ii) y^{-2}, $y^2 + x = cy$. (iii) $(xy)^{-3}$, $y^6 = c\exp(-x^{-2}y^{-2})$. (iv) $(x+y)$, $x^3y+2x^2y^2+xy^3 = c$, $x+y = 0$. (v) $(x^2+y^2)^{-2}$, $(c + 2x)(x^2 + y^2) = 1$. (vi) x^2y, $x^4y^2 + x^3y^5 = c$.

3.9. $xy + \ln(x/y) = c + 1/(xy)$.

3.10. (i) Note that $u_x + y_yy' = 0$ and $M + Ny' = 0$. (ii) $M_x = N_y$ and $M_y = -N_x$ imply $\frac{\partial}{\partial y}\left(\frac{M}{M^2+N^2}\right) = \frac{\partial}{\partial x}\left(\frac{N}{M^2+N^2}\right)$.

Lecture 4
Elementary First-Order Equations

Suppose in the DE of first order (3.1), $M(x,y) = X_1(x)Y_1(y)$ and $N(x,y) = X_2(x)Y_2(y)$, so that it takes the form

$$X_1(x)Y_1(y) + X_2(x)Y_2(y)y' = 0. \qquad (4.1)$$

If $Y_1(y)X_2(x) \neq 0$ for all $(x,y) \in S$, then (4.1) can be written as an exact DE

$$\frac{X_1(x)}{X_2(x)} + \frac{Y_2(y)}{Y_1(y)}y' = 0 \qquad (4.2)$$

in which the variables are separated. Such a DE (4.2) is said to be *separable*. The solution of this exact equation is given by

$$\int \frac{X_1(x)}{X_2(x)}dx + \int \frac{Y_2(y)}{Y_1(y)}dy = c. \qquad (4.3)$$

Here both the integrals are indefinite and constants of integration have been absorbed in c.

Equation (4.3) contains all the solutions of (4.1) for which $Y_1(y)X_2(x) \neq 0$. In fact, when we divide (4.1) by $Y_1 X_2$ we might have lost some solutions, and the ones which are not in (4.3) for some c must be coupled with (4.3) to obtain all solutions of (4.1).

Example 4.1. The DE in Example 3.2 may be written as

$$\frac{1}{x} + \frac{1}{y(1-y)}y' = 0, \quad xy(1-y) \neq 0$$

for which (4.3) gives the solution $y = (1-cx)^{-1}$. Other possible solutions for which $x(y-y^2) = 0$ are $x = 0$, $y = 0$, and $y = 1$. However, the solution $y = 1$ is already included in $y = (1-cx)^{-1}$ for $c = 0$, and $x = 0$ is not a solution, and hence all solutions of this DE are given by $y = 0$, $y = (1-cx)^{-1}$.

A function $f(x,y)$ defined in a domain D (an open connected set in $\mathbb{R}^2$) is said to be *homogeneous* of degree k if for all real λ and $(x,y) \in D$

$$f(\lambda x, \lambda y) = \lambda^k f(x,y). \qquad (4.4)$$

R.P. Agarwal and D. O'Regan, *An Introduction to Ordinary Differential Equations*, doi: 10.1007/978-0-387-71276-5_4, © Springer Science + Business Media, LLC 2008

For example, the functions $3x^2 - xy - y^2$, $\sin(x^2/(x^2 - y^2))$, $(x^4 + 7y^4)^{1/5}$, $(1/x^2)\sin(x/y) + (x/y^3)(\ln y - \ln x)$, and $(6e^{y/x}/x^{2/3}y^{1/3})$ are homogeneous of degree 2, 0, 4/5, -2, and -1, respectively.

In (4.4) if $\lambda = 1/x$, then it is the same as

$$x^k f\left(1, \frac{y}{x}\right) = f(x, y). \tag{4.5}$$

This implies that a homogeneous function of degree zero is a function of a single variable $v \ (= y/x)$.

The first-order DE (3.1) is said to be *homogeneous* if $M(x, y)$ and $N(x, y)$ are homogeneous functions of the same degree, say, n. If (3.1) is homogeneous, then in view of (4.5) it can be written as

$$x^n M\left(1, \frac{y}{x}\right) + x^n N\left(1, \frac{y}{x}\right) y' = 0. \tag{4.6}$$

In (4.6) we use the substitution $y(x) = xv(x)$, to obtain

$$x^n(M(1, v) + vN(1, v)) + x^{n+1}N(1, v)v' = 0. \tag{4.7}$$

Equation (4.7) is separable and admits the integrating factor

$$\mu = \frac{1}{x^{n+1}(M(1, v) + vN(1, v))} = \frac{1}{xM(x, y) + yN(x, y)} \tag{4.8}$$

provided $xM + yN \neq 0$.

The vanishing of $xM + yN$ implies that (4.7) is simply

$$x^{n+1}N(1, v)v' = xN(x, y)v' = 0$$

for which the integrating factor is $1/xN$. Thus, in this case the general solution of (3.1) is given by $y = cx$.

We summarize these results in the following theorem.

Theorem 4.1. The homogeneous DE (4.6) can be transformed to a separable DE by the transformation $y = xv$ which admits an integrating factor $1/(xM + yN)$ provided $xM + yN \neq 0$. Further, if $xM + yN = 0$, then its integrating factor is $1/xN$ and it has $y = cx$ as its general solution.

Example 4.2. The DE

$$y' = \frac{x^2 + xy + y^2}{x^2}, \quad x \neq 0$$

is homogeneous. The transformation $y = xv$ simplifies this DE to $xv' = 1 + v^2$. Thus, $\tan^{-1}(y/x) = \ln|cx|$ is the general solution of the given DE. Alternatively, the integrating factor

$$\left[x\left(\frac{x^2 + xy + y^2}{x^2}\right) - y\right]^{-1} = \left(\frac{x^2 + y^2}{x}\right)^{-1}$$

converts the given DE into the exact equation

$$\frac{1}{x} + \frac{y}{x^2 + y^2} - \frac{x}{x^2 + y^2}y' = 0$$

whose general solution using (3.7) is given by $-\tan^{-1}(y/x) + \ln x = c$, which is the same as obtained earlier.

Often, it is possible to introduce a new set of variables given by the equations

$$u = \phi(x, y), \quad v = \psi(x, y) \tag{4.9}$$

which convert a given DE (1.9) into a form that can be solved rather easily. Geometrically, relations (4.9) can be regarded as a mapping of a region in the xy-plane into the uv-plane. We wish to be able to solve these relations for x and y in terms of u and v. For this, we should assume that the mapping is one-to-one. In other words, we can assume that $\partial(u, v)/\partial(x, y) \neq 0$ over a region in $\mathbb{R}^2$, which implies that there is no functional relationship between u and v. Thus, if (u_0, v_0) is the image of (x_0, y_0) under the transformation (4.9), then it can be uniquely solved for x and y in a neighborhood of the point (x_0, y_0). This leads to the inverse transformation

$$x = x(u, v), \quad y = y(u, v). \tag{4.10}$$

The image of a curve in the xy-plane is a curve in the uv-plane; and the relation between slopes at the corresponding points of these curves is

$$y' = \frac{y_u + y_v\dfrac{dv}{du}}{x_u + x_v\dfrac{dv}{du}}. \tag{4.11}$$

Relations (4.10) and (4.11) can be used to convert the DE (1.9) in terms of u and v, which hopefully can be solved explicitly. Finally, replacement of u and v in terms of x and y by using (4.9) leads to an implicit solution of the DE (1.9).

Unfortunately, for a given nonlinear DE there is no way to predict a transformation which leads to a solution of (1.9). Finding such a transformation can only be learned by practice. We therefore illustrate this technique in the following examples.

Example 4.3. Consider the DE

$$3x^2ye^y + x^3e^y(y+1)y' = 0.$$

Setting $u = x^3$, $v = ye^y$ we obtain

$$3x^2\frac{dv}{dy} = e^y(y+1)\frac{du}{dx},$$

and this changes the given DE into the form $v + u(dv/du) = 0$ for which the solution is $uv = c$, equivalently, $x^3ye^y = c$ is the general solution of the given DE.

We now consider the DE

$$y' = f\left(\frac{a_1x + b_1y + c_1}{a_2x + b_2y + c_2}\right) \tag{4.12}$$

in which a_1, b_1, c_1, a_2, b_2 and c_2 are constants. If c_1 and c_2 are not both zero, then it can be converted to a homogeneous equation by means of the transformation

$$x = u + h, \quad y = v + k \tag{4.13}$$

where h and k are the solutions of the system of simultaneous linear equations

$$\begin{aligned} a_1h + b_1k + c_1 &= 0 \\ a_2h + b_2k + c_2 &= 0, \end{aligned} \tag{4.14}$$

and the resulting homogeneous DE

$$\frac{dv}{du} = f\left(\frac{a_1u + b_1v}{a_2u + b_2v}\right) \tag{4.15}$$

can be solved easily.

However, the system (4.14) can be solved for h and k provided $\Delta = a_1b_2 - a_2b_1 \neq 0$. If $\Delta = 0$, then $a_1x + b_1y$ is proportional to $a_2x + b_2y$, and hence (4.12) is of the form

$$y' = f(\alpha x + \beta y), \tag{4.16}$$

which can be solved easily by using the substitution $\alpha x + \beta y = z$.

Example 4.4. Consider the DE

$$y' = \frac{1}{2}\left(\frac{x+y-1}{x+2}\right)^2.$$

The straight lines $h + k - 1 = 0$ and $h + 2 = 0$ intersect at $(-2, 3)$, and hence the transformation $x = u - 2$, $y = v + 3$ changes the given DE to the form

$$\frac{dv}{du} = \frac{1}{2}\left(\frac{u+v}{u}\right)^2,$$

which has the solution $2\tan^{-1}(v/u) = \ln|u| + c$, and thus

$$2\tan^{-1}\frac{y-3}{x+2} = \ln|x+2| + c$$

is the general solution of the given DE.

Example 4.5. The DE

$$(x+y+1) + (2x+2y+1)y' = 0$$

suggests that we should use the substitution $x+y = z$. This converts the given DE to the form $(2z+1)z' = z$ for which the solution is $2z + \ln|z| = x+c$, $z \neq 0$; or equivalently, $x+2y+\ln|x+y| = c$, $x+y \neq 0$. If $z = x+y = 0$, then the given DE is satisfied by the relation $x+y = 0$. Thus, all solutions of the given DE are $x+2y+\ln|x+y| = c$, $x+y = 0$.

Let $f(x,y,\alpha) = 0$ and $g(x,y,\beta) = 0$ be the equations of two families of curves, each dependent on one parameter. When each member of the second family cuts each member of the first family according to a definite law, any curve of either of the families is said to be a *trajectory* of the family. The most important case is that in which curves of the families intersect at a constant angle. The *orthogonal trajectories* of a given family of curves are the curves that cut the given family at right angles. The slopes y_1' and y_2' of the tangents to the curves of the family and to the sought for orthogonal trajectories must at each point satisfy the orthogonality condition $y_1'y_2' = -1$.

Example 4.6. For the family of parabolas $y = ax^2$, we have $y' = 2ax$ or $y' = 2y/x$. Thus, the DE of the desired orthogonal trajectories is $y' = -(x/2y)$. Separating the variables, we find $2yy' + x = 0$, and on integrating this DE we obtain the family of ellipses $x^2 + 2y^2 = c$.

Problems

4.1. Show that if (3.1) is both homogeneous and exact, and $Mx + Ny$ is not a constant, then its general solution is given by $Mx + Ny = c$.

4.2. Show that if the DE (3.1) is homogeneous and M and N possess continuous partial derivatives in some domain D, then

$$\frac{xM_x + yM_y}{xN_x + yN_y} = \frac{M}{N}.$$

4.3. Solve the following DEs:

(i) $x\sin y + (x^2+1)\cos yy' = 0$.

(ii) $(x + 2y - 1) + 3(x + 2y)y' = 0.$

(iii) $xy' - y = xe^{y/x}.$

(iv) $xy' - y = x \sin[(y - x)/x].$

(v) $y' = (x + y + 1)^2 - 2.$

(vi) $y' = (3x - y - 5)/(-x + 3y + 7).$

4.4. Use the given transformation to solve the following DEs:

(i) $3x^5 - y(y^2 - x^3)y' = 0,$ $u = x^3,\ v = y^2.$

(ii) $(2x + y) + (x + 5y)y' = 0,$ $u = x - y,\ v = x + 2y.$

(iii) $(x + 2y) + (y - 2x)y' = 0,$ $x = r\cos\theta,\ y = r\sin\theta.$

(iv) $(2x^2 + 3y^2 - 7)x - (3x^2 + 2y^2 - 8)yy' = 0,$ $u = x^2,\ v = y^2.$

4.5. Show that the transformation $y(x) = xv(x)$ converts the DE $y^n f(x) + g(y/x)(y - xy') = 0$ into a separable equation.

4.6. Show that the change of variable $y = x^n v$ in the DE $y' = x^{n-1} f(y/x^n)$ leads to a separable equation. Hence, solve the DE

$$x^3 y' = 2y(x^2 - y).$$

4.7. Show that the introduction of polar coordinates $x = r\cos\theta,\ y = r\sin\theta$ leads to separation of variables in a homogeneous DE $y' = f(y/x)$. Hence, solve the DE

$$y' = \frac{ax + by}{bx - ay}.$$

4.8. Solve

$$y' = \frac{y - xy^2}{x + x^2 y}$$

by making the substitution $y = vx^n$ for an appropriate n.

4.9. Show that the families of parabolas $y^2 = 2cx + c^2,\ x^2 = 4a(y + a)$ are self-orthogonal.

4.10. Show that the circles $x^2 + y^2 = px$ intersect the circles $x^2 + y^2 = qy$ at right angles.

*4.11. Show that if the functions $f,\ f_x,\ f_y,$ and f_{xy} are continuous on some region D in the xy-plane, f is never zero on D, and $ff_{xy} = f_x f_y$ on D, then the DE (1.9) is separable. The requirement $f(x, y) \neq 0$ is essential, for this consider the function

$$f(x, y) = \begin{cases} x^2 e^{2y}, & y \leq 0 \\ x^2 e^y, & y > 0. \end{cases}$$

Answers or Hints

4.1. First use Theorem 4.1 and then Lemma 3.3.

4.2. Use Theorem 4.1.

4.3. (i) $(x^2+1)\sin^2 y = c$. (ii) $x+3y+c = 3\ln(x+2y+2)$, $x+2y+2 = 0$.
(iii) $\exp(-y/x) + \ln|x| = c$, $x \neq 0$. (iv) $\tan(y-x)/2x = cx$, $x \neq 0$.
(v) $(x+y+1)(1-ce^{2x}) = 1+ce^{2x}$. (vi) $(x+y+1)^2(y-x+3) = c$.

4.4. (i) $(y^2-2x^3)(y^2+x^3)^2 = c$. (ii) $2x^2+2xy+5y^2 = c$. (iii) $\sqrt{x^2+y^2} = c\exp(2\tan^{-1} y/x)$. (iv) $(x^2-y^2-1)^5 = c(x^2+y^2-3)$.

4.5. The given DE is reduced to $v^n x^n f(x) = x^2 g(v)v'$.

4.6. $y(2\ln|x|+c) = x^2$, $y = 0$.

4.7. $\sqrt{x^2+y^2} = c\exp\left(\frac{b}{a}\tan^{-1}\frac{y}{x}\right)$.

4.8. $ye^{xy} = cx$.

4.9. For $y^2 = 2cx + c^2$, $y' = c/y$ and hence the DE for the orthogonal trajectories is $y' = -y/c$, but it can be shown that it is the same as $y' = c/y$.

4.10. For the given families $y' = (y^2 - x^2)/2xy$ and $y' = 2xy/(x^2 - y^2)$, respectively.

Lecture 5
First-Order Linear Equations

Let in the DE (3.1) the functions M and N be $p_1(x)y - r(x)$ and $p_0(x)$, respectively, then it becomes

$$p_0(x)y' + p_1(x)y = r(x), \tag{5.1}$$

which is a first-order linear DE. In (5.1) we shall assume that the functions $p_0(x)$, $p_1(x)$, $r(x)$ are continuous and $p_0(x) \neq 0$ in J. With these assumptions the DE (5.1) can be written as

$$y' + p(x)y = q(x), \tag{5.2}$$

where $p(x) = p_1(x)/p_0(x)$ and $q(x) = r(x)/p_0(x)$ are continuous functions in J.

The corresponding homogeneous equation

$$y' + p(x)y = 0 \tag{5.3}$$

obtained by taking $q(x) \equiv 0$ in (5.2) can be solved by separating the variables, i.e., $(1/y)y' + p(x) = 0$, and now integrating it to obtain

$$y(x) = c \exp\left(-\int^x p(t)dt\right). \tag{5.4}$$

In dividing (5.3) by y we have lost the solution $y(x) \equiv 0$, which is called the *trivial solution* (for a linear homogeneous DE $y(x) \equiv 0$ is always a solution). However, it is included in (5.4) with $c = 0$.

If $x_0 \in J$, then the function

$$y(x) = y_0 \exp\left(-\int_{x_0}^x p(t)dt\right) \tag{5.5}$$

clearly satisfies the DE (5.3) in J and passes through the point (x_0, y_0). Thus, it is the solution of the initial value problem (5.3), (1.10).

To find the solution of the DE (5.2) we shall use *the method of variation of parameters* due to Lagrange. In (5.4) we assume that c is a function of x, i.e.,

$$y(x) = c(x) \exp\left(-\int^x p(t)dt\right) \tag{5.6}$$

R.P. Agarwal and D. O'Regan, *An Introduction to Ordinary Differential Equations*,
doi: 10.1007/978-0-387-71276-5_5, © Springer Science + Business Media, LLC 2008

and search for $c(x)$ so that (5.6) becomes a solution of the DE (5.2). For this, substituting (5.6) into (5.2), we find

$$c'(x)\exp\left(-\int^x p(t)dt\right) \quad -c(x)p(x)\exp\left(-\int^x p(t)dt\right)$$

$$+c(x)p(x)\exp\left(-\int^x p(t)dt\right) \;=\; q(x),$$

which is the same as

$$c'(x) \;=\; q(x)\exp\left(\int^x p(t)dt\right). \tag{5.7}$$

Integrating (5.7), we obtain the required function

$$c(x) \;=\; c_1 + \int^x q(t)\exp\left(\int^t p(s)ds\right)dt.$$

Now substituting this $c(x)$ in (5.6), we find the solution of (5.2) as

$$y(x) \;=\; c_1\exp\left(-\int^x p(t)dt\right) + \int^x q(t)\exp\left(-\int^x_t p(s)ds\right)dt. \tag{5.8}$$

This solution $y(x)$ is of the form $c_1 u(x) + v(x)$. It is to be noted that $c_1 u(x)$ is the general solution of (5.3) and $v(x)$ is a particular solution of (5.2). Hence, the general solution of (5.2) is obtained by adding any particular solution of (5.2) to the general solution of (5.3).

From (5.8) the solution of the initial value problem (5.2), (1.10) where $x_0 \in J$ is easily obtained as

$$y(x) \;=\; y_0\exp\left(-\int^x_{x_0} p(t)dt\right) + \int^x_{x_0} q(t)\exp\left(-\int^x_t p(s)ds\right)dt. \tag{5.9}$$

This solution in the particular case when $p(x) \equiv p$ and $q(x) \equiv q$ simply reduces to

$$y(x) \;=\; \left(y_0 - \frac{q}{p}\right)e^{-p(x-x_0)} + \frac{q}{p}.$$

Example 5.1. Consider the initial value problem

$$xy' - 4y + 2x^2 + 4 \;=\; 0, \quad x \neq 0, \quad y(1) = 1. \tag{5.10}$$

Since $x_0 = 1$, $y_0 = 1$, $p(x) = -4/x$ and $q(x) = -2x - 4/x$, from (5.9) the solution of (5.10) can be written as

$$
\begin{aligned}
y(x) &= \exp\left(\int_1^x \frac{4}{t}\,dt\right) + \int_1^x \left(-2t - \frac{4}{t}\right)\exp\left(\int_t^x \frac{4}{s}\,ds\right)dt \\
&= x^4 + \int_1^x \left(-2t - \frac{4}{t}\right)\frac{x^4}{t^4}\,dt \\
&= x^4 + x^4\left(\frac{1}{x^2} + \frac{1}{x^4} - 2\right) = -x^4 + x^2 + 1.
\end{aligned}
$$

Alternatively, instead of using (5.9), we can find the solution of (5.10) as follows. For the corresponding homogeneous DE $y' - (4/x)y = 0$ the general solution is cx^4, and a particular solution of the DE (5.10) is

$$
\int^x \left(-2t - \frac{4}{t}\right)\exp\left(\int_t^x \frac{4}{s}\,ds\right)dt = x^2 + 1,
$$

and hence the general solution of the DE (5.10) is $y(x) = cx^4 + x^2 + 1$. Now in order to satisfy the initial condition $y(1) = 1$ it is necessary that $1 = c+1+1$, or $c = -1$. The solution of (5.10) is therefore $y(x) = -x^4 + x^2 + 1$.

Suppose $y_1(x)$ and $y_2(x)$ are two particular solutions of (5.2), then

$$
\begin{aligned}
y_1'(x) - y_2'(x) &= -p(x)y_1(x) + q(x) + p(x)y_2(x) - q(x) \\
&= -p(x)(y_1(x) - y_2(x)),
\end{aligned}
$$

which implies that $y(x) = y_1(x) - y_2(x)$ is a solution of (5.3). Thus, if two particular solutions of (5.2) are known, then $y(x) = c(y_1(x) - y_2(x)) + y_1(x)$ as well as $y(x) = c(y_1(x) - y_2(x)) + y_2(x)$ represents the general solution of (5.2). For example, $x + 1/x$ and x are two solutions of the DE $xy' + y = 2x$ and $y(x) = c/x + x$ is its general solution.

The DE $(xf(y) + g(y))y' = h(y)$ may not be integrable as it is, but if the roles of x and y are interchanged, then it can be written as

$$
h(y)\frac{dx}{dy} - f(y)x = g(y),
$$

which is a linear DE in x and can be solved by the preceding procedure. In fact, the solutions of (1.9) and $dx/dy = 1/f(x,y)$ determine the same curve in a region in $\mathbb{R}^2$ provided the function f is defined, continuous, and nonzero. For this, if $y = y(x)$ is a solution of (1.9) in J and $y'(x) = f(x, y(x)) \neq 0$, then $y(x)$ is monotonic function in J and hence has an inverse $x = x(y)$. This function x is such that

$$
\frac{dx}{dy} = \frac{1}{y'(x)} = \frac{1}{f(x, y(x))} \quad \text{in} \quad J.
$$

Example 5.2. The DE

$$y' = \frac{1}{(e^{-y} - x)}$$

can be written as $dx/dy + x = e^{-y}$ which can be solved to obtain $x = e^{-y}(y + c)$.

Certain nonlinear first-order DEs can be reduced to linear equations by an appropriate change of variables. For example, it is always possible for the *Bernoulli equation*

$$p_0(x)y' + p_1(x)y = r(x)y^n, \quad n \neq 0, 1. \tag{5.11}$$

In (5.11), $n = 0$ and 1 are excluded because in these cases this equation is obviously linear.

The equation (5.11) is equivalent to the DE

$$p_0(x)y^{-n}y' + p_1(x)y^{1-n} = r(x) \tag{5.12}$$

and now the substitution $v = y^{1-n}$ leads to the first-order linear DE

$$\frac{1}{1-n}p_0(x)v' + p_1(x)v = r(x). \tag{5.13}$$

Example 5.3. The DE $xy' + y = x^2y^2$, $x \neq 0$ can be written as $xy^{-2}y' + y^{-1} = x^2$. The substitution $v = y^{-1}$ converts this DE into $-xv' + v = x^2$, which can be solved to get $v = (c - x)x$, and hence the general solution of the given DE is $y(x) = (cx - x^2)^{-1}$.

As we have remarked in Lecture 2, we shall show that if one solution $y_1(x)$ of the Riccati equation (2.14) is known, then the substitution $y = y_1 + z^{-1}$ converts it into a first-order linear DE in z. Indeed, we have

$$y_1' - \frac{1}{z^2}z' = p(x)\left(y_1 + \frac{1}{z}\right)^2 + q(x)\left(y_1 + \frac{1}{z}\right) + r(x)$$

$$= (p(x)y_1^2 + q(x)y_1 + r(x)) + p(x)\left(\frac{2y_1}{z} + \frac{1}{z^2}\right) + q(x)\frac{1}{z}$$

and hence

$$-\frac{1}{z^2}z' = (2p(x)y_1 + q(x))\frac{1}{z} + p(x)\frac{1}{z^2},$$

which is the first-order linear DE

$$z' + (2p(x)y_1 + q(x))z + p(x) = 0. \tag{5.14}$$

Example 5.4. It is easy to verify that $y_1 = x$ is a particular solution of the Riccati equation $y' = 1 + x^2 - 2xy + y^2$. The substitution $y = x + z^{-1}$

converts this DE to the first-order linear DE $z' + 1 = 0$, whose general solution is $z = (c - x)$, $x \neq c$. Thus, the general solution of the given Riccati equation is $y(x) = x + 1/(c - x)$, $x \neq c$.

In many physical problems the nonhomogeneous term $q(x)$ in (5.2) is specified by different formulas in different intervals. This is often the case when (5.2) is considered as an *input–output* relation, i.e., the function $q(x)$ is an *input* and the solution $y(x)$ is an *output* corresponding to the input $q(x)$. Usually, in such situations the solution $y(x)$ is not defined at certain points, so that it is not continuous throughout the interval of interest. To understand such a case, for simplicity, we consider the initial value problem (5.2), (1.10) in the interval $[x_0, x_2]$, where the function $p(x)$ is continuous, and

$$q(x) = \begin{cases} q_1(x), & x_0 \leq x < x_1 \\ q_2(x), & x_1 < x \leq x_2. \end{cases}$$

We assume that the functions $q_1(x)$ and $q_2(x)$ are continuous in the intervals $[x_0, x_1)$ and $(x_1, x_2]$, respectively. With these assumptions the "solution" $y(x)$ of (5.2), (1.10) in view of (5.9) can be written as

$$y(x) = \begin{cases} y_1(x) = y_0 \exp\left(-\int_{x_0}^{x} p(t)dt\right) + \int_{x_0}^{x} q_1(t) \exp\left(-\int_{t}^{x} p(s)ds\right) dt, \\ \qquad\qquad\qquad\qquad\qquad\qquad\qquad\qquad\qquad x_0 \leq x < x_1 \\ y_2(x) = c \exp\left(-\int_{x_1}^{x} p(t)dt\right) + \int_{x_1}^{x} q_2(t) \exp\left(-\int_{t}^{x} p(s)ds\right) dt, \\ \qquad\qquad\qquad\qquad\qquad\qquad\qquad\qquad\qquad x_1 < x \leq x_2. \end{cases}$$

Clearly, at the point x_1 we cannot say much about the solution $y(x)$, it may not even be defined. However, if the limits $\lim_{x \to x_1^-} y_1(x)$ and $\lim_{x \to x_1^+} y_2(x)$ exist (which are guaranteed if both the functions $q_1(x)$ and $q_2(x)$ are bounded at $x = x_1$), then the relation

$$\lim_{x \to x_1^-} y_1(x) = \lim_{x \to x_1^+} y_2(x) \tag{5.15}$$

determines the constant c, so that the solution $y(x)$ is continuous on $[x_0, x_2]$.

Example 5.5. Consider the initial value problem

$$y' - \frac{4}{x}y = \begin{cases} -2x - \dfrac{4}{x}, & x \in [1, 2) \\ x^2, & x \in (2, 4] \end{cases} \tag{5.16}$$

$$y(1) = 1.$$

In view of Example 5.1 the solution of (5.16) can be written as

$$y(x) = \begin{cases} -x^4 + x^2 + 1, & x \in [1, 2) \\ c\dfrac{x^4}{16} + \dfrac{x^4}{2} - x^3, & x \in (2, 4]. \end{cases}$$

Now the relation (5.15) gives $c = -11$. Thus, the continuous solution of (5.16) is

$$y(x) = \begin{cases} -x^4 + x^2 + 1, & x \in [1, 2) \\ -\dfrac{3}{16}x^4 - x^3, & x \in (2, 4]. \end{cases}$$

Clearly, this solution is not differentiable at $x = 2$.

Problems

5.1. Show that the DE (5.2) admits an integrating factor which is a function of x alone. Use this to obtain its general solution.

5.2. (Principle of Superposition). If $y_1(x)$ and $y_2(x)$ are solutions of $y' + p(x)y = q_i(x)$, $i = 1, 2$, respectively, then show that $c_1 y_1(x) + c_2 y_2(x)$ is a solution of the DE $y' + p(x)y = c_1 q_1(x) + c_2 q_2(x)$, where c_1 and c_2 are constants.

5.3. Find the general solution of the following DEs:

(i) $y' - (\cot x)y = 2x \sin x$.
(ii) $y' + y \mid x \mid x^2 + x^3 = 0$.
(iii) $(y^2 - 1) + 2(x - y(1 + y)^2)y' = 0$.
(iv) $(1 + y^2) = (\tan^{-1} y - x)y'$.

5.4. Solve the following initial value problems:

(i) $y' + 2y = \begin{cases} 1, & 0 \leq x \leq 1 \\ 0, & x > 1 \end{cases}$, $y(0) = 0$.

(ii) $y' + p(x)y = 0$, $y(0) = 1$, where $p(x) = \begin{cases} 2, & 0 \leq x \leq 1 \\ 1, & x > 1. \end{cases}$

5.5. Let $q(x)$ be continuous in $[0, \infty)$ and $\lim_{x \to \infty} q(x) = L$. For the DE $y' + ay = q(x)$, show the following:

(i)   If $a > 0$, every solution approaches L/a as $x \to \infty$.
(ii) If $a < 0$, there is one and only one solution which approaches L/a as $x \to \infty$.

5.6. Let $y(x)$ be the solution of the initial value problem (5.2), (1.10) in $[x_0, \infty)$, and let $z(x)$ be a continuously differentiable function in $[x_0, \infty)$ such that $z' + p(x)z \leq q(x)$, $z(x_0) \leq y_0$. Show that $z(x) \leq y(x)$ for all x in $[x_0, \infty)$. In particular, for the problem $y' + y = \cos x$, $y(0) = 1$ verify that $2e^{-x} - 1 \leq y(x) \leq 1$, $x \in [0, \infty)$.

5.7. Find the general solution of the following nonlinear DEs:

(i) $2(1 + y^3) + 3xy^2 y' = 0.$

(ii) $y + x(1 + xy^4)y' = 0.$

(iii) $(1 - x^2)y' + y^2 - 1 = 0.$

(iv) $y' - e^{-x}y^2 - y - e^x = 0.$

∗5.8. Let the functions p_0, p_1, and r be continuous in $J = [\alpha, \beta]$ such that $p_0(\alpha) = p_0(\beta) = 0$, $p_0(x) > 0$, $x \in (\alpha, \beta)$, $p_1(x) > 0$, $x \in J$, and

$$\int_\alpha^{\alpha+\epsilon} \frac{dx}{p_0(x)} = \int_{\beta-\epsilon}^\beta \frac{dx}{p_0(x)} = \infty, \quad 0 < \epsilon < \beta - \alpha.$$

Show that all solutions of the DE (5.1) which exist in (α, β) converge to $r(\beta)/p_1(\beta)$ as $x \to \beta$. Further, show that one of these solutions converges to $r(\alpha)/p_1(\alpha)$ as $x \to \alpha$, while all other solutions converge to ∞, or $-\infty$.

Answers or Hints

5.1. Since $M = p(x)y - q(x)$, $N = 1$, $[(M_y - N_x)/N] = p(x)$, and hence the integrating factor is $\exp(\int^x p(t)dt)$.

5.2. Use the definition of a solution.

5.3. (i) $c \sin x + x^2 \sin x$. (ii) $ce^{-x} - x^3 + 2x^2 - 5x + 5$. (iii) $x(y-1)/(y+1) = y^2 + c$. (iv) $x = \tan^{-1} y - 1 + ce^{-\tan^{-1} y}$.

5.4. (i) $y(x) = \begin{cases} \frac{1}{2}(1 - e^{-2x}), & 0 \le x \le 1 \\ \frac{1}{2}(e^2 - 1)e^{-2x}, & x > 1 \end{cases}$ (ii) $y(x) = \begin{cases} e^{-2x}, & 0 \le x \le 1 \\ e^{-(x+1)}, & x > 1. \end{cases}$

5.5. (i) In $y(x) = y(x_0)e^{-a(x-x_0)} + [\int_{x_0}^x e^{at}q(t)dt]/e^{ax}$ take the limit $x \to \infty$. (ii) In $y(x) = e^{-ax}\left[y(x_0)e^{ax_0} + \int_{x_0}^\infty e^{at}q(t)dt - \int_x^\infty e^{at}q(t)dt\right]$ choose $y(x_0)$ so that $y(x_0)e^{ax_0} + \int_{x_0}^\infty e^{at}q(t)dt = 0$ $(\lim_{x\to\infty} q(x) = L)$. Now in $y(x) = -[\int_x^\infty e^{at}q(t)dt]/e^{ax}$ take the limit $x \to \infty$.

5.6. There exists a continuous function $r(x) \ge 0$ such that $z' + p(x)z = q(x) - r(x)$, $z(x_0) \le y_0$. Thus, for the function $\phi(x) = y(x) - z(x)$, $\phi' + p(x)\phi = r(x) \ge 0$, $\phi(x_0) = y_0 - z(x_0) \ge 0$.

5.7. (i) $x^2(1 + y^3) = c$. (ii) $xy^4 = 3(1 + cxy)$, $y = 0$. (iii) $(y-1)(1+x) = c(1-x)(1+y)$. (iv) $e^x \tan(x + c)$.

Lecture 6
Second-Order Linear Equations

Consider the homogeneous linear second-order DE with variable coefficients

$$p_0(x)y'' + p_1(x)y' + p_2(x)y = 0, \tag{6.1}$$

where $p_0(x)$ (> 0), $p_1(x)$ and $p_2(x)$ are continuous in J. There does not exist any method to solve it except in a few rather restrictive cases. However, the results below follow immediately from the general theory of first-order linear systems, which we shall present in later lectures.

Theorem 6.1. There exist exactly two solutions $y_1(x)$ and $y_2(x)$ of (6.1) which are linearly independent (essentially different) in J; i.e., there does not exist a constant c such that $y_1(x) = cy_2(x)$ for all $x \in J$.

Theorem 6.2. Two solutions $y_1(x)$ and $y_2(x)$ of (6.1) are linearly independent in J if and only if their *Wronskian* defined by

$$W(x) = W(y_1, y_2)(x) = \begin{vmatrix} y_1(x) & y_2(x) \\ y_1'(x) & y_2'(x) \end{vmatrix} = y_1(x)y_2'(x) - y_2(x)y_1'(x) \tag{6.2}$$

is different from zero for some $x = x_0$ in J.

Theorem 6.3. For the Wronskian defined in (6.2) the following Abel's identity (also known as the Ostrogradsky–Liouville formula) holds:

$$W(x) = W(x_0)\exp\left(-\int_{x_0}^{x} \frac{p_1(t)}{p_0(t)}dt\right), \quad x_0 \in J. \tag{6.3}$$

Thus, if the Wronskian is zero at some $x_0 \in J$, then it is zero for all $x \in J$.

Theorem 6.4. If $y_1(x)$ and $y_2(x)$ are solutions of (6.1) and c_1 and c_2 are arbitrary constants, then $c_1y_1(x) + c_2y_2(x)$ is also a solution of (6.1). Further, if $y_1(x)$ and $y_2(x)$ are linearly independent, then any solution $y(x)$ of (6.1) can be written as $y(x) = \bar{c}_1y_1(x) + \bar{c}_2y_2(x)$, where $\bar{c}_1$ and $\bar{c}_2$ are suitable constants.

Now we shall show that, if one solution $y_1(x)$ of (6.1) is known (by some clever method), then we can employ variation of parameters to find the second solution of (6.1). For this we let $y(x) = u(x)y_1(x)$ and substitute this in (6.1) to get

$$p_0(uy_1)'' + p_1(uy_1)' + p_2(uy_1) = 0,$$

R.P. Agarwal and D. O'Regan, *An Introduction to Ordinary Differential Equations*, doi: 10.1007/978-0-387-71276-5_6, © Springer Science + Business Media, LLC 2008

or
$$p_0 u'' y_1 + 2p_0 u' y_1' + p_0 u y_1'' + p_1 u' y_1 + p_1 u y_1' + p_2 u y_1 = 0,$$

or
$$p_0 u'' y_1 + (2p_0 y_1' + p_1 y_1) u' + (p_0 y_1'' + p_1 y_1' + p_2 y_1) u = 0.$$

However, since y_1 is a solution of (6.1), the above equation with $v = u'$ is the same as
$$p_0 y_1 v' + (2p_0 y_1' + p_1 y_1) v = 0, \tag{6.4}$$

which is a first-order equation, and it can be solved easily provided $y_1 \neq 0$ in J. Indeed, multiplying (6.4) by y_1/p_0, we get
$$(y_1^2 v' + 2y_1' y_1 v) + \frac{p_1}{p_0} y_1^2 v = 0,$$

which is the same as
$$(y_1^2 v)' + \frac{p_1}{p_0}(y_1^2 v) = 0$$

and hence
$$y_1^2 v = c \exp\left(-\int^x \frac{p_1(t)}{p_0(t)} dt\right),$$

or, on taking $c = 1$,
$$v(x) = \frac{1}{y_1^2(x)} \exp\left(-\int^x \frac{p_1(t)}{p_0(t)} dt\right).$$

Hence, the second solution of (6.1) is
$$y_2(x) = y_1(x) \int^x \frac{1}{y_1^2(t)} \exp\left(-\int^t \frac{p_1(s)}{p_0(s)} ds\right) dt. \tag{6.5}$$

Example 6.1.   It is easy to verify that $y_1(x) = x^2$ is a solution of the DE
$$x^2 y'' - 2xy' + 2y = 0, \quad x \neq 0.$$

For the second solution we use (6.5), to obtain
$$y_2(x) = x^2 \int^x \frac{1}{t^4} \exp\left(-\int^t \left(-\frac{2s}{s^2}\right) ds\right) dt = x^2 \int^x \frac{1}{t^4} t^2 dt = -x.$$

Now we shall find a particular solution of the nonhomogeneous equation
$$p_0(x) y'' + p_1(x) y' + p_2(x) y = r(x). \tag{6.6}$$

For this also we shall apply the method of variation of parameters. Let $y_1(x)$ and $y_2(x)$ be two solutions of (6.1). We assume $y(x) = c_1(x) y_1(x) + c_2(x) y_2(x)$ is a solution of (6.6). Note that $c_1(x)$ and $c_2(x)$ are two unknown functions, so we can have two sets of conditions which determine $c_1(x)$ and $c_2(x)$. Since

$$y' = c_1 y_1' + c_2 y_2' + c_1' y_1 + c_2' y_2$$

as a first condition, we assume that

$$c_1' y_1 + c_2' y_2 = 0. \tag{6.7}$$

Thus, we have

$$y' = c_1 y_1' + c_2 y_2'$$

and on differentiation

$$y'' = c_1 y_1'' + c_2 y_2'' + c_1' y_1' + c_2' y_2'.$$

Substituting these in (6.6), we find

$$c_1(p_0 y_1'' + p_1 y_1' + p_2 y_1) + c_2(p_0 y_2'' + p_1 y_2' + p_2 y_2) + p_0(c_1' y_1' + c_2' y_2') = r(x).$$

Clearly, this equation in view of the fact that $y_1(x)$ and $y_2(x)$ are solutions of (6.1) is the same as

$$c_1' y_1' + c_2' y_2' = \frac{r(x)}{p_0(x)}. \tag{6.8}$$

Solving (6.7), (6.8) we find

$$c_1' = -\frac{y_2(x)r(x)/p_0(x)}{\begin{vmatrix} y_1(x) & y_2(x) \\ y_1'(x) & y_2'(x) \end{vmatrix}}, \quad c_2' = \frac{y_1(x)r(x)/p_0(x)}{\begin{vmatrix} y_1(x) & y_2(x) \\ y_1'(x) & y_2'(x) \end{vmatrix}};$$

and hence a particular solution of (6.6) is

$$\begin{aligned}
y_p(x) &= c_1(x)y_1(x) + c_2(x)y_2(x) \\
&= -y_1(x) \int^x \frac{y_2(t)r(t)/p_0(t)}{\begin{vmatrix} y_1(t) & y_2(t) \\ y_1'(t) & y_2'(t) \end{vmatrix}} dt + y_2(x) \int^x \frac{y_1(t)r(t)/p_0(t)}{\begin{vmatrix} y_1(t) & y_2(t) \\ y_1'(t) & y_2'(t) \end{vmatrix}} dt \\
&= \int^x H(x,t) \frac{r(t)}{p_0(t)} dt,
\end{aligned} \tag{6.9}$$

where

$$H(x,t) = \begin{vmatrix} y_1(t) & y_2(t) \\ y_1(x) & y_2(x) \end{vmatrix} \bigg/ \begin{vmatrix} y_1(t) & y_2(t) \\ y_1'(t) & y_2'(t) \end{vmatrix}. \tag{6.10}$$

The general solution of (6.6) which is obtained by adding this particular solution with the general solution of (6.1) appears as

$$y(x) = c_1 y_1(x) + c_2 y_2(x) + y_p(x). \tag{6.11}$$

The following properties of the function $H(x,t)$ are immediate:

(i) $H(x,t)$ is defined for all $(x,t) \in J \times J$;

(ii) $\partial^j H(x,t)/\partial x^j$, $j = 0, 1, 2$ are continuous for all $(x,t) \in J \times J$;

(iii)   for each fixed $t \in J$ the function $z(x) = H(x,t)$ is a solution of the homogeneous DE (6.1) satisfying $z(t) = 0$, $z'(t) = 1$; and

(iv)   the function

$$v(x) = \int_{x_0}^x H(x,t) \frac{r(t)}{p_0(t)} dt$$

is a particular solution of the nonhomogeneous DE (6.6) satisfying $y(x_0) = y'(x_0) = 0$.

Example 6.2.   Consider the DE

$$y'' + y = \cot x.$$

For the corresponding homogeneous DE $y'' + y = 0$, $\sin x$ and $\cos x$ are the solutions. Thus, its general solution can be written as

$$y(x) = c_1 \cos x + c_2 \sin x + \int^x \frac{\begin{vmatrix} \sin t & \cos t \\ \sin x & \cos x \end{vmatrix}}{\begin{vmatrix} \sin t & \cos t \\ \cos t & -\sin t \end{vmatrix}} \frac{\cos t}{\sin t} dt$$

$$= c_1 \cos x + c_2 \sin x - \int^x (\sin t \cos x - \sin x \cos t) \frac{\cos t}{\sin t} dt$$

$$= c_1 \cos x + c_2 \sin x - \cos x \int^x \cos t\, dt + \sin x \int^x \frac{\cos^2 t}{\sin t} dt$$

$$= c_1 \cos x + c_2 \sin x - \cos x \sin x + \sin x \int^x \frac{1 - \sin^2 t}{\sin t} dt$$

$$= c_1 \cos x + c_2 \sin x - \cos x \sin x - \sin x \int^x \sin t\, dt + \sin x \int^x \frac{1}{\sin t} dt$$

$$= c_1 \cos x + c_2 \sin x + \sin x \int^x \frac{\operatorname{cosec} t(\operatorname{cosec} t - \cot t)}{(\operatorname{cosec} t - \cot t)} dt$$

$$= c_1 \cos x + c_2 \sin x + \sin x \, \ln[\operatorname{cosec} x - \cot x].$$

From the general theory of first-order linear systems, which we shall present in later lectures, it also follows that if the functions $p_0(x)$ (> 0), $p_1(x)$, $p_2(x)$, and $r(x)$ are continuous on J and $x_0 \in J$, then the initial value problem: (6.6) together with the *initial conditions*

$$y(x_0) = y_0, \quad y'(x_0) = y_1 \tag{6.12}$$

has a unique solution.

Now we shall show that second-order DEs with constant coefficients can be solved explicitly. In fact, to find the solution of the equation

$$y'' + ay' + by = 0, \tag{6.13}$$

where a and b are constants, as a first step we look back at the equation $y' + ay = 0$ (a is a constant) for which all solutions are multiples of $y = e^{-ax}$. Thus, for (6.13) also some form of exponential function would be a reasonable choice and would utilize the property that the differentiation of an exponential function e^{rx} always yields a constant multiplied by e^{rx}.

Thus, we try $y = e^{rx}$ and find the value(s) of r. For this, we have

$$r^2 e^{rx} + are^{rx} + be^{rx} = (r^2 + ar + b)e^{rx} = 0,$$

which gives

$$r^2 + ar + b = 0. \tag{6.14}$$

Hence, e^{rx} is a solution of (6.13) if r is a solution of (6.14). Equation (6.14) is called the *characteristic equation*. For the roots of (6.14) we have the following three cases:

1. Distinct real roots. If r_1 and r_2 are real and distinct roots of (6.14), then $e^{r_1 x}$ and $e^{r_2 x}$ are two solutions of (6.13) and its general solution can be written as

$$y(x) = c_1 e^{r_1 x} + c_2 e^{r_2 x}.$$

In the particular case when $r_1 = r$, $r_2 = -r$ (then the DE (6.13) is $y'' - r^2 y = 0$) we have

$$y(x) = c_1 e^{rx} + c_2 e^{-rx} = \left(\frac{A+B}{2}\right) e^{rx} + \left(\frac{A-B}{2}\right) e^{-rx}$$

$$= A\left(\frac{e^{rx} + e^{-rx}}{2}\right) + B\left(\frac{e^{rx} - e^{-rx}}{2}\right) = A\cosh rx + B\sinh rx.$$

2. Repeated real roots. If $r_1 = r_2 = r$ is a repeated root of (6.14), then e^{rx} is a solution. To find the second solution, we let $y(x) = u(x)e^{rx}$ and substitute it in (6.13), to get

$$e^{rx}(u'' + 2ru' + r^2 u) + ae^{ru}(u' + ru) + bue^{rx} = 0,$$

or

$$u'' + (2r + a)u' + (r^2 + ar + b)u = u'' + (2r + a)u' = 0.$$

Now since r is a repeated root of (6.14) it follows that $2r + a = 0$ and hence $u'' = 0$, i.e., $u(x) = c_1 + c_2 x$. Thus,

$$y(x) = (c_1 + c_2 x)e^{rx} = c_1 e^{rx} + c_2 x e^{rx}.$$

Hence, the second solution of (6.13) is xe^{rx}.

3. Complex conjugate roots.

Let $r_1 = \mu + i\nu$ and $r_2 = \mu - i\nu$ where $i = \sqrt{-1}$, so that

$$e^{(\mu \pm i\nu)x} = e^{\mu x}(\cos \nu x \pm i \sin \nu x).$$

Since for the DE (6.13) real part, i.e., $e^{\mu x} \cos \nu x$ and the complex part, i.e., $e^{\mu x} \sin \nu x$ both are solutions, the general solution of (6.13) can be written as

$$y(x) = c_1 e^{\mu x} \cos \nu x + c_2 e^{\mu x} \sin \nu x.$$

In the particular case when $r_1 = i\nu$ and $r_2 = -i\nu$ (then the DE (6.13) is $y'' + \nu^2 y = 0$) we have $y(x) = c_1 \cos \nu x + c_2 \sin \nu x$.

Finally, in this lecture we shall find the solution of the *Cauchy–Euler equation*

$$x^2 y'' + axy' + by = 0, \quad x > 0. \tag{6.15}$$

We assume $y(x) = x^m$ to obtain

$$x^2 m(m-1)x^{m-2} + axmx^{m-1} + bx^m = 0,$$

or

$$m(m-1) + am + b = 0. \tag{6.16}$$

This is the characteristic equation for (6.15), and as earlier for (6.14) the nature of its roots determines the solution:

Real, distinct roots $m_1 \neq m_2$: $y(x) = c_1 x^{m_1} + c_2 x^{m_2}$,

Real, repeated roots $m = m_1 = m_2$: $y(x) = c_1 x^m + c_2 (\ln x) x^m$,

Complex conjugate roots $m_1 = \mu + i\nu$, $m_2 = \mu - i\nu$: $y(x) = c_1 x^\mu \cos(\nu \ln x) + c_2 x^\mu \sin(\nu \ln x)$.

In the particular case

$$x^2 y'' + xy' - \lambda^2 y = 0, \quad x > 0, \quad \lambda > 0 \tag{6.17}$$

the characteristic equation is $m(m-1) + m - \lambda^2 = 0$, or $m^2 - \lambda^2 = 0$. The roots are $m = \pm \lambda$ and hence the solution of (6.17) appears as

$$y(x) = c_1 x^\lambda + c_2 x^{-\lambda}. \tag{6.18}$$

Problems

6.1. Let $y_1(x), y_2(x), y_3(x)$ and $\lambda(x)$ be differentiable functions in J. Show that for all $x \in J$,

(i) $W(y_1, y_2 + y_3)(x) = W(y_1, y_2)(x) + W(y_1, y_3)(x);$

(ii) $W(\lambda y_1, \lambda y_2)(x) = \lambda^2(x)W(y_1, y_2)(x);$

(iii) $W(y_1, \lambda y_1)(x) = \lambda'(x)y_1^2(x).$

6.2. Show that the functions $y_1(x) = c \ (\neq 0)$ and $y_2(x) = 1/x^2$ satisfy the nonlinear DE $y'' + 3xyy' = 0$ in $(0, \infty)$, but $y_1(x) + y_2(x)$ does not satisfy the given DE. (This shows that Theorem 6.4 holds good only for the linear equations.)

6.3. Given the solution $y_1(x)$, find the second solution of the following DEs:

(i) $(x^2 - x)y'' + (3x - 1)y' + y = 0 \quad (x \neq 0, 1), \quad y_1(x) = (x - 1)^{-1}.$

(ii) $x(x - 2)y'' + 2(x - 1)y' - 2y = 0 \quad (x \neq 0, 2), \quad y_1(x) = (1 - x).$

(iii) $xy'' - y' - 4x^3y = 0 \quad (x \neq 0), \quad y_1(x) = \exp(x^2).$

(iv) $(1 - x^2)y'' - 2xy' + 2y = 0 \quad (|x| < 1), \quad y_1(x) = x.$

6.4. The differential equation

$$xy'' - (x + n)y' + ny = 0$$

is interesting because it has an exponential solution and a polynomial solution.

(i) Verify that one solution is $y_1(x) = e^x.$

(ii) Show that the second solution has the form $y_2(x) = ce^x \int^x t^n e^{-t} dt.$ Further, show that with $c = -1/n!,$

$$y_2(x) = 1 + \frac{x}{1!} + \frac{x^2}{2!} + \cdots + \frac{x^n}{n!}.$$

Note that $y_2(x)$ is the first $n + 1$ terms of the Taylor series about $x = 0$ for e^x, that is, for $y_1(x)$.

6.5. For the differential equation

$$y'' + \delta(xy' + y) = 0,$$

verify that $y_1(x) = \exp(-\delta x^2/2)$ is one solution. Find its second solution.

6.6. Let $y_1(x) \neq 0$ and $y_2(x)$ be two linearly independent solutions of the DE (6.1). Show that $y(x) = y_2(x)/y_1(x)$ is a nonconstant solution of the DE

$$y_1(x)y'' + \left(2y_1'(x) + \frac{p_1(x)}{p_0(x)}y_1(x)\right)y' = 0.$$

6.7. Let $y_1(x)$ and $y_2(x)$ be solutions of the DE

$$y'' + p_1(x)y' + p_2(x)y = 0 \tag{6.19}$$

in J. Show the following:

(i)   If $y_1(x)$ and $y_2(x)$ vanish at the same point in J, then $y_1(x)$ is a constant multiple of $y_2(x)$.

(ii)   If $y_1(x)$ and $y_2(x)$ have maxima or minima at the same point in the open interval J, then $y_1(x)$ and $y_2(x)$ are not the linearly independent solutions.

(iii) If $W(y_1, y_2)(x)$ is independent of x, then $p_1(x) = 0$ for all $x \in J$.

(iv)   If $y_1(x)$ and $y_2(x)$ are linearly independent, then $y_1(x)$ and $y_2(x)$ cannot have a common point of inflexion in J unless $p_1(x)$ and $p_2(x)$ vanish simultaneously there.

(v)   If $W(y_1, y_2)(x^*) = y_1(x^*) = 0$, then either $y_1(x) = 0$ for all $x \in J$, or $y_2(x) = (y_2'(x^*)/y_1'(x^*))y_1(x)$.

6.8. Let $y_1(x)$ and $y_2(x)$ be linearly independent solutions of (6.19), and $W(x)$ be their Wronskian. Show that

$$y'' + p_1(x)y' + p_2(x)y = \frac{W}{y_1} \frac{d}{dx} \left(\frac{y_1^2}{W} \frac{d}{dx} \left(\frac{y}{y_1} \right) \right).$$

6.9. Show that the DE (6.1) can be transformed into a first-order nonlinear DE by means of a change of dependent variable

$$y = \exp \left(\int^x f(t)w(t)dt \right),$$

where $f(x)$ is any nonvanishing differentiable function. In particular, if $f(x) = p_0(x)$, then show that (6.1) reduces to the Riccati equation,

$$w' + p_0(x)w^2 + \frac{p_0'(x) + p_1(x)}{p_0(x)}w + \frac{p_2(x)}{p_0^2(x)} = 0. \qquad (6.20)$$

6.10. If $w_1(x)$ and $w_2(x)$ are two different solutions of the DE (6.20) with $p_0(x) = 1$, i.e.,

$$w' + w^2 + p_1(x)w + p_2(x) = 0, \qquad (6.21)$$

then show that its general solution $w(x)$ is given by

$$\frac{w(x) - w_1(x)}{w(x) - w_2(x)} \exp \left(\int^x (w_1(t) - w_2(t))dt \right) = c_1.$$

Further, if $w_3(x)$ is another known solution of (6.21), then

$$\frac{w(x) - w_3(x)}{w(x) - w_2(x)} = c_2 \frac{w_1(x) - w_3(x)}{w_1(x) - w_2(x)}.$$

6.11. Find the general solution of the following homogeneous DEs:

(i) $y'' + 7y' + 10y = 0.$
(ii) $y'' - 8y' + 16y = 0.$
(iii) $y'' + 2y' + 3y = 0.$

6.12. Find the general solution of the following nonhomogeneous DEs:

(i) $y'' + 4y = \sin 2x.$
(ii) $y'' + 4y' + 3y = e^{-3x}.$
(iii) $y'' + 5y' + 4y = e^{-4x}.$

6.13. Show that if the real parts of all solutions of (6.14) are negative, then $\lim_{x \to \infty} y(x) = 0$ for every solution of (6.13).

6.14. Show that the solution of the initial value problem

$$y'' - 2(r + \beta)y' + r^2 y = 0, \quad y(0) = 0, \quad y'(0) = 1$$

can be written as

$$y_\beta(x) = \frac{1}{2\sqrt{\beta(2r + \beta)}} \left[e^{[r+\beta+\sqrt{\beta(2r+\beta)}]x} - e^{[r+\beta-\sqrt{\beta(2r+\beta)}]x} \right].$$

Further, show that $\lim_{\beta \to 0} y_\beta(x) = xe^{rx}.$

6.15. Verify that $y_1(x) = x$ and $y_2(x) = 1/x$ are solutions of

$$x^3 y'' + x^2 y' - xy = 0.$$

Use this information and the variation of parameters method to find the general solution of

$$x^3 y'' + x^2 y' - xy = x/(1 + x).$$

Answers or Hints

6.1. Use the definition of Wronskian.

6.2. Verify directly.

6.3. (i) $\ln x/(x-1)$. (ii) $(1/2)(1-x) \ln[(x-2)/x]-1$. (iii) e^{-x^2}. (iv) $(x/2) \times \ln[(1 + x)/(1 - x)] - 1$.

6.4. (i) Verify directly. (ii) Use (6.5).

6.5. $e^{-\delta x^2/2} \int^x e^{\delta t^2/2} dt.$

6.6. Use $y_2(x) = y_1(x)y(x)$ and the fact that $y_1(x)$ and $y_2(x)$ are solutions.

6.7. (i) Use Abel's identity. (ii) If both attain maxima or minima at x_0, then $\phi_1'(x_0) = \phi_2'(x_0) = 0$. (iii) Use Abel's identity. (iv) If x_0 is a common point of inflexion, then $\phi_1''(x_0) = \phi_2''(x_0) = 0$. (v) $W(x^*) = 0$ implies $\phi_2(x) = c\phi_1(x)$. If $\phi_1'(x^*) = 0$, then $\phi_1(x) \equiv 0$, and if $\phi_1'(x^*) \neq 0$ then $c = \phi_2'(x^*)/\phi_1'(x^*)$.

6.8. Directly show right-hand side is the same as left-hand side.

6.9. Verify directly.

6.10. Use the substitution $w = z + w_1$ to obtain $z' + (2w_1 + p_1(x))z + z^2 = 0$, which is a Bernoulli equation whose multiplier is $z^{-2}\exp(-\int^x (2u_1 + p_1)dt)$. Hence, if w_1 is a solution of (6.21), then its integrating factor is $(w - w_1)^{-2}\exp(-\int^x (2u_1 + p_1)dt)$. Now use Theorem 3.4.

6.11. (i) $c_1 e^{-2x} + c_2 e^{-5x}$. (ii) $(c_1 + c_2 x)e^{4x}$. (iii) $c_1 e^{-x}\cos\sqrt{2}x + c_2 e^{-x}\times \sin\sqrt{2}x$.

6.12. (i) $c_1\cos 2x + c_2\sin 2x - \frac{1}{4}x\cos 2x$. (ii) $c_1 e^{-x} + c_2 e^{-3x} - \frac{1}{2}xe^{-3x}$ (iii) $c_1 e^{-x} + c_2 e^{-4x} - \frac{1}{3}xe^{-4x}$.

6.13. Use explicit forms of the solution.

6.14. Note that $\sqrt{\beta(\beta + 2r)} \to 0$ as $\beta \to 0$.

6.15. $c_1 x + (c_2/x) + (1/2)[(x - (1/x))\ln(1 + x) - x\ln x - 1]$.

Lecture 7

Preliminaries to Existence and Uniqueness of Solutions

So far, mostly we have engaged ourselves in solving DEs, tacitly assuming that there always exists a solution. However, the theory of existence and uniqueness of solutions of the initial value problems is quite complex. We begin to develop this theory for the initial value problem

$$y' = f(x, y), \quad y(x_0) = y_0, \tag{7.1}$$

where $f(x, y)$ will be assumed to be continuous in a domain D containing the point (x_0, y_0). By a solution of (7.1) in an interval J containing x_0, we mean a function $y(x)$ satisfying (i) $y(x_0) = y_0$, (ii) $y'(x)$ exists for all $x \in J$, (iii) for all $x \in J$ the points $(x, y(x)) \in D$, and (iv) $y'(x) = f(x, y(x))$ for all $x \in J$.

For the initial value problem (7.1) later we shall prove that the continuity of the function $f(x, y)$ alone is sufficient for the existence of at least one solution in a sufficiently small neighborhood of the point (x_0, y_0). However, if $f(x, y)$ is not continuous, then the nature of the solutions of (7.1) is quite arbitrary. For example, the initial value problem

$$y' = \frac{2}{x}(y - 1), \quad y(0) = 0$$

has no solution, while the problem

$$y' = \frac{2}{x}(y - 1), \quad y(0) = 1$$

has an infinite number of solutions $y(x) = 1 + cx^2$, where c is an arbitrary constant.

The use of integral equations to establish existence theorems is a standard device in the theory of DEs. It owes its efficiency to the smoothening properties of integration as contrasted with coarsening properties of differentiation. If two functions are close enough, their integrals must be close enough, whereas their derivatives may be far apart and may not even exist. We shall need the following result to prove the existence, uniqueness, and several other properties of the solutions of the initial value problem (7.1).

R.P. Agarwal and D. O'Regan, *An Introduction to Ordinary Differential Equations*, doi: 10.1007/978-0-387-71276-5_7, © Springer Science + Business Media, LLC 2008

Theorem 7.1. Let $f(x, y)$ be continuous in the domain D, then any solution of (7.1) is also a solution of the integral equation

$$y(x) = y_0 + \int_{x_0}^{x} f(t, y(t))dt \tag{7.2}$$

and conversely.

Proof. Any solution $y(x)$ of the DE $y' = f(x, y)$ converts it into an identity in x, i.e., $y'(x) = f(x, y(x))$. An integration of this equality yields

$$y(x) - y(x_0) = \int_{x_0}^{x} f(t, y(t))dt.$$

Conversely, if $y(x)$ is any solution of (7.2) then $y(x_0) = y_0$ and since $f(x, y)$ is continuous, differentiating (7.2) we find $y'(x) = f(x, y(x))$. ∎

While continuity of the function $f(x, y)$ is sufficient for the existence of a solution of (7.1), it does not imply uniqueness. For example, the function $f(x, y) = y^{2/3}$ is continuous in the entire xy-plane, but the problem $y' = y^{2/3}$, $y(0) = 0$ has at least two solutions $y(x) \equiv 0$ and $y(x) = x^3/27$. To ensure the uniqueness we shall begin with the assumption that the variation of the function $f(x, y)$ relative to y remains bounded, i.e.,

$$|f(x, y_1) - f(x, y_2)| \leq L|y_1 - y_2| \tag{7.3}$$

for all (x, y_1), (x, y_2) in the domain D. The function $f(x, y)$ is said to satisfy a uniform *Lipschitz condition* in any domain D if the inequality (7.3) holds for all point-pairs (x, y_1), (x, y_2) in D having the same x. The nonnegative constant L is called the *Lipschitz constant*.

The function $y^{2/3}$ violates the Lipschitz condition in any domain containing $y = 0$, whereas the function $f(x, y) = x - y$ satisfies the Lipschitz condition in $D = \mathbb{R}^2$ with $L = 1$. As an another example, the function $f(x, y) = e^y$ satisfies the Lipschitz condition in $D = \{(x, y) : x \in \mathbb{R}, |y| \leq c\}$ with $L = e^c$, where c is some positive constant.

Obviously, if inequality (7.3) is satisfied in D, then the function $f(x, y)$ is continuous with respect to y in D; however, it is not necessarily differentiable with respect to y, e.g., the function $f(x, y) = |y|$ is not differentiable in $\mathbb{R}^2$ but satisfies (7.3) with $L = 1$.

If the function $f(x, y)$ is differentiable with respect to y, then it is easy to compute the Lipschitz constant. In fact, we shall prove the following theorem.

Theorem 7.2. Let the domain D be convex and the function $f(x, y)$ be differentiable with respect to y in D. Then for the Lipschitz condition

(7.3) to be satisfied, it is necessary and sufficient that

$$\sup_D \left| \frac{\partial f(x,y)}{\partial y} \right| \leq L. \tag{7.4}$$

Proof. Since $f(x,y)$ is differentiable with respect to y and the domain D is convex, for all (x,y_1), $(x,y_2) \in D$ the mean value theorem provides

$$f(x,y_1) - f(x,y_2) = \frac{\partial f(x,y^*)}{\partial y}(y_1 - y_2),$$

where y^* lies between y_1 and y_2. Thus, in view of (7.4) the inequality (7.3) is immediate.

Conversely, inequality (7.3) implies that

$$\left| \frac{\partial f(x,y_1)}{\partial y_1} \right| = \lim_{y_2 \to y_1} \left| \frac{f(x,y_1) - f(x,y_2)}{y_1 - y_2} \right| \leq L. \quad \blacksquare$$

To prove the existence, uniqueness, and several other properties of the solutions of (7.1), we shall also need a *Gronwall's-type integral inequality*, which is contained in the following result.

Theorem 7.3. Let $u(x)$, $p(x)$ and $q(x)$ be nonnegative continuous functions in the interval $|x - x_0| \leq a$ and

$$u(x) \leq p(x) + \left| \int_{x_0}^{x} q(t)u(t)dt \right| \quad \text{for} \quad |x - x_0| \leq a. \tag{7.5}$$

Then the following inequality holds:

$$u(x) \leq p(x) + \left| \int_{x_0}^{x} p(t)q(t) \exp\left(\left| \int_{t}^{x} q(s)ds \right| \right) dt \right| \quad \text{for} \quad |x - x_0| \leq a. \tag{7.6}$$

Proof. We shall prove (7.6) for $x_0 \leq x \leq x_0 + a$ whereas for $x_0 - a \leq x \leq x_0$ the proof is similar. We define

$$r(x) = \int_{x_0}^{x} q(t)u(t)dt$$

so that $r(x_0) = 0$, and

$$r'(x) = q(x)u(x).$$

Since from (7.5), $u(x) \leq p(x) + r(x)$, it follows that

$$r'(x) \leq p(x)q(x) + q(x)r(x),$$

which on multiplying by $\exp\left(-\int_{x_0}^x q(s)ds\right)$ is the same as

$$\left(\exp\left(-\int_{x_0}^x q(s)ds\right)r(x)\right)' \leq p(x)q(x)\exp\left(-\int_{x_0}^x q(s)ds\right).$$

Integrating the above inequality, we obtain

$$r(x) \leq \int_{x_0}^x p(t)q(t)\exp\left(\int_t^x q(s)ds\right)dt$$

and now (7.6) follows from $u(x) \leq p(x) + r(x)$. ∎

Corollary 7.4. If in Theorem 7.3 the function $p(x) \equiv 0$, then $u(x) \equiv 0$.

Corollary 7.5. If in Theorem 7.3 the function $p(x)$ is nondecreasing in $[x_0, x_0 + a]$ and nonincreasing in $[x_0 - a, x_0]$, then

$$u(x) \leq p(x)\exp\left(\left|\int_{x_0}^x q(t)dt\right|\right) \quad \text{for} \quad |x - x_0| \leq a. \tag{7.7}$$

Proof. Once again we shall prove (7.7) for $x_0 \leq x \leq x_0 + a$ and for $x_0 - a \leq x \leq x_0$ the proof is similar. Since $p(x)$ is nondecreasing from (7.6) we find

$$\begin{aligned}
u(x) &\leq p(x)\left[1 + \int_{x_0}^x q(t)\exp\left(\int_t^x q(s)ds\right)dt\right] \\
&= p(x)\left[1 - \int_{x_0}^x \frac{d}{dt}\exp\left(\int_t^x q(s)ds\right)dt\right] \\
&= p(x)\exp\left(\int_{x_0}^x q(t)dt\right). \quad ∎
\end{aligned}$$

Corollary 7.6. If in Theorem 7.3 functions $p(x) = c_0 + c_1|x - x_0|$ and $q(x) = c_2$, where c_0, c_1 and c_2 are nonnegative constants, then

$$u(x) \leq \left(c_0 + \frac{c_1}{c_2}\right)\exp(c_2|x - x_0|) - \frac{c_1}{c_2}. \tag{7.8}$$

Proof. For the given functions $p(x)$ and $q(x)$, in the interval $[x_0, x_0 + a]$ inequality (7.6) is the same as

$$\begin{aligned}
u(x) &\leq c_0 + c_1(x - x_0) + \int_{x_0}^x [c_0 + c_1(t - x_0)]c_2 e^{c_2(x-t)}dt \\
&= c_0 + c_1(x - x_0) + \left\{-[c_0 + c_1(t - x_0)e^{c_2(x-t)}\Big|_{x_0}^x - \frac{c_1}{c_2}e^{c_2(x-t)}\Big|_{x_0}^x\right\} \\
&= c_0 + c_1(x - x_0) - c_0 - c_1(x - x_0) + c_0 e^{c_2(x-x_0)} - \frac{c_1}{c_2} + \frac{c_1}{c_2}e^{c_2(x-x_0)} \\
&= \left(c_0 + \frac{c_1}{c_2}\right)\exp(c_2(x - x_0)) - \frac{c_1}{c_2}. \quad ∎
\end{aligned}$$

Finally, in this lecture we recall several definitions and theorems from real analysis which will be needed in Lectures 8 and 9.

Definition 7.1. The sequence of functions $\{y_m(x)\}$ is said to converge uniformly to a function $y(x)$ in the interval $[\alpha, \beta]$ if for every real number $\epsilon > 0$ there exists an integer N such that whenever $m \geq N$, $|y_m(x) - y(x)| \leq \epsilon$ for all x in $[\alpha, \beta]$.

Theorem 7.7. Let $\{y_m(x)\}$ be a sequence of continuous functions in $[\alpha, \beta]$ that converges uniformly to $y(x)$. Then $y(x)$ is continuous in $[\alpha, \beta]$.

Theorem 7.8. Let $\{y_m(x)\}$ be a sequence converging uniformly to $y(x)$ in $[\alpha, \beta]$, and let $f(x, y)$ be a continuous function in the domain D such that for all m and x in $[\alpha, \beta]$ the points $(x, y_m(x))$ are in D. Then

$$\lim_{m \to \infty} \int_\alpha^\beta f(t, y_m(t))dt = \int_\alpha^\beta \lim_{m \to \infty} f(t, y_m(t))dt = \int_\alpha^\beta f(t, y(t))dt.$$

Theorem 7.9 (Weierstrass' M-Test). Let $\{y_m(x)\}$ be a sequence of functions with $|y_m(x)| \leq M_m$ for all x in $[\alpha, \beta]$ with $\sum_{m=0}^\infty M_m < \infty$. Then $\sum_{m=0}^\infty y_m(x)$ converges uniformly in $[\alpha, \beta]$ to a unique function $y(x)$.

Definition 7.2. A set S of functions is said to be equicontinuous in an interval $[\alpha, \beta]$ if for every given $\epsilon > 0$ there exists a $\delta > 0$ such that if x_1, $x_2 \in [\alpha, \beta]$, $|x_1 - x_2| \leq \delta$ then $|y(x_1) - y(x_2)| \leq \epsilon$ for all $y(x)$ in S.

Definition 7.3. A set S of functions is said to be uniformly bounded in an interval $[\alpha, \beta]$ if there exists a number M such that $|y(x)| \leq M$ for all $y(x)$ in S.

Theorem 7.10 (Ascoli–Arzela Theorem). An infinite set S of functions uniformly bounded and equicontinuous in $[\alpha, \beta]$ contains a sequence which converges uniformly in $[\alpha, \beta]$.

Theorem 7.11 (Implicit Function Theorem). Let $f(x, y)$ be defined in the strip $T = [\alpha, \beta] \times \mathbb{R}$, and continuous in x and differentiable in y, also $0 < m \leq f_y(x, y) \leq M < \infty$ for all $(x, y) \in T$. Then the equation $f(x, y) = 0$ has a unique continuous solution $y(x)$ in $[\alpha, \beta]$.

Problems

7.1. Show that the initial value problem

$$y'' = f(x, y), \quad y(x_0) = y_0, \quad y'(x_0) = y_1, \tag{7.9}$$

where $f(x, y)$ is continuous in a domain D containing the point (x_0, y_0), is equivalent to the integral equation

$$y(x) = y_0 + (x - x_0)y_1 + \int_{x_0}^{x} (x - t)f(t, y(t))dt.$$

7.2. Find the domains in which the following functions satisfy the Lipschitz condition (7.3), also find the Lipschitz constants:

(i) $\dfrac{y}{(1 + x^2)}$.   (ii) $\dfrac{x}{(1 + y^2)}$.   (iii) $x^2 \cos^2 y + y \sin^2 x$.

(iv) $|xy|$.   (v) $y + [x]$.   (vi) $x^2 y^2 + xy + 1$.

7.3. By computing appropriate Lipschitz constants, show that the following functions satisfy the Lipschitz condition in the given domains:

(i) $x \sin y + y \cos x$, $|x| \leq a$, $|y| \leq b$.

(ii) $x^3 e^{-xy^2}$, $0 \leq x \leq a$, $|y| < \infty$.

(iii) $x^2 e^{x+y}$, $|x| \leq a$, $|y| \leq b$.

(iv) $p(x)y + q(x)$, $|x| \leq 1$, $|y| < \infty$ where $p(x)$ and $q(x)$ are continuous functions in the interval $|x| \leq 1$.

7.4. Show that the following functions do not satisfy the Lipschitz condition (7.3) in the given domains:

(i) $f(x, y) = \begin{cases} \dfrac{x^3 y}{x^4 + y^2}, & (x, y) \neq (0,0) \\ 0, & (x, y) = (0,0) \end{cases}$, $|x| \leq 1$, $|y| \leq 2$.

(ii) $f(x, y) = \begin{cases} \dfrac{\sin y}{x}, & x \neq 0 \\ 0, & x = 0 \end{cases}$, $|x| \leq 1$, $|y| < \infty$.

7.5. Let $u(x)$ be a nonnegative continuous function in the interval $|x - x_0| \leq a$, and $C \geq 0$ be a given constant, and

$$u(x) \leq \left| \int_{x_0}^{x} Cu^\alpha(t)dt \right|, \quad 0 < \alpha < 1.$$

Show that for all x in $|x - x_0| \leq a$,

$$u(x) \leq [C(1 - \alpha)|x - x_0|]^{(1-\alpha)^{-1}}.$$

7.6. Let c_0 and c_1 be nonnegative constants, and $u(x)$ and $q(x)$ be nonnegative continuous functions for all $x \geq 0$ satisfying

$$u(x) \leq c_0 + c_1 \int_{0}^{x} q(t)u^2(t)dt.$$

Show that for all $x \geq 0$ for which $c_0 c_1 \int_0^x q(t)dt < 1$,

$$u(x) \leq c_0 \left[1 - c_0 c_1 \int_0^x q(t)dt \right]^{-1}.$$

7.7. Suppose that $y = y(x)$ is a solution of the initial value problem $y' = yg(x, y)$, $y(0) = 1$ on the interval $[0, \beta]$, where $g(x, y)$ is a bounded and continuous function in the (x, y) plane. Show that there exists a constant C such that $|y(x)| \leq e^{Cx}$ for all $x \in [0, \beta]$.

∗7.8. Suppose $\alpha > 0$, $\gamma > 0$, c_0, c_1, c_2 are nonnegative constants and $u(x)$ is a nonnegative bounded continuous solution of either the inequality

$$u(x) \leq c_0 e^{-\alpha x} + c_1 \int_0^x e^{-\alpha(x-t)} u(t)dt + c_2 \int_0^\infty e^{-\gamma t} u(x+t)dt, \quad x \geq 0,$$

or the inequality

$$u(x) \leq c_0 e^{\alpha x} + c_1 \int_x^0 e^{\alpha(x-t)} u(t)dt + c_2 \int_{-\infty}^0 e^{\gamma t} u(x+t)dt, \quad x \leq 0.$$

If

$$\beta = \frac{c_1}{\alpha} + \frac{c_2}{\gamma} < 1,$$

then in either case, show that

$$u(x) \leq (1-\beta)^{-1} c_0 e^{-[\alpha-(1-\beta)^{-1}c_1]|x|}.$$

∗7.9. Suppose a, b, c are nonnegative continuous functions on $[0, \infty)$ and $u(x)$ is a nonnegative bounded continuous solution of the inequality

$$u(x) \leq a(x) + \int_0^x b(x-t)u(t)dt + \int_0^\infty c(t)u(x+t)dt, \quad x \geq 0,$$

where $a(x) \to 0$, $b(x) \to 0$ as $x \to \infty$. If

$$\int_0^\infty [b(t) + c(t)]dt < 1,$$

then show that $u(x) \to 0$ as $x \to \infty$.

7.10. Show that the sequence $\{nx/(nx+1)\}$, $0 \leq x \leq 1$ converges pointwise to the function $f(x) = \begin{cases} 0, & x = 0 \\ 1, & 0 < x \leq 1. \end{cases}$

7.11. Show that the sequence $\{nx^2/(nx+1)\}$, $0 \leq x \leq 1$ converges uniformly to the function $f(x) = x$. Further, verify that

$$\lim_{n\to\infty} \int_0^1 \frac{nx^2}{nx+1} dx = \int_0^1 \lim_{n\to\infty} \frac{nx^2}{nx+1} dx = \frac{1}{2}.$$

*7.12. Show the following:

(i)   In the Ascoli–Arzela theorem (Theorem 7.10), the interval $[\alpha, \beta]$ can be replaced by any finite open interval (α, β).

(ii)   The Ascoli–Arzela theorem remains true if instead of uniform boundedness on the whole interval (α, β), we have $|f(x_0)| < M$ for every $f \in S$ and some $x_0 \in (\alpha, \beta)$.

Answers or Hints

7.1. $y(x) = y_0 + (x - x_0)y_1 + \int_{x_0}^{x} \left[\int_{x_0}^{t} f(s, y(s)) ds \right] dt$

$\qquad = y_0 + (x - x_0)y_1 + \left[t \int_{x_0}^{t} f(s, y(s)) ds \Big|_{x_0}^{x} - \int_{x_0}^{x} tf(t, y(t)) dt \right]$

$\qquad = y_0 + (x - x_0)y_1 + \int_{x_0}^{x} (x - t) f(t, y(t)) dt.$

7.2. (i) $\mathbb{R}^2$, 1. (ii) $|x| \le a$, $|y| < \infty$, $(3\sqrt{3}/8)a$. (iii) $|x| \le a$, $|y| < \infty$, $a^2 + 1$. (iv) $|x| \le a$, $|y| < \infty$, a. (v) $\mathbb{R}^2$, 1. (vi) $|x| \le a$, $|y| \le b$, $2a^2b + a$.

7.3. (i) $a + 1$. (ii) $\max\{2a^3, 2a^4\}$. (iii) $a^2 e^{a+b}$. (iv) $\max_{-1 \le x \le 1} |p(x)|$.

7.4. (i) $|f(x, y) - f(x, 0)| = |x^3 y/(x^4 + y^2)| \le L|y|$, i.e., $|x^3/(x^4 + y^2)| \le L$; however, along the curve $y = x^2$ this is impossible. (ii) $|f(x, y) - f(x, 0)| = |x^{-1} \sin y| \le L|y|$; but, this is impossible.

7.5. For $x \in [x_0, x_0 + a]$ let $r(x) = \int_{x_0}^{x} Cu^\alpha(t) dt$ so that $r'(x) < C(r(x) + \epsilon)^\alpha$, where $\epsilon > 0$ and $r(x_0) = 0$. Integrate this inequality and then let $\epsilon \to 0$.

7.6. Let $r(x) = c_0 + c_1 \int_{0}^{x} q(t)u^2(t) dt$ so that $r'(x) < c_1 q(x)(r(x) + \epsilon)^2$, where $\epsilon > 0$ and $r(0) = c_0$. Integrate this inequality and then let $\epsilon \to 0$.

7.7. Use Corollary 7.6.

7.10. Verify directly.

7.11. Verify directly.

Lecture 8

Picard's Method
of Successive Approximations

We shall solve the integral equation (7.2) by using the method of successive approximations due to Picard. For this, let $y_0(x)$ be any continuous function (we often pick $y_0(x) \equiv y_0$) which we assume to be the initial approximation of the unknown solution of (7.2), then we define $y_1(x)$ as

$$y_1(x) = y_0 + \int_{x_0}^{x} f(t, y_0(t))dt.$$

We take this $y_1(x)$ as our next approximation and substitute this for $y(x)$ on the right side of (7.2) and call it $y_2(x)$. Continuing in this way, the $(m+1)$st approximation $y_{m+1}(x)$ is obtained from $y_m(x)$ by means of the relation

$$y_{m+1}(x) = y_0 + \int_{x_0}^{x} f(t, y_m(t))dt, \quad m = 0, 1, 2, \ldots. \tag{8.1}$$

If the sequence $\{y_m(x)\}$ converges uniformly to a continuous function $y(x)$ in some interval J containing x_0 and for all $x \in J$ the points $(x, y_m(x)) \in D$, then using Theorem 7.8 we may pass to the limit in both sides of (8.1), to obtain

$$y(x) = \lim_{m \to \infty} y_{m+1}(x) = y_0 + \lim_{m \to \infty} \int_{x_0}^{x} f(t, y_m(t))dt = y_0 + \int_{x_0}^{x} f(t, y(t))dt,$$

so that $y(x)$ is the desired solution.

Example 8.1. The initial value problem $y' = -y$, $y(0) = 1$ is equivalent to solving the integral equation

$$y(x) = 1 - \int_0^x y(t)dt.$$

R.P. Agarwal and D. O'Regan, *An Introduction to Ordinary Differential Equations*,
doi: 10.1007/978-0-387-71276-5_8, © Springer Science + Business Media, LLC 2008

Let $y_0(x) = 1$, to obtain

$$y_1(x) \;=\; 1 - \int_0^x dt \;=\; 1 - x$$

$$y_2(x) \;=\; 1 - \int_0^x (1-t)dt \;=\; 1 - x + \frac{x^2}{2!}$$

$$\cdots$$

$$y_m(x) \;=\; \sum_{i=0}^m (-1)^i \frac{x^i}{i!}.$$

Recalling Taylor's series expansion of e^{-x}, we see that $\lim_{m\to\infty} y_m(x) = e^{-x}$. The function $y(x) = e^{-x}$ is indeed the solution of the given initial value problem in $J = \mathbb{R}$.

An important characteristic of this method is that it is constructive, moreover bounds on the difference between iterates and the solution are easily available. Such bounds are useful for the approximation of solutions and also in the study of qualitative properties of solutions. The following result provides sufficient conditions for the uniform convergence of the sequence $\{y_m(x)\}$ to the unique solution $y(x)$ of the integral equation (7.2), or equivalently of the initial value problem (7.1).

Theorem 8.1. Let the following conditions be satisfied

(i) $f(x, y)$ is continuous in the closed rectangle $\overline{S} : |x - x_0| \le a,\ |y - y_0| \le b$ and hence there exists a $M > 0$ such that $|f(x, y)| \le M$ for all $(x, y) \in \overline{S}$,

(ii) $f(x, y)$ satisfies a uniform Lipschitz condition (7.3) in $\overline{S}$,

(iii) $y_0(x)$ is continuous in $|x - x_0| \le a$, and $|y_0(x) - y_0| \le b$.

Then the sequence $\{y_m(x)\}$ generated by the Picard iterative scheme (8.1) converges to the unique solution $y(x)$ of the initial value problem (7.1). This solution is valid in the interval $J_h : |x - x_0| \le h = \min\{a, b/M\}$. Further, for all $x \in J_h$ the following error estimate holds:

$$|y(x) - y_m(x)| \;\le\; N e^{Lh} \min\left\{1, \frac{(Lh)^m}{m!}\right\}, \quad m = 0, 1, \ldots \quad (8.2)$$

where $\max_{x \in J_h} |y_1(x) - y_0(x)| \le N$.

Proof.   First we shall show that the successive approximations $y_m(x)$ defined by (8.1) exist as continuous functions in J_h and $(x, y_m(x)) \in \overline{S}$ for all $x \in J_h$. Since $y_0(x)$ is continuous for all $x : |x - x_0| \le a$, the function $F_0(x) = f(x, y_0(x))$ is continuous in J_h, and hence $y_1(x)$ is continuous in J_h. Also,

$$|y_1(x) - y_0| \;\le\; \left| \int_{x_0}^x |f(t, y_0(t))| dt \right| \;\le\; M|x - x_0| \;\le\; Mh \;\le\; b.$$

Assuming that the assertion is true for $y_{m-1}(x)$ ($m \geq 2$), then it is sufficient to prove that it is also true for $y_m(x)$. For this, since $y_{m-1}(x)$ is continuous in J_h, the function $F_{m-1}(x) = f(x, y_{m-1}(x))$ is also continuous in J_h. Moreover,

$$|y_m(x) - y_0| \leq \left| \int_{x_0}^{x} |f(t, y_{m-1}(t))| dt \right| \leq M|x - x_0| \leq b.$$

Next we shall show that the sequence $\{y_m(x)\}$ converges uniformly in J_h. Since $y_1(x)$ and $y_0(x)$ are continuous in J_h, there exists a constant $N > 0$ such that $|y_1(x) - y_0(x)| \leq N$. We need to show that for all $x \in J_h$ the following inequality holds:

$$|y_m(x) - y_{m-1}(x)| \leq N \frac{(L|x - x_0|)^{m-1}}{(m-1)!}, \quad m = 1, 2, \ldots . \qquad (8.3)$$

For $m = 1$, the inequality (8.3) is obvious, further if it is true for $m = k \geq 1$, then (8.1) and hypothesis (ii) give

$$
\begin{aligned}
|y_{k+1}(x) - y_k(x)| &\leq \left| \int_{x_0}^{x} |f(t, y_k(t)) - f(t, y_{k-1}(t))| dt \right| \\
&\leq L \left| \int_{x_0}^{x} |y_k(t) - y_{k-1}(t)| dt \right| \\
&\leq L \left| \int_{x_0}^{x} N \frac{(L|t - x_0|)^{k-1}}{(k-1)!} dt \right| = N \frac{(L|x - x_0|)^k}{k!}.
\end{aligned}
$$

Thus, the inequality (8.3) is true for all m.

Next since

$$N \sum_{m-1}^{\infty} \frac{(L|x - x_0|)^{m-1}}{(m-1)!} \leq N \sum_{m=0}^{\infty} \frac{(Lh)^m}{m!} = N e^{Lh} < \infty,$$

from Theorem 7.9 it follows that the series

$$y_0(x) + \sum_{m=1}^{\infty} (y_m(x) - y_{m-1}(x))$$

converges absolutely and uniformly in the interval J_h, and hence its partial sums $y_1(x), y_2(x), \ldots$ converge to a continuous function in this interval, i.e., $y(x) = \lim_{m \to \infty} y_m(x)$. As we have seen earlier this $y(x)$ is a solution of (7.2).

To show that this $y(x)$ is the only solution, we assume that $z(x)$ is also a solution of (7.2) which exists in the interval J_h and $(x, z(x)) \in \bar{S}$ for all $x \in J_h$. Then hypothesis (ii) is applicable and we have

$$|y(x) - z(x)| \leq \left| \int_{x_0}^{x} |f(t, y(t)) - f(t, z(t))| dt \right| \leq L \left| \int_{x_0}^{x} |y(t) - z(t)| dt \right|.$$

However, for the above integral inequality Corollary 7.4 implies that $|y(x) - z(x)| = 0$ for all $x \in J_h$, and hence $y(x) = z(x)$ for all $x \in J_h$.

Finally, we shall obtain the error bound (8.2). For $n > m$ the inequality (8.3) gives

$$
\begin{aligned}
|y_n(x) - y_m(x)| &\leq \sum_{k=m}^{n-1} |y_{k+1}(x) - y_k(x)| \leq \sum_{k=m}^{n-1} N \frac{(L|x - x_0|)^k}{k!} \\
&\leq N \sum_{k=m}^{n-1} \frac{(Lh)^k}{k!} = N(Lh)^m \sum_{k=0}^{n-m-1} \frac{(Lh)^k}{(m+k)!}.
\end{aligned}
$$

(8.4)

However, since $1/(m+k)! \leq 1/(m! \, k!)$ it follows that

$$|y_n(x) - y_m(x)| \leq N \frac{(Lh)^m}{m!} \sum_{k=0}^{n-m-1} \frac{(Lh)^k}{k!} \leq N \frac{(Lh)^m}{m!} e^{Lh}$$

and hence as $n \to \infty$, we get

$$|y(x) - y_m(x)| \leq N \frac{(Lh)^m}{m!} e^{Lh}. \tag{8.5}$$

Inequality (8.4) also provides

$$|y_n(x) - y_m(x)| \leq N \sum_{k=m}^{n-1} \frac{(Lh)^k}{k!} \leq N e^{Lh}$$

and as $n \to \infty$, we find

$$|y(x) - y_m(x)| \leq N e^{Lh}. \tag{8.6}$$

Combining (8.5) and (8.6) we obtain the required error bound (8.2). ∎

Theorem 8.1 is called a *local existence theorem* since it guarantees a solution only in the neighborhood of the point (x_0, y_0).

Example 8.2. Consider the initial value problem

$$y' = 1 + y^2, \quad y(0) = 0 \tag{8.7}$$

for which the unique solution $y(x) = \tan x$ exists in the interval $(-\pi/2, \pi/2)$. To apply Theorem 8.1 we note that (i) the function $1 + y^2$ is continuous in the rectangle $\overline{S}: |x| \leq a, |y| \leq b$, and $1 + y^2 \leq 1 + b^2 = M$; (ii) in the rectangle $\overline{S}$ the function $1 + y^2$ satisfies (7.3) with $L = 2b$; and (iii) $y_0(x) \equiv$

0 is continuous in $|x| \leq a$ and $|y_0(x)| \leq b$. Thus, there exists a unique solution of (8.7) in the interval $|x| \leq h = \min\{a, b/(1+b^2)\}$. However, since $b/(1+b^2) \leq 1/2$ (with equality for $b=1$) the optimum interval which Theorem 8.1 can give is $|x| \leq 1/2$. Further, the iterative scheme (8.1) for the problem (8.7) takes the form

$$y_{m+1}(x) = x + \int_0^x y_m^2(t)dt, \quad y_0(x) \equiv 0, \quad m = 0, 1, \ldots . \quad (8.8)$$

From (8.8) it is easy to obtain $y_1(x) = x$, $y_2(x) = x + x^3/3$. Thus, the error bound (8.2) with $b = 1$, $h = 1/2$ and $m = 2$ gives

$$\left| \tan x - x - \frac{x^3}{3} \right| \leq \frac{1}{2} e \min\left\{ 1, \frac{1}{2} \right\} = \frac{1}{4}e, \quad -\frac{1}{2} \leq x \leq \frac{1}{2}. \quad (8.9)$$

Obviously, in (8.9) the right side is too crude.

If the solution of the initial value problem (7.1) exists in the entire interval $|x - x_0| \leq a$, we say that the solution exists globally. The following result is called a *global existence theorem*.

Theorem 8.2. Let the following conditions be satisfied:

(i) $f(x, y)$ is continuous in the strip $T : |x - x_0| \leq a$, $|y| < \infty$,

(ii) $f(x, y)$ satisfies a uniform Lipschitz condition (7.3) in T,

(iii) $y_0(x)$ is continuous in $|x - x_0| \leq a$.

Then the sequence $\{y_m(x)\}$ generated by the Picard iterative scheme (8.1) exists in the entire interval $|x - x_0| \leq a$, and converges to the unique solution $y(x)$ of the initial value problem (7.1).

Proof. For any continuous function $y_0(x)$ in $|x - x_0| \leq a$ an easy inductive argument establishes the existence of each $y_m(x)$ in $|x - x_0| \leq a$ satisfying $|y_m(x)| < \infty$. Also, as in the proof of Theorem 8.1 it is easy to verify that the sequence $\{y_m(x)\}$ converges to $y(x)$ in $|x - x_0| \leq a$ (replacing h by a throughout the proof and recalling that the function $f(x, y)$ satisfies the Lipschitz condition in the strip T). ∎

Corollary 8.3. Let $f(x, y)$ be continuous in $\mathbb{R}^2$ and satisfy a uniform Lipschitz condition (7.3) in each strip $T_a : |x| \leq a$, $|y| < \infty$ with the Lipschitz constant L_a. Then the initial value problem (7.1) has a unique solution which exists for all x.

Proof. For any x there exists an $a > 0$ such that $|x - x_0| \leq a$. Since, the strip T is contained in the strip $T_{a+|x_0|}$ the function $f(x, y)$ satisfies the conditions of Theorem 8.2 in the strip T. Hence, the result follows for any x. ∎

Problems

8.1. Compute the first few Picard's iterates with $y_0(x) \equiv 0$ for the initial value problem $y' = xy + 2x - x^3$, $y(0) = 0$ and show that they converge to the solution $y(x) = x^2$ for all x.

8.2. For the following initial value problems compute the first three iterates with the initial approximation $y_0(x) \equiv x$:

(i) $y' = x^2 - y^2 - 1$, $y(0) = 0$.

(ii) $y' = (x + 2y)/(2x + y)$, $y(1) = 1$.

(iii) $y' = x^2 + y^2$, $y(0) = 0$.

8.3. Discuss the existence and uniqueness of the solutions of the following initial value problems:

(i) $y' = 1 + y^{2/3}$, $y(0) = 0$.

(ii) $y' = \sin(xy)$, $y(0) = 1$.

(iii) $y' = (x + y)x^2 y^2$, $y(0) = 1$.

(iv) $y' = e^x + x/y$, $y(0) = 1$.

8.4. Show that the following initial value problems possess a unique solution for all real x:

(i) $y' + p(x)y = q(x)$, $y(x_0) = y_0$, where $p(x)$ and $q(x)$ are continuous in $\mathbb{R}$.

(ii) $y' = p(x)f(\cos y) + q(x)g(\sin y)$, $y(x_0) = y_0$, where $p(x)$ and $q(x)$ are continuous in $\mathbb{R}$, and f and g are polynomials of degree m and n, respectively.

(iii) $y' = y^3 e^x (1 + y^2)^{-1} + x^2 \cos y$, $y(x_0) = y_0$.

(iv) $y' = (\cos x)e^{-y^2} + \sin y$, $y(x_0) = y_0$.

8.5. Show that the initial value problem

$$y' = (x^2 - y^2)\sin y + y^2 \cos y, \quad y(0) = 0$$

has a unique solution $y(x) \equiv 0$ in the closed rectangle $\overline{S}: |x| \le a$, $|y| \le b$.

8.6. Show that Theorem 8.1 guarantees the existence of a unique solution of the initial value problem $y' = e^{2y}$, $y(0) = 0$ in the interval $(-1/2e, 1/2e)$. Also, solve this problem and verify that the solution exists in a larger interval.

8.7. The function $f(x, y) = (\tan x)y + 1$ is continuous in the open strip $|x| < \pi/2$, $|y| < \infty$. Solve the initial value problem $y' = (\tan x)y + 1$, $y(0) = 1$ and verify that the solution exists in a larger interval $(-3\pi/2, \pi/2)$.

8.8. Let $f_1(x)$ and $f_2(y) \neq 0$ be continuous in $|x - x_0| \leq a$, $|y - y_0| \leq b$, respectively. Show that the initial value problem $y' = f_1(x)f_2(y)$, $y(x_0) = y_0$ has a unique solution in the interval $|x - x_0| \leq h$ where $h \leq a$.

8.9. Consider the DE (3.1), where the functions M, N are continuous and having continuous partial derivatives M_y, N_x in the rectangle S : $|x - x_0| < a$, $|y - y_0| < b$ $(0 < a, b < \infty)$. Suppose that $N(x, y) \neq 0$ for all $(x, y) \in S$ and the condition (3.4) holds. Show that the initial value problem (3.1), (1.10) has a unique solution in the interval $|x - x_0| \leq h$ where $h \leq a$.

8.10. If $f(x, y)$ has continuous partial derivatives of all orders in a domain D, then show that the mth Picard's approximation $y_m(x)$ of the initial value problem (7.1) has the same value and the same derivatives up to order m at x_0 as the true solution.

8.11. Let $f(x, y)$ be continuously $p > 0$ times differentiable with respect to x and y. Show that every solution of the DE (1.9) is continuously $p + 1$ times differentiable with respect to x.

8.12. Consider the initial value problem

$$y' = f(x, y) = \begin{cases} y(1 - 2x), & x > 0 \\ y(2x - 1), & x < 0 \end{cases} \tag{8.10}$$

$$y(1) = 1.$$

Clearly, the function $f(x, y)$ is discontinuous at all $(0, y)$, $y \neq 0$. Show that

$$y(x) = \begin{cases} e^{x - x^2}, & x \geq 0 \\ e^{x^2 - x}, & x \leq 0 \end{cases}$$

is the unique continuous (but not differentiable at $x = 0$) solution of (8.10) which is valid for all x.

8.13. Let the conditions of Theorem 8.1 be satisfied. Show that the successive approximations

$$y_{m+1}(x) = y_0 + (x - x_0)y_1 + \int_{x_0}^{x} (x - t)f(t, y_m(t))dt, \quad m = 0, 1, \ldots \tag{8.11}$$

with the initial approximation $y_0(x) = y_0$ converge to the unique solution of the initial value problem (7.9) in the interval J_h : $|x - x_0| \leq h = \min\{a, b/M_1\}$, where $M_1 = |y_1| + Ma/2$.

Answers or Hints

8.1. $y_m(x) = x^2 - x^{2(m+1)}/[4.6.8\cdots 2(m+1)]$.

8.2. (i) $-x, -x, -x$. (ii) x, x, x. (iii) $2x^3/3$, $(x^3/3) + (4x^7/63)$, $(x^3/3) + (x^7/63) + (8x^{11}/2079) + (16x^{15}/59535)$.

8.3. (i) Unique solution $3(y^{1/3} - \tan^{-1} y^{1/3}) = x$. (ii) Global unique solution. (iii) Local unique solution. (iv) Local unique solution.

8.4. Apply Corollary 8.3.

8.5. The function $(x^2 - y^2)\sin y + y^2 \cos y$ satisfies the Lipschitz condition (7.3) for all $(x, y) \in \overline{S}$.

8.6. The solution $y(x) = \ln 1/\sqrt{(1 - 2x)}$ exists for all $x \in [-1/2, 1/2]$.

8.7. The solution $y(x) = \tan[(x/2) + (\pi/4)]$ exists in the interval $(-3\pi/2, \pi/2)$.

8.8. The function $G(y) = \int_{y_0}^{y} dt/f_2(t) = \int_{x_0}^{x} f_1(s)ds$ exists and it is continuous and monotonic as long as $|y - y_0| \leq b$.

8.9. From (3.3), $u(x, y) = u(x_0, y_0)$. Now use implicit function theorem.

8.10. $y_0(x_0) = y_0 = y(x_0)$, $y_1'(x) = f(x, y_0(x))$, $y_1'(x_0) = f(x_0, y_0(x_0)) = f(x_0, y_0) = y'(x_0)$, $y_2''(x_0) = \frac{\partial f}{\partial x}(x_0, y_1(x_0)) + \frac{\partial f}{\partial y}(x_0, y_1(x_0))y_1'(x_0) = y''(x_0)$.

8.11. Use $y'(x) = f(x, y(x))$.

8.12. Verify directly.

8.13. The proof is similar to that of Theorem 8.1.

Lecture 9
Existence Theorems

As promised in Lecture 7, here we shall prove that the continuity of the function $f(x, y)$ alone is sufficient for the existence of a solution of the initial value problem (7.1).

Theorem 9.1 (Peano's Existence Theorem). Let $f(x, y)$ be continuous and bounded in the strip $T: |x - x_0| \leq a$, $|y| < \infty$. Then the initial value problem (7.1) has at least one solution in $|x - x_0| \leq a$.

Proof. We shall give the existence proof in the interval $[x_0, x + a]$, and its extension to $[x_0 - a, x_0]$ is immediate. We define a sequence of functions $\{y_m(x)\}$ by the scheme

$$y_m(x) = y_0, \quad x_0 \leq x \leq x_0 + \frac{a}{m}$$

$$y_m(x) = y_0 + \int_{x_0}^{x-(a/m)} f(t, y_m(t))dt, \quad x_0 + k\frac{a}{m} \leq x \leq x_0 + (k+1)\frac{a}{m},$$
$$k = 1, 2, \ldots, m-1.$$
$$(9.1)$$

The first equation defines $y_m(x)$ in $[x_0, x_0 + a/m]$; then the second equation defines $y_m(x)$ at first in $[x_0 + a/m, x_0 + 2a/m]$ and then in $[x_0 + 2a/m, x_0 + 3a/m]$ and so on. Since $f(x, y)$ is bounded in T, we can assume that $|f(x, y)| \leq M$ for all $(x, y) \in T$. Now for any two points x_1, x_2 in $[x_0, x_0 + a]$, we have

$$|y_m(x_2) - y_m(x_1)|$$
$$= 0 \quad \text{if } x_1, x_2 \in \left[x_0, x_0 + \frac{a}{m}\right]$$
$$= \left|\int_{x_0}^{x_2-(a/m)} f(t, y_m(t))dt\right| \leq M\left|x_2 - \frac{a}{m} - x_0\right| \leq M|x_2 - x_1|$$
$$\quad \text{if } x_1 \in \left[x_0, x_0 + \frac{a}{m}\right], \quad x_2 \in \left[x_0 + k\frac{a}{m}, x_0 + (k+1)\frac{a}{m}\right]$$
$$= \left|\int_{x_1-(a/m)}^{x_2-(a/m)} f(t, y_m(t))dt\right| \leq M|x_2 - x_1| \quad \text{otherwise.}$$

Thus, it follows that

$$|y_m(x_2) - y_m(x_1)| \leq M|x_2 - x_1|, \quad x_1, x_2 \in [x_0, x_0 + a].$$

R.P. Agarwal and D. O'Regan, *An Introduction to Ordinary Differential Equations*, doi: 10.1007/978-0-387-71276-5_9, © Springer Science + Business Media, LLC 2008

Hence, $|y_m(x_2) - y_m(x_1)| \leq \epsilon$ provided $|x_2 - x_1| \leq \epsilon/M = \delta$; i.e., the sequence $\{y_m(x)\}$ is equicontinuous. Moreover, for all $x \in [x_0, x_0 + a]$, we have

$$|y_m(x)| \leq |y_0| + M \left| x - \frac{a}{m} - x_0 \right| \leq |y_0| + Ma,$$

i.e., the sequence $\{y_m(x)\}$ is uniformly bounded in $[x_0, x_0 + a]$. Therefore, from Theorem 7.10 the sequence $\{y_m(x)\}$ contains a subsequence $\{y_{m_p}(x)\}$ which converges uniformly in $[x_0, x_0 + a]$ to a continuous function $y(x)$. To show that the function $y(x)$ is a solution of the initial value problem (7.1), we let $p \to \infty$ in the relation

$$y_{m_p}(x) = y_0 + \int_{x_0}^{x} f(t, y_{m_p}(t))dt - \int_{x-(a/m_p)}^{x} f(t, y_{m_p}(t))dt.$$

Since $f(x, y)$ is continuous and the convergence is uniform, in the first integral we can take the limit inside the integral sign to obtain $\int_{x_0}^{x} f(t, y(t))dt$. The second integral does not exceed $M(a/m_p)$ and hence tends to zero. Thus, $y(x)$ is a solution of the integral equation (7.2). ∎

Corollary 9.2. Let $f(x, y)$ be continuous in $\overline{S}$, and hence there exists a $M > 0$ such that $|f(x, y)| \leq M$ for all $(x, y) \in \overline{S}$. Then the initial value problem (7.1) has at least one solution in J_h.

Proof. The proof is the same as that of Theorem 9.1 with some obvious changes. ∎

Example 9.1. The function $f(x, y) = y^{2/3}$ is continuous for all (x, y) in $\mathbb{R}^2$. Thus, from Corollary 9.2 the initial value problem $y' = y^{2/3}$, $y(0) = 0$ has at least one solution in the interval $|x| \leq h = \min\{a, b^{1/3}\}$. However, we can choose b sufficiently large so that $h = a$. Hence, the given problem in fact has at least one solution for all x in $\mathbb{R}$.

Next for a given continuous function $f(x, y)$ in a domain D, we need the following definition.

Definition 9.1. A function $y(x)$ defined in J is said to be an ϵ-*approximate solution* of the DE $y' = f(x, y)$ if (i) $y(x)$ is continuous for all x in J, (ii) for all $x \in J$ the points $(x, y(x)) \in D$, (iii) $y(x)$ has a piecewise continuous derivative in J which may fail to be defined only for a finite number of points, say, $x_1, x_2, \ldots, x_k$, and (iv) $|y'(x) - f(x, y(x))| \leq \epsilon$ for all $x \in J$, $x \neq x_i$, $i = 1, 2, \ldots, k$.

The existence of an ϵ-approximate solution is proved in the following theorem.

Theorem 9.3. Let $f(x, y)$ be continuous in $\overline{S}$ and hence there exists a $M > 0$ such that $|f(x, y)| \leq M$ for all $(x, y) \in \overline{S}$. Then for any $\epsilon > 0$, there exists an ϵ-approximate solution $y(x)$ of the DE $y' = f(x, y)$ in the interval J_h such that $y(x_0) = y_0$.

Proof. Since $f(x, y)$ is continuous in the closed rectangle $\overline{S}$, it is uniformly continuous in this rectangle. Thus, for a given $\epsilon > 0$ there exists a $\delta > 0$ such that

$$|f(x, y) - f(x_1, y_1)| \leq \epsilon \qquad (9.2)$$

for all (x, y), (x_1, y_1) in $\overline{S}$ whenever $|x - x_1| \leq \delta$ and $|y - y_1| \leq \delta$.

We shall construct an ϵ-approximate solution in the interval $x_0 \leq x \leq x_0 + h$ and a similar process will define it in the interval $x_0 - h \leq x \leq x_0$. For this, we divide the interval $x_0 \leq x \leq x_0 + h$ into m parts $x_0 < x_1 < \cdots < x_m = x_0 + h$ such that

$$x_i - x_{i-1} \leq \min\left\{\delta, \frac{\delta}{M}\right\}, \quad i = 1, 2, \ldots, m. \qquad (9.3)$$

Next we define a function $y(x)$ in the interval $x_0 \leq x \leq x_0 + h$ by the recursive formula

$$y(x) = y(x_{i-1}) + (x - x_{i-1})f(x_{i-1}, y(x_{i-1})), \quad x_{i-1} \leq x \leq x_i, \ i = 1, 2, \ldots, m. \qquad (9.4)$$

Obviously, this function $y(x)$ is continuous and has a piecewise continuous derivative $y'(x) = f(x_{i-1}, y(x_{i-1}))$, $x_{i-1} < x < x_i$, $i = 1, 2, \ldots, m$ which fails to be defined only at the points x_i, $i = 1, 2, \ldots, m-1$. Since in each subinterval $[x_{i-1}, x_i]$, $i = 1, 2, \ldots, m$ the function $y(x)$ is a straight line, to prove $(x, y(x)) \in \overline{S}$ it suffices to show that $|y(x_i) - y_0| \leq b$ for all $i = 1, 2, \ldots, m$. For this, in (9.4) let $i = 1$ and $x = x_1$ to obtain

$$|y(x_1) - y_0| = (x_1 - x_0)|f(x_0, y_0)| \leq Mh \leq b.$$

Now let the assertion be true for $i = 1, 2, \ldots, k-1 < m-1$, then from (9.4), we find

$$\begin{aligned}
y(x_1) - y_0 &= (x_1 - x_0)f(x_0, y_0) \\
y(x_2) - y(x_1) &= (x_2 - x_1)f(x_1, y(x_1)) \\
&\cdots \\
y(x_k) - y(x_{k-1}) &= (x_k - x_{k-1})f(x_{k-1}, y(x_{k-1}))
\end{aligned}$$

and hence,

$$y(x_k) - y_0 = \sum_{\ell=1}^{k}(x_\ell - x_{\ell-1})f(x_{\ell-1}, y(x_{\ell-1})),$$

which gives

$$|y(x_k) - y_0| \leq \sum_{\ell=1}^{k}(x_\ell - x_{\ell-1})M = M(x_k - x_0) \leq Mh \leq b.$$

Finally, if $x_{i-1} < x < x_i$ then from (9.4) and (9.3), we have

$$|y(x) - y(x_{i-1})| \le M|x - x_{i-1}| \le M\frac{\delta}{M} = \delta$$

and hence from (9.2), we find

$$|y'(x) - f(x, y(x))| = |f(x_{i-1}, y(x_{i-1})) - f(x, y(x))| \le \epsilon$$

for all $x \in J_h$, $x \ne x_i$, $i = 1, 2, \ldots, m-1$. This completes the proof that $y(x)$ is an ϵ-approximate solution of the DE $y' = f(x, y)$. This method of constructing an approximate solution is known as *Cauchy–Euler method.* ∎

Now we restate Corollary 9.2, and prove it as a consequence of Theorem 9.3.

Theorem 9.4 (Cauchy–Peano's Existence Theorem). Let the conditions of Theorem 9.3 be satisfied. Then the initial value problem (7.1) has at least one solution in J_h.

Proof.   Once again we shall give the proof only in the interval $x_0 \le x \le x_0 + h$. Let $\{\epsilon_m\}$ be a monotonically decreasing sequence of positive numbers such that $\epsilon_m \to 0$. For each ϵ_m we use Theorem 9.3 to construct an ϵ-approximate solution $y_m(x)$. Now as in Theorem 9.1, for any two points x and x^* in $[x_0, x_0 + h]$ it is easy to prove that

$$|y_m(x) - y_m(x^*)| \le M|x - x^*|$$

and from this it follows that the sequence $\{y_m(x)\}$ is equicontinuous. Further, as in Theorem 9.3 for each x in $[x_0, x_0 + h]$, we have $|y_m(x)| \le |y_0| + b$, and hence the sequence $\{y_m(x)\}$ is also uniformly bounded. Therefore, again Theorem 7.10 is applicable and the sequence $\{y_m(x)\}$ contains a subsequence $\{y_{m_p}(x)\}$ which converges uniformly in $[x_0, x_0 + h]$ to a continuous function $y(x)$. To show that the function $y(x)$ is a solution of (7.1), we define

$$
\begin{aligned}
e_m(x) &= y'_m(x) - f(x, y_m(x)), \quad \text{at the points where } y'_m(x) \text{ exists} \\
&= 0, \quad \text{otherwise.}
\end{aligned}
$$

Thus, it follows that

$$y_m(x) = y_0 + \int_{x_0}^x [f(t, y_m(t)) + e_m(t)]dt \tag{9.5}$$

and $|e_m(x)| \le \epsilon_m$. Since $f(x, y)$ is continuous in $\overline{S}$ and $y_{m_p}(x)$ converges to $y(x)$ uniformly in $[x_0, x_0 + h]$, the function $f(x, y_{m_p}(x))$ converges to $f(x, y(x))$ uniformly in $[x_0, x_0 + h]$. Further, since $\epsilon_{m_p} \to 0$ we find that $|\epsilon_{m_p}(x)|$ converges to zero uniformly in $[x_0, x_0 + h]$. Thus, by replacing m

by m_p in (9.5) and letting $p \to \infty$, we find that $y(x)$ is a solution of the integral equation (7.2).   ∎

Corollary 9.2 essentially states the following: If in a domain D the function $f(x, y)$ is continuous, then for every point (x_0, y_0) in D there is a rectangle $\overline{S}$ such that (7.1) has a solution $y(x)$ in J_h. Since $\overline{S}$ lies in D, by applying Corollary 9.2 to the point at which the solution goes out of $\overline{S}$, we can extend the region in which the solution exists. For example, the function $y(x) = 1/(1 - x)$ is the solution of the problem $y' = y^2$, $y(0) = 1$. Clearly, this solution exists in $(-\infty, 1)$. For this problem

$$\overline{S}: \ |x| \leq a, \ |y - 1| \leq b, \quad M = \max_{\overline{S}} y^2 = (1 + b)^2$$

and $h = \min\{a, b/(1 + b)^2\}$. Since $b/(1 + b)^2 \leq 1/4$ we can (independent of the choice of a) take $h = 1/4$. Thus, Corollary 9.2 gives the existence of a solution $y_1(x)$ only in the interval $|x| \leq 1/4$. Now consider the continuation of $y_1(x)$ to the right obtained by finding a solution $y_2(x)$ of the problem $y' = y^2$, $y(1/4) = 4/3$. For this new problem $\overline{S}: \ |x - 1/4| \leq a, \ |y - 4/3| \leq b$, and $\max_{\overline{S}} y^2 = (4/3 + b)^2$. Since $b/(4/3 + b)^2 \leq 3/16$ we can take $h = 3/16$. Thus, $y_2(x)$ exists in the interval $|x - 1/4| < 3/16$. This ensures the existence of a solution

$$y(x) = \begin{cases} y_1(x), & -1/4 \leq x \leq 1/4 \\ y_2(x), & 1/4 \leq x \leq 7/10 \end{cases}$$

in the interval $-1/4 \leq x \leq 7/16$. This process of continuation of the solution can be used further to the right of the point $(7/16, 16/9)$, or to the left of the point $(-1/4, 4/5)$. In order to establish how far the solution can be continued, we need the following lemma.

Lemma 9.5.   Let $f(x, y)$ be continuous in the domain D and let $\sup_D |f(x, y)| < M$. Further, let the initial value problem (7.1) has a solution $y(x)$ in an interval $J = (\alpha, \beta)$. Then the limits $\lim_{x \to \alpha^+} y(x) = y(\alpha + 0)$ and $\lim_{x \to \beta^-} y(x) = y(\beta - 0)$ exist.

Proof.   For $\alpha < x_1 < x_2 < \beta$, integral equation (7.2) gives that

$$|y(x_2) - y(x_1)| \leq \int_{x_1}^{x_2} |f(t, y(t))| dt \leq M|x_2 - x_1|.$$

Therefore, $y(x_2) - y(x_1) \to 0$ as $x_1, x_2 \to \alpha^+$. Thus, by the Cauchy criterion of convergence $\lim_{x \to \alpha^+} y(x)$ exists. A similar argument holds for $\lim_{x \to \beta^-} y(x)$.   ∎

Theorem 9.6.   Let the conditions of Lemma 9.5 be satisfied and let $(\beta, y(\beta - 0)) \in D$ $((\alpha, y(\alpha + 0)) \in D)$. Then the solution $y(x)$ of the initial value problem (7.1) in (α, β) can be extended over the interval $(\alpha, \beta + \gamma]$ $([\alpha - \gamma, \beta))$ for some $\gamma > 0$.

Proof. We define the function $y_1(x)$ as follows: $y_1(x) = y(x)$ for $x \in (\alpha, \beta)$ and $y_1(\beta) = y(\beta - 0)$. Then since for all $x \in (\alpha, \beta]$

$$
\begin{aligned}
y_1(x) &= y(\beta - 0) + \int_\beta^x f(t, y_1(t)) dt \\
&= y_0 + \int_{x_0}^\beta f(t, y_1(t)) dt + \int_\beta^x f(t, y_1(t)) dt \\
&= y_0 + \int_{x_0}^x f(t, y_1(t)) dt,
\end{aligned}
$$

the left-hand derivative $y_1'(\beta - 0)$ exists and $y_1'(\beta - 0) = f(\beta, y_1(\beta))$. Thus, $y_1(x)$ is a continuation of $y(x)$ in the interval $(\alpha, \beta]$. Next let $y_2(x)$ be a solution of the problem $y' = f(x, y)$, $y(\beta) = y_1(\beta)$ existing in the interval $[\beta, \beta + \gamma]$, then the function

$$
y_3(x) = \begin{cases} y_1(x), & x \in (\alpha, \beta] \\ y_2(x), & x \in [\beta, \beta + \gamma] \end{cases}
$$

is a continuation of $y(x)$ in the interval $(\alpha, \beta + \gamma]$. For this, it suffices to note that

$$
y_3(x) = y_0 + \int_{x_0}^x f(t, y_3(t)) dt \tag{9.6}
$$

for all $x \in (\alpha, \beta + \gamma]$. In fact (9.6) is obvious for all $x \in (\alpha, \beta]$ from the definition of $y_3(x)$ and for $x \in [\beta, \beta + \gamma]$, we have

$$
\begin{aligned}
y_3(x) &= y(\beta - 0) + \int_\beta^x f(t, y_3(t)) dt \\
&= y_0 + \int_{x_0}^\beta f(t, y_3(t)) dt + \int_\beta^x f(t, y_3(t)) dt \\
&= y_0 + \int_{x_0}^x f(t, y_3(t)) dt.
\end{aligned}
$$

Problems

9.1. Let $f(x, y)$ be continuous and $|f(x, y)| \leq c_1 + c_2|y|^\alpha$ for all $(x, y) \in T : |x - x_0| \leq a$, $|y| < \infty$ where c_1 and c_2 are nonnegative constants and $0 \leq \alpha < 1$. Show that the initial value problem (7.1) has at least one solution in the interval $|x - x_0| \leq a$.

9.2. Let $f(x, y)$ be continuous and satisfy the Lipschitz condition (7.3) in a domain D. Further, let $y_i(x)$, $i = 1, 2$ be ϵ_i-approximate solutions of the DE $y' = f(x, y)$ in J and $x_0 \in J$. Show that for all $x \in J$

$$
|y_1(x) - y_2(x)| \leq \left(|y_1(x_0) - y_2(x_0)| + \frac{\epsilon_1 + \epsilon_2}{L} \right) \exp(L|x - x_0|) - \frac{\epsilon_1 + \epsilon_2}{L}.
$$

9.3. Show that the solution of the problem $y' = -x/y$, $y(0) = 1$ cannot be extended beyond the interval $-1 < x < 1$.

9.4. Show that the solution of the problem $y' = y^2$, $y(0) = 2$ is extendable only to the interval $-\infty < x < 1/2$.

9.5. Show that the solution of the problem $y' = 2xy^2$, $y(0) = 1$ exists only in the interval $|x| < 1$.

9.6. Show that the solution of the problem $y' = 1+y^2$, $y(0) = 1$ cannot be extended beyond the interval $-3\pi/4 < x < \pi/4$.

9.7. Find the maximum interval in which the solution of the problem $y' + (\sin x)y^2 = 3(xy)^2$, $y(0) = 2$ can be extended.

9.8. Find the maximum interval of existence of solutions of $y' + 3y^{4/3} \sin x = 0$ satisfying (i) $y(\pi/2) = 0$, (ii) $y(\pi/2) = 1/8$, and (iii) $y(\pi/2) = 8$.

9.9. Solve the initial value problem

$$yy' - 3x^2(1 + y^2) = 0, \quad y(0) = 1.$$

Find also the largest interval on which the solution is defined.

Answers or Hints

9.1. In the rectangle $\overline{S} : |x - x_0| \le a$, $|y - y_0| \le b$, $|f(x,y)| \le c_1 + c_2(|y_0| + b)^\alpha = K$. Note that $b/K \to \infty$ as $b \to \infty$.

9.2. For $x \ge x_0$, $|y_i'(x) - f(x, y_i(x))| \le \epsilon_i$, $i = 1, 2$ yields $|y_i(x) - y_i(x_0) - \int_{x_0}^x f(t, y_i(t))dt| \le \epsilon_i(x - x_0)$. Now use $|p - q| \le |p| + |q|$ and Corollary 7.6.

9.3. Although the solution $y(x) = \sqrt{1 - x^2}$ is defined on $[-1, 1]$ its derivative is not defined at $x = \pm 1$.

9.4. The solution is $y(x) = 2/(1 - 2x)$.

9.5. The solution is $y(x) = 1/(1 - x^2)$.

9.6. The solution is $y(x) = \tan(x + \pi/4)$.

9.7. The solution is $y(x) = 1/[(3/2) - \cos x - x^3]$, which is defined in $(-\infty, 0.9808696 \cdots)$.

9.8. (i) The solution is $y(x) \equiv 0$, which is defined on $\mathbb{R}$. (ii) The solution is $y(x) = 1/(2 - \cos x)^3$, which is defined on $\mathbb{R}$. (iii) The solution is $y(x) = 1/[(1/2) - \cos x]^3$, which is defined in $(\pi/3, 5\pi/3)$.

9.9. $y = \left(2e^{2x^3} - 1\right)^{1/2}$, $x > \left(\frac{1}{2} \ln \frac{1}{2}\right)^{1/3}$.

Lecture 10
Uniqueness Theorems

In our previous lectures we have proved that the continuity of the function $f(x, y)$ in the closed rectangle $\overline{S}$ is sufficient for the existence of at least one solution of the initial value problem (7.1) in the interval J_h, and to achieve the uniqueness (i.e., existence of at most one solution) some additional condition on $f(x, y)$ is required. In fact, continuous functions $f(x, y)$ have been constructed (see Lavrentev [30], Hartman [20]) so that from any given point (x_0, y_0) the equation $y' = f(x, y)$ has at least two solutions in every neighborhood of (x_0, y_0). In Theorem 8.1 this additional condition was assumed to be the Lipschitz continuity. In the following, we shall provide several such conditions which are sufficient for the uniqueness of the solutions of (7.1).

Theorem 10.1 (Lipschitz Uniqueness Theorem). Let $f(x, y)$ be continuous and satisfy a uniform Lipschitz condition (7.3) in $\overline{S}$. Then (7.1) has at most one solution in $|x - x_0| \leq a$.

Proof. In Theorem 8.1 the uniqueness of the solutions of (7.1) is proved in the interval J_h; however, it is clear that J_h can be replaced by the interval $|x - x_0| \leq a$. ∎

Theorem 10.2 (Peano's Uniqueness Theorem). Let $f(x, y)$ be continuous in $\overline{S}_+ : x_0 \leq x \leq x_0 + a$, $|y - y_0| \leq b$ and nonincreasing in y for each fixed x in $x_0 \leq x \leq x_0 + a$. Then (7.1) has at most one solution in $x_0 \leq x \leq x_0 + a$.

Proof. Suppose $y_1(x)$ and $y_2(x)$ are two solutions of (7.1) in $x_0 \leq x \leq x_0 + a$ which differ somewhere in $x_0 \leq x \leq x_0 + a$. We assume that $y_2(x) > y_1(x)$ in $x_1 < x < x_1 + \epsilon \leq x_0 + a$, while $y_1(x) = y_2(x)$ in $x_0 \leq x \leq x_1$, i.e., x_1 is the greatest lower bound of the set A consisting of those x for which $y_2(x) > y_1(x)$. This greatest lower bound exists because the set A is bounded below by x_0 at least. Thus, for all $x \in (x_1, x_1 + \epsilon)$ we have $f(x, y_1(x)) \geq f(x, y_2(x))$; i.e., $y_1'(x) \geq y_2'(x)$. Hence, the function $z(x) = y_2(x) - y_1(x)$ is nonincreasing, since if $z(x_1) = 0$ we should have $z(x) \leq 0$ in $(x_1, x_1 + \epsilon)$. This contradiction proves that $y_1(x) = y_2(x)$ in $x_0 \leq x \leq x_0 + a$. ∎

Example 10.1. The function $|y|^{1/2}\text{sgn } y$, where sgn $y = 1$ if $y \geq 0$, and -1 if $y < 0$ is continuous, nondecreasing, and the initial value problem $y' = |y|^{1/2}\text{sgn } y$, $y(0) = 0$ has two solutions $y(x) \equiv 0$, $y(x) = x^2/4$ in the

R.P. Agarwal and D. O'Regan, *An Introduction to Ordinary Differential Equations*,
doi: 10.1007/978-0-387-71276-5_10, © Springer Science + Business Media, LLC 2008

interval $[0, \infty)$. Thus, in Theorem 10.2 "nonincreasing" cannot be replaced by "nondecreasing."

For our next result, we need the following lemma.

Lemma 10.3. Let $w(z)$ be continuous and increasing function in the interval $[0, \infty)$, and $w(0) = 0$, $w(z) > 0$ for $z > 0$, with also

$$\lim_{\epsilon \to 0+} \int_\epsilon \frac{dz}{w(z)} = \infty. \tag{10.1}$$

Let $u(x)$ be a nonnegative continuous function in $[0, a]$. Then the inequality

$$u(x) \le \int_0^x w(u(t))dt, \quad 0 < x \le a \tag{10.2}$$

implies that $u(x) \equiv 0$ in $[0, a]$.

Proof. Define $v(x) = \max_{0 \le t \le x} u(t)$ and assume that $v(x) > 0$ for $0 < x \le a$. Then $u(x) \le v(x)$ and for each x there is an $x_1 \le x$ such that $u(x_1) = v(x)$. From this, we have

$$v(x) = u(x_1) \le \int_0^{x_1} w(u(t))dt \le \int_0^x w(v(t))dt;$$

i.e., the nondecreasing function $v(x)$ satisfies the same inequality as $u(x)$ does. Let us set

$$\overline{v}(x) = \int_0^x w(v(t))dt,$$

then $\overline{v}(0) = 0$, $v(x) \le \overline{v}(x)$, $\overline{v}'(x) = w(v(x)) \le w(\overline{v}(x))$. Hence, for $0 < \delta < a$, we have

$$\int_\delta^a \frac{\overline{v}'(x)}{w(\overline{v}(x))}dx \le a - \delta < a.$$

However, from (10.1) it follows that

$$\int_\delta^a \frac{\overline{v}'(x)}{w(\overline{v}(x))}dx = \int_\epsilon^\alpha \frac{dz}{w(z)}, \quad \overline{v}(\delta) = \epsilon, \quad \overline{v}(a) = \alpha$$

becomes infinite when $\epsilon \to 0$ ($\delta \to 0$). This contradiction shows that $v(x)$ cannot be positive, so $v(x) \equiv 0$, and hence $u(x) = 0$ in $[0, a]$. ∎

Theorem 10.4 (Osgood's Uniqueness Theorem). Let $f(x, y)$ be continuous in $\overline{S}$ and for all (x, y_1), $(x, y_2) \in \overline{S}$ it satisfies

$$|f(x, y_1) - f(x, y_2)| \le w(|y_1 - y_2|), \tag{10.3}$$

where $w(z)$ is the same as in Lemma 10.3. Then (7.1) has at most one solution in $|x - x_0| \le a$.

Proof. Suppose $y_1(x)$ and $y_2(x)$ are two solutions of (7.1) in $|x-x_0| \leq a$. Then from (10.3) it follows that

$$|y_1(x) - y_2(x)| \leq \left| \int_{x_0}^{x} w(|y_1(t) - y_2(t)|) dt \right|.$$

For any x in $[x_0, x_0 + a]$, we set $u(x) = |y_1(x_0 + x) - y_2(x_0 + x)|$. Then the nonnegative continuous function $u(x)$ satisfies the inequality (10.2), and therefore, Lemma 10.3 implies that $u(x) = 0$ in $[0, a]$, i.e., $y_1(x) = y_2(x)$ in $[x_0, x_0 + a]$. If x is in $[x_0 - a, x_0]$, then the proof remains the same except that we need to define the function $u(x) = |y_1(x_0 - x) - y_2(x_0 - x)|$ in $[0, a]$. ∎

For our next result, we shall prove the following lemma.

Lemma 10.5. Let $u(x)$ be nonnegative continuous function in $|x - x_0| \leq a$, and $u(x_0) = 0$, and let $u(x)$ be differentiable at $x = x_0$ with $u'(x_0) = 0$. Then the inequality

$$u(x) \leq \left| \int_{x_0}^{x} \frac{u(t)}{t - x_0} dt \right| \tag{10.4}$$

implies that $u(x) = 0$ in $|x - x_0| \leq a$.

Proof. It suffices to prove the lemma only for $x_0 \leq x \leq x_0 + a$. We define

$$v(x) = \int_{x_0}^{x} \frac{u(t)}{t - x_0} dt.$$

This integral exists since

$$\lim_{x \to x_0} \frac{u(x)}{x - x_0} = u'(x_0) = 0.$$

Further, we have

$$v'(x) = \frac{u(x)}{x - x_0} \leq \frac{v(x)}{x - x_0}$$

and hence $d/dx[v(x)/(x - x_0)] \leq 0$, which implies that $v(x)/(x - x_0)$ is nonincreasing. Since $v(x_0) = 0$, this gives $v(x) \leq 0$, which is a contradiction to $v(x) \geq 0$. So, $v(x) \equiv 0$, and hence $u(x) = 0$ in $[x_0, x_0 + a]$. ∎

Theorem 10.6 (Nagumo's Uniqueness Theorem). Let $f(x, y)$ be continuous in $\overline{S}$ and for all (x, y_1), $(x, y_2) \in \overline{S}$ it satisfies

$$|f(x, y_1) - f(x, y_2)| \leq k|x - x_0|^{-1}|y_1 - y_2|, \quad x \neq x_0, \quad k \leq 1. \tag{10.5}$$

Then (7.1) has at most one solution in $|x - x_0| \leq a$.

Proof. Suppose $y_1(x)$ and $y_2(x)$ are two solutions of (7.1) in $|x-x_0| \leq a$. Then from (10.5) it follows that

$$|y_1(x) - y_2(x)| \leq \left| \int_{x_0}^{x} |t - x_0|^{-1}|y_1(t) - y_2(t)| dt \right|.$$

We set $u(x) = |y_1(x) - y_2(x)|$; then the nonnegative function $u(x)$ satisfies the inequality (10.4). Further, since $u(x)$ is continuous in $|x - x_0| \leq a$, and $u(x_0) = 0$, from the mean value theorem we have

$$
\begin{aligned}
u'(x_0) &= \lim_{h \to 0} \frac{u(x_0 + h) - u(x_0)}{h} \\
&= \lim_{h \to 0} \frac{|y_1(x_0) + h y_1'(x_0 + \theta_1 h) - y_2(x_0) - h y_2'(x_0 + \theta_2 h)|}{h}, \\
&\qquad\qquad\qquad\qquad\qquad\qquad\qquad\qquad 0 < \theta_1,\ \theta_2 < 1 \\
&= (\operatorname{sgn} h) \lim_{h \to 0} |y_1'(x_0 + \theta_1 h) - y_2'(x_0 + \theta_2 h)| = 0.
\end{aligned}
$$

Thus, the conditions of Lemma 10.5 are satisfied and $u(x) \equiv 0$, i.e., $y_1(x) = y_2(x)$ in $|x - x_0| \leq a$. ∎

Example 10.2. It is easy to verify that the function

$$
f(x, y) = \begin{cases} 0 & 0 \leq x \leq 1,\ y \leq 0 \\[2mm] \dfrac{(1 + \epsilon)y}{x} & 0 \leq x \leq 1,\ 0 < y < x^{1+\epsilon},\ \epsilon > 0 \\[2mm] (1 + \epsilon)x^\epsilon & 0 \leq x \leq 1,\ x^{1+\epsilon} \leq y \end{cases}
$$

is continuous and satisfies the condition (10.5) (except $k = 1 + \epsilon > 1$) in $\overline{S} : [0, 1] \times \mathbb{R}$. For this function the initial value problem (7.1) with $(x_0, y_0) = (0, 0)$ has an infinite number of solutions $y(x) = cx^{1+\epsilon}$, where c is an arbitrary constant such that $0 < c < 1$. Thus, in condition (10.5) the constant $k \leq 1$ is the best possible, i.e., it cannot be replaced by $k > 1$.

Theorem 10.7 (Krasnoselski–Krein Uniqueness Theorem).

Let $f(x, y)$ be continuous in $\overline{S}$ and for all (x, y_1), $(x, y_2) \in \overline{S}$ it satisfies

$$
|f(x, y_1) - f(x, y_2)| \leq k|x - x_0|^{-1}|y_1 - y_2|, \quad x \neq x_0,\ k > 0 \qquad (10.6)
$$

$$
|f(x, y_1) - f(x, y_2)| \leq C|y_1 - y_2|^\alpha, \quad C > 0,\ 0 < \alpha < 1,\ k(1 - \alpha) < 1. \tag{10.7}
$$

Then (7.1) has at most one solution in $|x - x_0| \leq a$.

Proof. Suppose $y_1(x)$ and $y_2(x)$ are two solutions of (7.1) in $|x - x_0| \leq a$. We shall show that $y_1(x) = y_2(x)$ only in the interval $[x_0, x_0 + a]$. For this, from (10.7) we have

$$
u(x) = |y_1(x) - y_2(x)| \leq \int_{x_0}^x C u^\alpha(t)\,dt
$$

and hence Problem 7.5 gives that

$$
u(x) \leq [C(1 - \alpha)(x - x_0)]^{(1-\alpha)^{-1}} \leq [C(x - x_0)]^{(1-\alpha)^{-1}}.
$$

Thus, the function $v(x) = u(x)/(x - x_0)^k$ satisfies the inequality

$$0 \leq v(x) \leq C^{(1-\alpha)^{-1}}(x - x_0)^{(1-\alpha)^{-1}-k}. \tag{10.8}$$

Since $k(1-\alpha) < 1$, it is immediate that $\lim_{x \to x_0} v(x) = 0$. Hence, if we define $v(x_0) = 0$, then the function $v(x)$ is continuous in $[x_0, x_0 + a]$. We wish to show that $v(x) = 0$ in $[x_0, x_0+a]$. If $v(x) > 0$ at any point in $[x_0, x_0+a]$, then there exists a point $x_1 > x_0$ such that $0 < m = v(x_1) = \max_{x_0 \leq x \leq x_0+a} v(x)$. However, from (10.6) we obtain

$$
\begin{aligned}
m = v(x_1) &\leq (x_1 - x_0)^{-k} \int_{x_0}^{x_1} k(t - x_0)^{-1}u(t)dt \\
&\leq (x_1 - x_0)^{-k} \int_{x_0}^{x_1} k(t - x_0)^{k-1}v(t)dt \\
&< m(x_1 - x_0)^{-k} \int_{x_0}^{x_1} k(t - x_0)^{k-1}dt \\
&= m(x_1 - x_0)^{-k}(x_1 - x_0)^k = m,
\end{aligned}
$$

which is the desired contradiction. So, $v(x) \equiv 0$, and hence $u(x) = 0$ in $[x_0, x_0 + a]$. ∎

Theorem 10.8 (Van Kampen Uniqueness Theorem). Let

$f(x, y)$ be continuous in $\overline{S}$ and for all $(x, y) \in \overline{S}$ it satisfies

$$|f(x, y)| \leq A|x - x_0|^p, \quad p > -1, \quad A > 0. \tag{10.9}$$

Further, let for all (x, y_1), $(x, y_2) \in \overline{S}$ it satisfies

$$|f(x, y_1) - f(x, y_2)| \leq \frac{C}{|x - x_0|^r}|y_1 - y_2|^q, \quad q \geq 1, \quad C > 0 \tag{10.10}$$

with $q(1 + p) - r = p$, $\rho = C(2A)^{q-1}/(p+1)^q < 1$. Then (7.1) has at most one solution in $|x - x_0| \leq a$.

Proof. Suppose $y_1(x)$ and $y_2(x)$ are two solutions of (7.1) in $|x - x_0| \leq a$. We shall show that $y_1(x) = y_2(x)$ only in the interval $[x_0 - a, x_0]$. For this, from (10.9) we have

$$
\begin{aligned}
u(x) = |y_1(x) - y_2(x)| &\leq \int_x^{x_0} |f(t, y_1(t)) - f(t, y_2(t))|dt \\
&\leq 2A \int_x^{x_0} (x_0 - t)^p dt = \frac{2A}{p+1}(x_0 - x)^{p+1}.
\end{aligned}
$$

Using this estimate and (10.10), we obtain

$$
\begin{aligned}
u(x) &\leq C \int_x^{x_0} \frac{1}{(x_0 - t)^r} u^q(t)dt \\
&\leq C \left(\frac{2A}{p+1}\right)^q \int_x^{x_0} (x_0 - t)^{q(p+1)-r}dt = \rho \left(\frac{2A}{p+1}\right)(x_0 - x)^{p+1}.
\end{aligned}
$$

Now using this new estimate and (10.10), we get

$$u(x) \leq \rho^{1+q} \left(\frac{2A}{p+1} \right) (x_0 - x)^{p+1}.$$

Continuing in this way, we find

$$u(x) \leq \rho^{1+q+q^2+\cdots+q^m} \left(\frac{2A}{p+1} \right) (x_0 - x)^{p+1}, \quad m = 1, 2, \ldots.$$

Since $q \geq 1$ and $\rho < 1$, it follows that $u(x) = 0$ in $|x_0 - a, x_0]$.   ∎

Problems

10.1. Consider the initial value problem

$$y' = f(x, y) = \begin{cases} \dfrac{4x^3 y}{x^4 + y^2}, & (x, y) \neq (0, 0) \\ 0, & (x, y) = (0, 0) \end{cases} \tag{10.11}$$

$$y(0) = 0.$$

Show that the function $f(x, y)$ is continuous but does not satisfy the Lipschitz condition in any region containing the origin (see Problem 7.4). Further, show that (10.11) has an infinite number of solutions.

10.2. Given the equation $y' = xg(x, y)$, suppose that g and $\partial g / \partial y$ are defined and continuous for all (x, y). Show the following:

(i) $y(x) \equiv 0$ is a solution.

(ii)   If $y = y(x)$, $x \in (\alpha, \beta)$ is a solution and if $y(x_0) > 0$, $x_0 \in (\alpha, \beta)$, then $y(x) > 0$ for all $x \in (\alpha, \beta)$.

(iii)   If $y = y(x)$, $x \in (\alpha, \beta)$ is a solution and if $y(x_0) < 0$, $x_0 \in (\alpha, \beta)$, then $y(x) < 0$ for all $x \in (\alpha, \beta)$.

10.3. Let $f(x, y)$ be continuous and satisfy the generalized Lipschitz condition

$$|f(x, y_1) - f(x, y_2)| \leq L(x)|y_1 - y_2|$$

for all (x, y_1), (x, y_2) in $\overline{S}$, where the function $L(x)$ is such that the integral $\int_{x_0-a}^{x_0+a} L(t)dt$ exists. Show that (7.1) has at most one solution in $|x-x_0| \leq a$.

10.4. Give some examples to show that the Lipschitz condition in Theorem 10.1 is just a sufficient condition for proving the uniqueness of the solutions of (7.1) but not the necessary condition.

10.5. Let $f(x, y)$ be continuous in $\overline{S}_+$ and for all (x, y_1), (x, y_2) in $\overline{S}_+$ with $y_2 \geq y_1$ satisfy one sided Lipschitz condition

$$f(x, y_2) - f(x, y_1) \leq L(y_2 - y_1).$$

Show that (7.1) has at most one solution in $x_0 \leq x \leq x_0 + a$.

10.6. Let $f(x, y)$ be continuous in $\overline{S}_- : x_0 - a \leq x \leq x_0$, $|y - y_0| \leq b$ and nondecreasing in y for each fixed x in $x_0 - a \leq x \leq x_0$. Show that (7.1) has at most one solution in $x_0 - a \leq x \leq x_0$.

10.7. Show that the functions $w(z) = Lz^\alpha$ $(\alpha \geq 1)$, and

$$w(z) = \begin{cases} -z \ln z, & 0 \leq z \leq e^{-1} \\ e^{-1}, & z > e^{-1} \end{cases}$$

satisfy the conditions of Lemma 10.3.

10.8. Consider the function $f(x, y)$ in the strip $T : \ -\infty < x \leq 1$, $-\infty < y < \infty$ defined by

$$f(x, y) = \begin{cases} 0 & -\infty < x \leq 0, \ -\infty < y < \infty \\ 2x & 0 < x \leq 1, \ -\infty < y < 0 \\ 2x - \dfrac{4y}{x} & 0 < x \leq 1, \ 0 \leq y \leq x^2 \\ -2x & 0 < x \leq 1, \ x^2 < y < \infty. \end{cases}$$

Show that the problem $y' = f(x, y)$, $y(0) = 0$ has a unique solution in the interval $-\infty < x \leq 1$. Further, show that the Picard iterates with $y_0(x) \equiv 0$ for this problem do not converge.

10.9. Consider the function $f(x, y)$ in the strip $T : 0 \leq x \leq 1$, $-\infty < y < \infty$ defined by

$$f(x, y) = \begin{cases} 0 & 0 \leq x \leq 1, \ x^{1/(1-\alpha)} < y < \infty, \ 0 < \alpha < 1 \\ kx^{\alpha/(1-\alpha)} - k\dfrac{y}{x} & 0 \leq x \leq 1, \ 0 \leq y \leq x^{1/(1-\alpha)}, \ k > 0 \\ kx^{\alpha/(1-\alpha)} & 0 \leq x \leq 1, \ -\infty < y < 0, \ k(1-\alpha) < 1. \end{cases}$$

Show that the problem $y' = f(x, y)$, $y(0) = 0$ has a unique solution in $[0, 1]$.

*10.10 (Rogers' Uniqueness Theorem). Let $f(x, y)$ be continuous in the strip $T : 0 \leq x \leq 1$, $-\infty < y < \infty$ and satisfy the condition

$$f(x, y) = o\left(e^{-1/x}x^{-2}\right)$$

uniformly for $0 \leq y \leq \delta$, $\delta > 0$ arbitrary. Further, let for all (x, y_1), (x, y_2) $\in T$ it satisfy

$$|f(x, y_1) - f(x, y_2)| \leq \frac{1}{x^2}|y_1 - y_2|.$$

Show that the problem $y' = f(x, y)$, $y(0) = 0$ has at most one solution in $[0, 1]$.

10.11. Consider the function $f(x, y)$ in the strip $T : 0 \leq x \leq 1$, $-\infty < y < \infty$ defined by

$$f(x, y) = \begin{cases} \left(1 + \dfrac{1}{x}\right) e^{-1/x} & 0 \leq x \leq 1, \quad xe^{-1/x} \leq y < \infty \\[2mm] \dfrac{y}{x^2} + e^{-1/x} & 0 \leq x \leq 1, \quad 0 \leq y \leq xe^{-1/x} \\[2mm] e^{-1/x} & 0 \leq x \leq 1, \quad -\infty < y \leq 0. \end{cases}$$

Show that the problem $y' = f(x, y)$, $y(0) = 0$ has a unique solution in $[0, 1]$.

10.12. Consider the function $f(x, y)$ in the strip $T : 0 \leq x \leq 1$, $-\infty < y < \infty$ defined by

$$f(x, y) = \begin{cases} 0 & 0 \leq x \leq 1, \quad -\infty < y \leq 0 \\[2mm] \dfrac{y}{x^2} & 0 \leq x \leq 1, \quad 0 \leq y \leq e^{-1/x} \\[2mm] \dfrac{e^{-1/x}}{x^2} & 0 \leq x \leq 1, \quad e^{-1/x} \leq y < \infty. \end{cases}$$

Show that the problem $y' = f(x, y)$, $y(0) = 0$ has an infinite number of solutions in $[0, 1]$.

Answers or Hints

10.1. $y = c^2 - \sqrt{x^4 + c^4}$, where c is arbitrary.

10.2. (i) Verify directly. (ii) Use Theorem 10.1. (iii) Use Theorem 10.1.

10.3. Since $\int_{x_0-a}^{x_0+a} L(t)dt$ exists, Corollary 7.4 is applicable.

10.4. Consider the Problem 8.3(i), or $y' = y \ln(1/y)$, $y(0) = \alpha \geq 0$.

10.5. Suppose two solutions $y_1(x)$ and $y_2(x)$ are such that $y_2(x) > y_1(x)$, $x_1 < x < x_1 + \epsilon \leq x_0 + a$ and $y_1(x) = y_2(x)$, $x_0 \leq x \leq x_1$. Now apply Corollary 7.4.

10.6. The proof is similar to that of Theorem 10.2.

10.7. Verify directly.

10.8. The given function is continuous and bounded by 2 in T. Also it satisfies the conditions of Theorem 10.2 (also, see Problem 10.6). The only solution is $y(x) = \begin{cases} 0, & -\infty < x \leq 0 \\ x^2/3, & 0 < x \leq 1. \end{cases}$ The successive approximations are $y_{2m-1}(x) = x^2$, $y_{2m}(x) = -x^2$, $m = 1, 2, \ldots$.

10.9. Show that conditions of Theorem 10.7 are satisfied. The only solution is $y(x) = k(1 - \alpha)x^{1/(1-\alpha)}/[k(1 - \alpha) + 1]$.

10.11. Show that conditions of Problem 10.10 are satisfied. The only solution is $y(x) = xe^{-1/x}$.

10.12. Show that the condition $f(x, y) = o\left(e^{-1/x}x^{-2}\right)$ of Problem 10.10 is not satisfied. For each $0 \le c \le 1$, $y(x) = ce^{-1/x}$ is a solution.

Lecture 11
Differential Inequalities

Let the function $f(x, y)$ be continuous in a given domain D. A function $y(x)$ is said to be a solution of the differential inequality $y' > f(x, y)$ in $J = [x_0, x_0 + a)$ if (i) $y'(x)$ exists for all $x \in J$, (ii) for all $x \in J$ the points $(x, y(x)) \in D$, and (iii) $y'(x) > f(x, y(x))$ for all $x \in J$. The solutions of the differential inequalities $y' \geq f(x, y)$, $y' < f(x, y)$, and $y' \leq f(x, y)$ are defined analogously. For example, $y(x) = \cot x$ is a solution of the differential inequality $y' < -y^2$ in the interval $(0, \pi)$.

Our first basic result for differential inequalities is stated in the following theorem.

Theorem 11.1. Let $f(x, y)$ be continuous in the domain D and $y_1(x)$ and $y_2(x)$ be the solutions of the differential inequalities

$$y_1' \leq f(x, y_1), \quad y_2' > f(x, y_2) \tag{11.1}$$

on J. Then $y_1(x_0) < y_2(x_0)$ implies that

$$y_1(x) < y_2(x) \quad \text{for all} \quad x \in J. \tag{11.2}$$

Proof. If (11.2) is not true, then the set $A = \{x : x \in J, \ y_1(x) \geq y_2(x)\}$ is nonempty. Let x^* be the greatest lower bound of A, then $x_0 < x^*$ and $y_1(x^*) = y_2(x^*)$. Now for $h < 0$ we have $y_1(x^* + h) < y_2(x^* + h)$, and hence

$$
\begin{aligned}
y_1'(x^* - 0) &= \lim_{h \to 0} \frac{y_1(x^* + h) - y_1(x^*)}{h} \\
&\geq \lim_{h \to 0} \frac{y_2(x^* + h) - y_2(x^*)}{h} = y_2'(x^* - 0).
\end{aligned}
$$

Therefore, from (11.1) we obtain $f(x^*, y_1(x^*)) > f(x^*, y_2(x^*))$, which is a contradiction to $y_1(x^*) = y_2(x^*)$. Hence, the set A is empty and (11.2) follows. ∎

Obviously, Theorem 11.1 holds even when we replace $\leq$ by $<$ and $>$ by $\geq$ in (11.1).

Corollary 11.2. Let $f(x, y)$ be continuous in the domain D. Further, we assume the following:

(i) $y(x)$ is a solution of the initial value problem (7.1) in $J = [x_0, x_0 + a)$.

R.P. Agarwal and D. O'Regan, *An Introduction to Ordinary Differential Equations*, doi: 10.1007/978-0-387-71276-5_11, © Springer Science + Business Media, LLC 2008

(ii) $y_1(x)$ and $y_2(x)$ are the solutions of the differential inequalities $y_1' < f(x, y_1)$, $y_2' > f(x, y_2)$ in J.

(iii) $y_1(x_0) \leq y_0 \leq y_2(x_0)$.

Then $y_1(x) < y(x) < y_2(x)$ for all $x \in (x_0, x_0 + a)$.

Proof.   We shall prove only $y(x) < y_2(x)$ in the interval $(x_0, x_0 + a)$. If $y_0 < y_2(x_0)$, then the result follows from Theorem 11.1. Thus, we shall assume that $y_0 = y_2(x_0)$. Let $z(x) = y_2(x) - y(x)$, then $z'(x_0) = y_2'(x_0) - y'(x_0) > f(x_0, y_2(x_0)) - f(x_0, y(x_0)) = 0$, i.e., $z(x)$ is increasing to the right of x_0 in a sufficiently small interval $[x_0, x_0 + \delta]$. Therefore, we have $y(x_0 + \delta) < y_2(x_0 + \delta)$. Now an application of Theorem 11.1 gives that $y(x) < y_2(x)$ for all $x \in [x_0 + \delta, x_0 + a)$. Since δ can be chosen sufficiently small, the conclusion follows.   ∎

In Theorem 11.1, and consequently in Corollary 11.2, several refinements are possible, e.g., it is enough if the inequalities (11.1) hold in J except at a countable subset of J.

Example 11.1.   Consider the initial value problem

$$y' = y^2 + x^2, \quad y(0) = 1, \quad x \in [0, 1). \tag{11.3}$$

For the function $y_1(x) = 1 + x^3/3$, $y_1(0) = 1$ and for $x \in (0, 1)$, we have

$$y_1'(x) = x^2 < \left(1 + \frac{x^3}{3}\right)^2 + x^2 = y_1^2(x) + x^2.$$

Similarly, for the function $y_2(x) = \tan(x + \pi/4)$, $y_2(0) = 1$ and for $x \in (0, 1)$, we find

$$y_2'(x) = \sec^2\left(x + \frac{\pi}{4}\right) = \tan^2\left(x + \frac{\pi}{4}\right) + 1 > y_2^2(x) + x^2.$$

Thus, from Corollary 11.2 the solution $y(x)$ of the problem (11.3) can be bracketed between $y_1(x)$ and $y_2(x)$, i.e.,

$$1 + \frac{x^3}{3} < y(x) < \tan\left(x + \frac{\pi}{4}\right), \quad x \in (0, 1).$$

As the first application of Theorem 11.1, we shall prove the following result.

Theorem 11.3.   Let $f(x, y)$ be continuous in the domain D and for all (x, y), (x, z) in D with $x \geq x_0$, $y \geq z$

$$f(x, y) - f(x, z) \leq L(y - z). \tag{11.4}$$

Further, we assume that conditions (i)–(iii) of Corollary 11.2 with strict inequalities in (ii) replaced by with equalities are satisfied. Then $y_1(x) \leq y(x) \leq y_2(x)$ for all $x \in J$.

Proof.   We define $z_1(x) = y_1(x) - \epsilon e^{\lambda(x-x_0)}$, where $\epsilon > 0$ and $\lambda > L$. Then from the above assumptions, we obtain

$$
\begin{aligned}
z_1'(x) = y_1'(x) - \epsilon \lambda e^{\lambda(x-x_0)} &\leq f(x, y_1(x)) - \epsilon \lambda e^{\lambda(x-x_0)} \\
&\leq f(x, z_1(x)) + \epsilon(L - \lambda)e^{\lambda(x-x_0)} \\
&< f(x, z_1(x)).
\end{aligned}
$$

Similarly, for the function $z_2(x) = y_2(x) + \epsilon e^{\lambda(x-x_0)}$, we find

$$z_2'(x) > f(x, z_2(x)).$$

Also, $z_1(x_0) < y_1(x_0) \leq y_0 \leq y_2(x_0) < z_2(x_0)$ is obvious. Hence, the conditions of Theorem 11.1 for the functions $z_1(x)$ and $z_2(x)$ are satisfied, and we get

$$z_1(x) < y(x) < z_2(x) \tag{11.5}$$

for all $x \in J$. The desired conclusion now follows by letting $\epsilon \to 0$ in (11.5). ∎

Corollary 11.4.   Let the conditions of Theorem 11.3 with (11.4) replaced by the Lipschitz condition (7.3) be satisfied for all $x \geq x_0$, and let (iii) of Corollary 11.2 be replaced by $y_1(x_0) = y_0 = y_2(x_0)$. Then for any $x_1 \in J$ such that $x_1 > x_0$, either $y_1(x_1) < y(x_1)$ $(y(x_1) < y_2(x_1))$ or $y_1(x) = y(x)$ $(y(x) = y_2(x))$ for all $x \in [x_0, x_1]$.

Proof.   For $x \geq x_0$ and $y \geq z$ the Lipschitz condition (7.3) is equivalent to the following

$$-L(y - z) \leq f(x,y) - f(x,z) \leq L(y - z) \tag{11.6}$$

and hence from Theorem 11.3 it follows that $y_1(x) \leq y(x) \leq y_2(x)$. Now since $y_1(x_0) = y(x_0) = y_2(x_0)$, unless $y(x) = y_1(x)$ $(y(x) = y_2(x))$, there is some $x_1 > x_0$ at which $y_1(x_1) < y(x_1)$ $(y(x_1) < y_2(x_1))$. However, from (11.6) we find

$$y_1'(x) - y'(x) \leq f(x, y_1(x)) - f(x, y(x)) \leq L(y(x) - y_1(x)),$$

which is the same as

$$\frac{d}{dx}\left(e^{Lx}(y_1(x) - y(x))\right) \leq 0.$$

Hence, the function $e^{Lx}(y_1(x) - y(x))$ cannot increase and for any $x > x_1$

$$e^{Lx}(y_1(x) - y(x)) \leq e^{Lx_1}(y_1(x_1) - y(x_1)) < 0.$$

Thus, $y_1(x) < y(x)$ for all $x > x_1$. Consequently, if $y(x_1) = y_1(x_1)$ at any point x_1, then $y(x) = y_1(x)$ in $[x_0, x_1]$. ∎

For our next application of Theorem 11.1, we need the following definition.

Definition 11.1. A solution $r(x)$ $(\rho(x))$ of the initial value problem (7.1) which exists in an interval J is said to be a *maximal* (*minimal*) *solution* if for an arbitrary solution $y(x)$ of (7.1) existing in J, the inequality $y(x) \leq r(x)$ $(\rho(x) \leq y(x))$ holds for all $x \in J$.

Obviously, if the maximal and minimal solutions exist, then these are unique. The existence of these solutions is proved in the following theorem.

Theorem 11.5. Let $f(x, y)$ be continuous in $\overline{S}_+ : x_0 \leq x \leq x_0 + a$, $|y - y_0| \leq b$ and hence there exists a $M > 0$ such that $|f(x, y)| \leq M$ for all $(x, y) \in \overline{S}_+$. Then there exist a maximal solution $r(x)$ and a minimal solution $\rho(x)$ of (7.1) in the interval $[x_0, x_0 + \alpha]$, where $\alpha = \min\{a, b/(2M + b)\}$.

Proof. We shall prove the existence of the maximal solution $r(x)$ only. Let $0 < \epsilon \leq b/2$, and consider the initial value problem

$$y' = f(x, y) + \epsilon, \quad y(x_0) = y_0 + \epsilon. \tag{11.7}$$

Since the function $f_\epsilon(x, y) = f(x, y) + \epsilon$ is continuous in $\overline{S}_\epsilon : x_0 \leq x \leq x_0 + a$, $|y - (y_0 + \epsilon)| \leq b/2$, and $\overline{S}_\epsilon \subseteq \overline{S}_+$ we find that $|f_\epsilon(x, y)| \leq M + b/2$ in $\overline{S}_\epsilon$. Hence, from Corollary 9.2 it follows that the problem (11.7) has a solution $y(x, \epsilon)$ in the interval $[x_0, x_0 + \alpha]$, where $\alpha = \min\{a, b/(2M + b)\}$. For $0 < \epsilon_2 < \epsilon_1 \leq \epsilon$, we have $y(x_0, \epsilon_2) < y(x_0, \epsilon_1)$ and $y'(x, \epsilon_2) = f(x, y(x, \epsilon_2)) + \epsilon_2$, $y'(x, \epsilon_1) > f(x, y(x, \epsilon_1)) + \epsilon_2$, $x \in [x_0, x_0 + \alpha]$. Thus, Theorem 11.1 is applicable and we have $y(x, \epsilon_2) < y(x, \epsilon_1)$ for all $x \in [x_0, x_0 + \alpha]$. Now as in Theorem 9.1 it is easy to see that the family of functions $y(x, \epsilon)$ is equicontinuous and uniformly bounded in $[x_0, x_0 + \alpha]$, therefore, from Theorem 7.10 there exists a decreasing sequence $\{\epsilon_n\}$ such that $\epsilon_n \to 0$ as $n \to \infty$, and $\lim_{n \to \infty} y(x, \epsilon_n)$ exists uniformly in $[x_0, x_0 + \alpha]$. We denote this limiting function by $r(x)$. Obviously, $r(x_0) = y_0$, and the uniform continuity of f, with

$$y(x, \epsilon_n) = y_0 + \epsilon_n + \int_{x_0}^{x} [f(t, y(t, \epsilon_n)) + \epsilon_n] dt$$

yields $r(x)$ as a solution of (7.1).

Finally, we shall show that $r(x)$ is the maximal solution of (7.1) in $[x_0, x_0 + \alpha]$. For this, let $y(x)$ be any solution of (7.1) in $[x_0, x_0 + \alpha]$. Then $y(x_0) = y_0 < y_0 + \epsilon = y(x_0, \epsilon)$, and $y'(x) < f(x, y(x)) + \epsilon$, $y'(x, \epsilon) = f(x, y(x, \epsilon)) + \epsilon$ for all $x \in [x_0, x_0 + \alpha]$ and $0 < \epsilon \leq b/2$. Thus, from Theorem 11.1 it follows that $y(x) < y(x, \epsilon)$, $x \in [x_0, x_0 + \alpha]$. Now the uniqueness of the maximal solution shows that $y(x, \epsilon)$ tends uniformly to $r(x)$ in $[x_0, x_0 + \alpha]$ as $\epsilon \to 0$. ■

Obviously, the process of continuation of the solutions of (7.1) discussed in Lecture 9 can be employed for the maximal solution $r(x)$ as well as for the minimal solution $\rho(x)$.

Example 11.2. For the initial value problem $y' = |y|^{1/2}$, $y(0) = 0$ it is clear that $r(x) = x^2/4$, $\rho(x) = 0$ if $x \geq 0$; $r(x) = 0$, $\rho(x) = -x^2/4$ if $x \leq 0$; and

$$
r(x) = \begin{cases} \dfrac{x^2}{4} & \text{if } x \geq 0 \\ 0 & \text{if } x \leq 0, \end{cases} \qquad
\rho(x) = \begin{cases} 0 & \text{if } x \geq 0 \\ -\dfrac{x^2}{4} & \text{if } x \leq 0. \end{cases}
$$

Finally, as an application of maximal solution $r(x)$ we shall prove the following theorem.

Theorem 11.6. Let $f(x, y)$ be continuous in the domain D, and let $r(x)$ be the maximal solution of (7.1) in the interval $J = [x_0, x_0 + a)$. Also, let $y(x)$ be a solution of the differential inequality

$$
y'(x) \leq f(x, y(x)) \tag{11.8}
$$

in J. Then $y(x_0) \leq y_0$ implies that

$$
y(x) \leq r(x) \quad \text{for all} \quad x \subset J. \tag{11.9}
$$

Proof. For $x_1 \in (x_0, x_0+a)$ an argument similar to that for Theorem 11.5 shows that there exists a maximal solution $r(x, \epsilon)$ of (11.7) in $[x_0, x_1]$ for all sufficiently small $\epsilon > 0$ and $\lim_{\epsilon \to 0} r(x, \epsilon) = r(x)$ uniformly in $x \in [x_0, x_1]$. Now for (11.7) and (11.8) together with $y(x_0) \leq y_0 < r(x_0, \epsilon)$, Theorem 11.1 gives

$$
y(x) < r(x, \epsilon) \tag{11.10}
$$

in $[x_0, x_1]$. The inequality (11.9) follows by taking $\epsilon \to 0$ in (11.10). ∎

Problems

11.1. Give an example to show that in Theorem 11.1 strict inequalities cannot be replaced with equalities.

11.2. Let $y(x)$ be a solution of the initial value problem $y' = y - y^2$, $y(0) = y_0$, $0 < y_0 < 1$. Show that $y_0 < y(x) \leq 1$ for all $x \in (0, \infty)$.

11.3. Let $y(x)$ be a solution of the initial value problem $y' = y^2 - x$, $y(0) = 1$. Show that $1 + x < y(x) < 1/(1 - x)$ for all $x \in (0, 1)$.

11.4. Let $f_1(x, y)$ and $f_2(x, y)$ be continuous in the domain D, and $f_1(x, y) < f_2(x, y)$ for all $(x, y) \in D$. Further, let $y_1(x)$ and $y_2(x)$ be the solutions of the DEs $y_1' = f_1(x, y_1)$ and $y_2' = f_2(x, y_2)$, respectively, existing in $J = [x_0, x_0 + a)$ such that $y_1(x_0) < y_2(x_0)$. Show that $y_1(x) < y_2(x)$ for all $x \in J$.

*11.5. For a given function $y(x)$ the *Dini derivatives* are defined as follows:

$$D^+ y(x) = \limsup_{h \to 0^+} \left(\frac{y(x+h) - y(x)}{h} \right) \quad \text{(upper-right derivative)}$$

$$D_+ y(x) = \liminf_{h \to 0^+} \left(\frac{y(x+h) - y(x)}{h} \right) \quad \text{(lower-right derivative)}$$

$$D^- y(x) = \limsup_{h \to 0^-} \left(\frac{y(x+h) - y(x)}{h} \right) \quad \text{(upper-left derivative)}$$

$$D_- y(x) = \liminf_{h \to 0^-} \left(\frac{y(x+h) - y(x)}{h} \right) \quad \text{(lower-left derivative)}.$$

In case $D^+ y(x) = D_+ y(x)$, the right-hand derivative denoted by $y_+'(x)$ exists and $y_+'(x) = D^+ y(x) = D_+ y(x)$.

In case $D^- y(x) = D_- y(x)$, the left-hand derivative denoted by $y_-'(x)$ exists and $y_-'(x) = D^- y(x) = D_- y(x)$.

In case $y_+'(x) = y_-'(x)$, the derivative exists and $y'(x) = y_+'(x) = y_-'(x)$. Conversely, if $y'(x)$ exists, then all the four Dini derivatives are equal.

Clearly, $D_+ y(x) \leq D^+ y(x)$ and $D_- y(x) \leq D^- y(x)$. Show the following:

(i) If $y(x) \in C[x_0, x_0 + a)$, then a necessary and sufficient condition for $y(x)$ to be nonincreasing in $[x_0, x_0 + a)$ is that $Dy(x) \leq 0$, where D is a fixed Dini derivative.

(ii) If $y(x), z(x) \in C[x_0, x_0 + a)$, and $Dy(x) \leq z(x)$, where D is a fixed Dini derivative, then $D_- y(x) \leq z(x)$.

(iii) Theorem 11.1 remains true when in the inequalities (11.1) the derivative is replaced by any fixed Dini derivative.

(iv) If $y(x) \in C^{(1)}[x_0, x_0+a)$, then $z(x) = |y(x)|$ has a right-hand derivative $z_+'(x)$ and $z_+'(x) \leq |y'(x)|$.

*11.6. Let $f(x, y)$ be continuous in the domain D, and $y(x)$ be a solution of the differential inequality $D_- y \leq f(x, y)$ in $(x_0 - a, x_0]$ and $y(x_0) \geq y_0$. Show that $\rho(x) \leq y(x)$ as far as the minimal solution $\rho(x)$ of (7.1) exists to the left of x_0.

Answers or Hints

11.1. The problem $y' = y^{2/3}$, $y(0) = 0$ has solutions $y_1(x) = x^3/27$, $y_2(x)$

$\equiv 0$ in $[0, \infty)$. For these functions equalities hold everywhere in Theorem 11.1. However, $x^3/27 \not\leq 0$ in $(0, \infty)$.

11.2. Let $y_1(x) = y_0$, $y_2(x) = 1 + \epsilon$, $\epsilon > 0$ and use Theorem 11.1 and Corollary 11.2.

11.3. Let $y_1(x) = 1 + x$, $y_2(x) = 1/(1 - x)$ and use Corollary 11.2.

11.4. Define $f(x, y) = (f_1(x, y) + f_2(x, y))/2$ so that $y'_1 < f(x, y_1)$, $y'_2 > f(x, y_2)$ and $y_1(x_0) < y_2(x_0)$. Now use Theorem 11.1.

Lecture 12
Continuous Dependence on Initial Conditions

The initial value problem (7.1) describes a model of a physical problem in which often some parameters such as lengths, masses, temperatures, etc., are involved. The values of these parameters can be measured only up to certain degree of accuracy. Thus, in (7.1) the initial condition (x_0, y_0) as well as the function $f(x, y)$ may be subject to some errors either by necessity or for convenience. Hence, it is important to know how the solution of (7.1) changes when (x_0, y_0) and $f(x, y)$ are slightly altered. We shall answer this question quantitatively in the following theorem.

Theorem 12.1. Let the following conditions be satisfied:

(i) $f(x, y)$ is continuous and bounded by M in a domain D containing the points (x_0, y_0) and (x_1, y_1).

(ii) $f(x, y)$ satisfies a uniform Lipschitz condition (7.3) in D.

(iii) $g(x, y)$ is continuous and bounded by M_1 in D.

(iv) $y(x)$ and $z(x)$, the solutions of the initial value problems (7.1) and

$$z' = f(x, z) + g(x, z), \quad z(x_1) = y_1,$$

respectively, exist in an interval J containing x_0 and x_1.

Then for all $x \in J$, the following inequality holds:

$$
\begin{aligned}
|y(x) - z(x)| \;\leq\; &\left(|y_0 - y_1| + (M + M_1)|x_1 - x_0| + \frac{1}{L}M_1 \right) \\
&\times \exp\left(L|x - x_0|\right) - \frac{1}{L}M_1.
\end{aligned}
\tag{12.1}
$$

Proof. From Theorem 7.1 for all $x \in J$ it follows that

$$
\begin{aligned}
z(x) \;=\; & y_1 + \int_{x_1}^{x} [f(t, z(t)) + g(t, z(t))]dt \\
\;=\; & y_1 + \int_{x_0}^{x} f(t, z(t))dt + \int_{x_1}^{x_0} f(t, z(t))dt + \int_{x_1}^{x} g(t, z(t))dt
\end{aligned}
$$

R.P. Agarwal and D. O'Regan, *An Introduction to Ordinary Differential Equations,*
doi: 10.1007/978-0-387-71276-5_12, © Springer Science + Business Media, LLC 2008

and hence, we find

$$
\begin{aligned}
y(x) - z(x) \;=\; & y_0 - y_1 + \int_{x_0}^{x} [f(t, y(t)) - f(t, z(t))] dt \\
& + \int_{x_0}^{x_1} f(t, z(t)) dt - \int_{x_1}^{x} g(t, z(t)) dt.
\end{aligned}
\tag{12.2}
$$

Now taking absolute values on both sides of (12.2) and using the hypotheses, we get

$$
\begin{aligned}
|y(x) - z(x)| \;\leq\; & |y_0 - y_1| + (M + M_1)|x_1 - x_0| + M_1|x - x_0| \\
& + L \left| \int_{x_0}^{x} |y(t) - z(t)| dt \right|.
\end{aligned}
\tag{12.3}
$$

Inequality (12.3) is exactly the same as that considered in Corollary 7.6 with $c_0 = |y_0 - y_1| + (M + M_1)|x_1 - x_0|$, $c_1 = M_1$, $c_2 = L$ and $u(x) = |y(x) - z(x)|$, and hence the inequality (12.1) follows. ∎

From the inequality (12.1) it is apparent that the difference between the solutions $y(x)$ and $z(x)$ in the interval J is small provided the changes in the initial point (x_0, y_0) as well as in the function $f(x, y)$ do not exceed prescribed amounts. Thus, the statement "if the function $f(x, y)$ and the initial point (x_0, y_0) vary continuously, then the solutions of (7.1) vary continuously" holds. It is also clear that the solution $z(x)$ of the initial value problem $z' = f(x, z) + g(x, z)$, $z(x_1) = y_1$ need not be unique in J.

Example 12.1. Consider the initial value problem

$$
y' = \sin(xy), \quad y(0) = 1
\tag{12.4}
$$

in the rectangle $\overline{S} : |x| \leq 1/2$, $|y - 1| \leq 1/2$. To apply Theorem 8.1 we note that $a = 1/2$, $b = 1/2$ and $\max_{\overline{S}} |\sin(xy)| \leq 1 \leq M$, and from Theorem 7.2 the function $\sin(xy)$ satisfies the Lipschitz condition (7.3) in $\overline{S}$, and $\max_{\overline{S}} |x \cos(xy)| = 1/2 = L$. Thus, the problem (12.4) has a unique solution in the interval $|x| \leq h \leq 1/2$.

As an approximation of the initial value problem (12.4), we consider

$$
z' = xz, \quad z(0) = 1.1,
\tag{12.5}
$$

which also has a unique solution $z(x) = 1.1 \exp(x^2/2)$ in the interval $|x| \leq 1/2$. Now by Taylor's formula, we find

$$
|g(x, y)| \;=\; |\sin(xy) - xy| \leq \frac{1}{6}|xy|^3 \leq \frac{1}{6}\left(\frac{1}{2}\right)^3 \left(\frac{3}{2}\right)^3 = \frac{9}{128} = M_1.
$$

Using Theorem 12.1 for the above initial value problems, we obtain an upper error bound for the difference between the solutions $y(x)$ and $z(x)$

$$|y(x) - z(x)| \leq \left(0.1 + \frac{9}{64}\right) \exp\left(\frac{|x|}{2}\right) - \frac{9}{64} \quad \text{for all} \quad |x| \leq \frac{1}{2}.$$

To emphasize the dependence of the initial point (x_0, y_0), we shall denote the solution of the initial value problem (7.1) as $y(x, x_0, y_0)$. In our next result we shall show that $y(x, x_0, y_0)$ is differentiable with respect to y_0.

Theorem 12.2. Let the following conditions be satisfied:

(i) $f(x, y)$ is continuous and bounded by M in a domain D containing the point (x_0, y_0).

(ii) $\partial f(x, y)/\partial y$ exists, continuous and bounded by L in D.

(iii) The solution $y(x, x_0, y_0)$ of the initial value problem (7.1) exists in an interval J containing x_0.

Then we have that $y(x, x_0, y_0)$ is differentiable with respect to y_0 and $z(x) = \partial y(x, x_0, y_0)/\partial y_0$ is the solution of the initial value problem

$$z' = \frac{\partial f}{\partial y}(x, y(x, x_0, y_0))z \tag{12.6}$$

$$z(x_0) = 1. \tag{12.7}$$

The DE (12.6) is called the *variational equation* corresponding to the solution $y(x, x_0, y_0)$.

Proof. Let $(x_0, y_1) \in D$ be such that the solution $y(x, x_0, y_1)$ of the initial value problem $y' = f(x, y)$, $y(x_0) = y_1$ exists in an interval J_1. Then for all $x \in J_2 = J \cap J_1$, Theorem 12.1 implies that

$$|y(x, x_0, y_0) - y(x, x_0, y_1)| \leq |y_0 - y_1|e^{L|x-x_0|},$$

i.e., $|y(x, x_0, y_0) - y(x, x_0, y_1)| \to 0$ as $|y_0 - y_1| \to 0$.

Now for all $x \in J_2$ it is easy to verify that

$$y(x, x_0, y_0) - y(x, x_0, y_1) - z(x)(y_0 - y_1)$$
$$= \int_{x_0}^{x} \left[f(t, y(t, x_0, y_0)) - f(t, y(t, x_0, y_1)) - \frac{\partial f}{\partial y}(t, y(t, x_0, y_0))z(t)(y_0 - y_1) \right] dt$$
$$= \int_{x_0}^{x} \frac{\partial f}{\partial y}(t, y(t, x_0, y_0))[y(t, x_0, y_0) - y(t, x_0, y_1) - z(t)(y_0 - y_1)]dt$$
$$+ \int_{x_0}^{x} \delta\{y(t, x_0, y_0), y(t, x_0, y_1)\}dt,$$

where $\delta\{y(x, x_0, y_0), y(x, x_0, y_1)\} \to 0$ as $|y(x, x_0, y_0) - y(x, x_0, y_1)| \to 0$, i.e., as $|y_0 - y_1| \to 0$.

Hence, we find that

$$|y(x, x_0, y_0) - y(x, x_0, y_1) - z(x)(y_0 - y_1)|$$

$$\leq L \left| \int_{x_0}^x |y(t, x_0, y_0) - y(t, x_0, y_1) - z(t)(y_0 - y_1)| dt \right| + o(|y_0 - y_1|).$$

Now applying Corollary 7.6, we get

$$|y(x, x_0, y_0) - y(x, x_0, y_1) - z(x)(y_0 - y_1)| \leq o(|y_0 - y_1|) \exp(L|x - x_0|).$$

Thus, $|y(x, x_0, y_0) - y(x, x_0, y_1) - z(x)(y_0 - y_1)| \to 0$ as $|y_0 - y_1| \to 0$. This completes the proof. ∎

In our next result we shall show that the conditions of Theorem 12.2 are also sufficient for the solution $y(x, x_0, y_0)$ to be differentiable with respect to x_0.

Theorem 12.3. Let the conditions of Theorem 12.2 be satisfied. Then the solution $y(x, x_0, y_0)$ is differentiable with respect to x_0 and $z(x) = \partial y(x, x_0, y_0)/\partial x_0$ is the solution of the variational equation (12.6), satisfying the initial condition

$$z(x_0) = -f(x_0, y_0). \tag{12.8}$$

Proof. The proof is similar to that of Theorem 12.2. ∎

We note that the variational equation (12.6) can be obtained directly by differentiating the relation $y'(x, x_0, y_0) = f(x, y(x, x_0, y_0))$ with respect to y_0 (or x_0). Further, since $y(x_0, x_0, y_0) = y_0$, differentiation with respect to y_0 gives the initial condition (12.7). To obtain (12.8), we begin with the integral equation

$$y(x, x_0, y_0) = y_0 + \int_{x_0}^x f(t, y(t, x_0, y_0))dt$$

and differentiate it with respect to x_0, to obtain

$$\frac{\partial y(x, x_0, y_0)}{\partial x_0} \bigg|_{x=x_0} = -f(x_0, y_0).$$

Finally, in this lecture we shall consider the initial value problem

$$y' = f(x, y, \lambda), \quad y(x_0) = y_0, \tag{12.9}$$

where $\lambda \in \mathbb{R}$ is a parameter. The proof of the following theorem is very similar to earlier results.

Theorem 12.4. Let the following conditions be satisfied:

(i) $f(x, y, \lambda)$ is continuous and bounded by M in a domain $D \subset \mathbb{R}^3$ containing the point (x_0, y_0, λ_0).

(ii) $\partial f(x, y, \lambda)/\partial y$, $\partial f(x, y, \lambda)/\partial \lambda$ exist, continuous and bounded, respectively, by L and L_1 in D.

Then the following hold:

(1) There exist positive numbers h and ϵ such that given any λ in the interval $|\lambda - \lambda_0| \leq \epsilon$, there exits a unique solution $y(x, \lambda)$ of the initial value problem (12.9) in the interval $|x - x_0| \leq h$.

(2) For all λ_1, λ_2 in the interval $|\lambda - \lambda_0| \leq \epsilon$, and x in $|x - x_0| \leq h$ the following inequality holds:

$$|y(x, \lambda_1) - y(x, \lambda_2)| \leq \frac{L_1 |\lambda_1 - \lambda_2|}{L} \left(\exp(L|x - x_0|) - 1 \right). \qquad (12.10)$$

(3) The solution $y(x, \lambda)$ is differentiable with respect to λ and $z(x, \lambda) = \partial y(x, \lambda)/\partial \lambda$ is the solution of the initial value problem

$$z'(x, \lambda) = \frac{\partial f}{\partial y}(x, y(x, \lambda), \lambda) z(x, \lambda) + \frac{\partial f}{\partial \lambda}(x, y(x, \lambda), \lambda), \qquad (12.11)$$

$$z(x_0, \lambda) = 0. \qquad (12.12)$$

If λ is such that $|\lambda - \lambda_0|$ is sufficiently small, then we have a *first-order approximation* of the solution $y(x, \lambda)$ given by

$$y(x, \lambda) \simeq y(x, \lambda_0) + (\lambda - \lambda_0) \left[\frac{\partial y}{\partial \lambda}(x, \lambda) \right]_{\lambda = \lambda_0} \qquad (12.13)$$
$$= y(x, \lambda_0) + (\lambda - \lambda_0) z(x, \lambda_0).$$

We illustrate this important idea in the following example.

Example 12.2. Consider the initial value problem

$$y' = \lambda y^2 + 1, \quad y(0) = 0 \quad (\lambda \geq 0) \qquad (12.14)$$

for which the solution $y(x, \lambda) = (1/\sqrt{\lambda}) \tan(\sqrt{\lambda} x)$ exists in the interval $(-\pi/(2\sqrt{\lambda}), \pi/(2\sqrt{\lambda}))$. Let in (12.14) the parameter $\lambda = 0$, so that $y(x, 0) = x$. Since $\partial f/\partial y = 2\lambda y$ and $\partial f/\partial \lambda = y^2$, the initial value problem corresponding to (12.11), (12.12) is

$$z'(x, 0) = x^2, \quad z(0, 0) = 0,$$

whose solution is $z(x,0) = x^3/3$. Thus, for λ near zero, (12.13) gives the approximation

$$y(x, \lambda) = \frac{1}{\sqrt{\lambda}} \tan(\sqrt{\lambda}\, x) \simeq x + \lambda \frac{x^3}{3}.$$

Problems

12.1. Let $y(x, \lambda)$ denote the solution of the initial value problem

$$y' + p(x)y = q(x), \quad y(x_0) = \lambda$$

in the interval $x_0 \le x < \infty$. Show that for each fixed $x > x_0$ and for each positive number ϵ there exists a positive number δ such that $|y(x, \lambda + \Delta\lambda) - y(x, \lambda)| \le \epsilon$, whenever $|\Delta\lambda| \le \delta$, i.e., the solution $y(x, \lambda)$ is continuous with respect to the parameter λ.

12.2. Prove Theorem 12.3.

12.3. Prove Theorem 12.4.

12.4. For the initial value problem

$$y' = x + e^x \sin(xy), \quad y(0) = 0 = y_0$$

estimate the variation of the solution in the interval $[0, 1]$ if y_0 is perturbed by 0.01.

12.5. For the initial value problem $y' = \lambda + \cos y$, $y(0) = 0$ obtain an upper estimate for $|y(x, \lambda_1) - y(x, \lambda_2)|$ in the interval $[0, 1]$.

12.6. For sufficiently small λ find a first-order approximation of the solution $y(x, \lambda)$ of the initial value problem $y' = y + \lambda(x + y^2)$, $y(0) = 1$.

12.7. State and prove an analog of Theorem 12.1 for the initial value problem (7.9).

12.8. Find the error in using the approximate solution

$$y(x) = \exp(-x^3/6)$$

for the initial value problem $y'' + xy = 0$, $y(0) = 1$, $y'(0) = 0$ in the interval $|x| \le 1/2$.

Answers or Hints

12.1. Let $z(x) = y(x, \lambda + \Delta\lambda) - y(x, \lambda)$. Then $z'(x) + p(x)z(x) = 0$, $z(x_0) = \Delta\lambda$, whose solution is $z(x) = \Delta\lambda \exp\left(-\int_{x_0}^{x} p(t)dt\right)$.

12.2. The proof is similar to that of Theorem 12.2.

12.3. For the existence and uniqueness of solutions of (12.9) see the remark following Theorem 16.7. To prove inequality (12.10) use Corollary 7.6.

12.4. $0.01e^e$.

12.5. $|\lambda_1 - \lambda_2|(e - 1)$.

12.6. $e^x + \lambda(e^{2x} - x - 1)$.

12.7. The result corresponding to Theorem 12.1 for (7.9) can be stated as follows: Let the conditions (i)–(iii) of Theorem 12.1 be satisfied and (iv)' $y(x)$ and $z(x)$, the solutions of (7.9) and $z'' = f(x, z) + g(x, z)$, $z(x_1) = z_0$, $z'(x_1) = z_1$, respectively, exist in $J = (\alpha, \beta)$ containing x_0 and x_1. Then for all $x \in J$, the following inequality holds:

$$|y(x) - z(x)| \leq \{[|y_0 - z_0| + |x_1 - x_0|(|z_1| + (\beta - \alpha)(M + M_1))] + \gamma\}$$
$$\times \exp(L(\beta - \alpha)|x - x_0|) - \gamma,$$

where $\gamma = [|y_1 - z_1| + M_1(\beta - \alpha)]/L(\beta - \alpha)$.

12.8. $(1/512)e^{1/12}$.

Lecture 13
Preliminary Results from Algebra and Analysis

For future reference we collect here several fundamental concepts and results from algebra and analysis.

A function $P_n(x)$ defined by

$$P_n(x) = a_0 + a_1 x + \cdots + a_n x^n = \sum_{i=0}^{n} a_i x^i, \quad a_n \neq 0$$

where $a_i \in \mathbb{R}$, $0 \leq i \leq n$, is called a *polynomial* of degree n in x. If $P_n(x_1) = 0$, then the number $x = x_1$ is called a *zero* of $P_n(x)$. The following *fundamental theorem of algebra* of complex numbers is valid.

Theorem 13.1. Every polynomial of degree $n \geq 1$ has at least one zero.

Thus, $P_n(x)$ has exactly n zeros; however, some of these may be the same, i.e., $P_n(x)$ can be written as

$$P_n(x) = a_n(x - x_1)^{r_1}(x - x_2)^{r_2} \cdots (x - x_k)^{r_k}, \quad r_i \geq 1, \quad 1 \leq i \leq k,$$

where $\sum_{i=1}^{k} r_i = n$. If $r_i = 1$, then x_i is called a *simple zero*, and if $r_i > 1$, then *multiple zero* of multiplicity r_i. Thus, if x_i is a multiple zero of multiplicity r_i, then $P^{(j)}(x_i) = 0$, $0 \leq j \leq r_i - 1$ and $P^{(r_i)}(x_i) \neq 0$.

A rectangular table of $m \times n$ elements arranged in m rows and n columns

$$\begin{bmatrix} a_{11} & a_{12} & \cdots & a_{1n} \\ a_{21} & a_{22} & \cdots & a_{2n} \\ \cdots & & & \\ a_{m1} & a_{m2} & \cdots & a_{mn} \end{bmatrix}$$

is called an $m \times n$ *matrix* and in short represented as $A = (a_{ij})$. We shall mainly deal with square matrices $(m = n)$, row matrices or row vectors $(m = 1)$, and column matrices or column vectors $(n = 1)$.

A matrix with $a_{ij} = 0$, $1 \leq i, j \leq n$, is called *null* or *zero matrix*, which we shall denote by 0.

R.P. Agarwal and D. O'Regan, *An Introduction to Ordinary Differential Equations*, doi: 10.1007/978-0-387-71276-5_13, © Springer Science + Business Media, LLC 2008

A matrix with $a_{ij} = 0$, $1 \le i \ne j \le n$ is called a *diagonal matrix*; if, in addition, $a_{ii} = 1$, $1 \le i \le n$, then it is an *identity matrix* which we shall denote by I.

The *transpose* of a matrix denoted by A^T is a matrix with elements a_{ji}. A matrix is called *symmetric* if $A = A^T$.

The sum of two matrices $A = (a_{ij})$ and $B = (b_{ij})$ is a matrix $C = (c_{ij})$ with elements $c_{ij} = a_{ij} + b_{ij}$.

Let α be a constant and A be a matrix; then αA is a matrix $C = (c_{ij})$ with elements $c_{ij} = \alpha a_{ij}$.

Let A and B be two matrices; then the product AB is a matrix $C = (c_{ij})$ with elements $c_{ij} = \sum_{k=1}^{n} a_{ik} b_{kj}$. Note that in general $AB \ne BA$, but $(AB)^T = B^T A^T$.

The *trace* of a matrix A is denoted by $\text{Tr}\, A$ and it is the sum of the diagonal elements, i.e.,

$$\text{Tr}\, A \;=\; \sum_{i=1}^{n} a_{ii}.$$

Associated with an $n \times n$ matrix $A = (a_{ij})$ there is a scalar called the *determinant* of A

$$\det A \;=\; \begin{vmatrix} a_{11} & a_{12} & \cdots & a_{1n} \\ a_{21} & a_{22} & \cdots & a_{2n} \\ \cdots & & & \\ a_{n1} & a_{n2} & \cdots & a_{nn} \end{vmatrix}.$$

An $(n-1) \times (n-1)$ determinant obtained by deleting ith row and jth column of the matrix A is called the *minor* $\tilde{a}_{ij}$ of the element a_{ij}. We define the *cofactor* of a_{ij} as $\alpha_{ij} = (-1)^{i+j} \tilde{a}_{ij}$. In terms of cofactors the determinant of A is defined as

$$\det A \;=\; \sum_{j=1}^{n} a_{ij} \alpha_{ij} \;=\; \sum_{i=1}^{n} a_{ij} \alpha_{ij}. \tag{13.1}$$

Further,

$$\sum_{j=1}^{n} a_{ij} \alpha_{kj} \;=\; 0 \quad \text{if} \quad i \ne k. \tag{13.2}$$

The following properties of determinants are fundamental:

(i)   If two rows (or columns) of A are equal or have a constant ratio, then $\det A = 0$.

(ii)   If any two consecutive rows (or columns) of A are interchanged, then the determinant of the new matrix A_1 is $-\det A$.

(iii)   If a row (or column) of A is multiplied by a constant α, then the determinant of the new matrix A_1 is $\alpha \det A$.

(iv) If a constant multiple of one row (or column) of A is added to another, then the determinant of the new matrix A_1 is unchanged.

(v) $\det A^T = \det A$.

(vi) $\det AB = (\det A)(\det B)$.

A linear system of n equations in n unknowns is a set of equations of the following form:

$$
\begin{aligned}
a_{11}u_1 + \cdots + a_{1n}u_n &= b_1 \\
a_{21}u_1 + \cdots + a_{2n}u_n &= b_2 \\
&\cdots \\
a_{n1}u_1 + \cdots + a_{nn}u_n &= b_n,
\end{aligned}
\tag{13.3}
$$

where a_{ij} and b_i, $1 \le i, j \le n$, are given real numbers and n unknowns are u_i, $1 \le i \le n$.

The system (13.3) can be written in a compact form as

$$
Au = b, \tag{13.4}
$$

where A is an $n \times n$ matrix (a_{ij}), b is an $n \times 1$ vector (b_i), and u is an $n \times 1$ unknown vector (u_i). If $b = 0$, the system (13.4) is called *homogeneous*, otherwise it is called *nonhomogeneous*.

The following result provides a necessary and sufficient condition for the system (13.4) to have a unique solution.

Theorem 13.2. The system (13.4) has a unique solution if and only if $\det A \ne 0$. Alternatively, if the homogeneous system has only the trivial solution ($u = 0$), then $\det A \ne 0$.

If $\det A = 0$ then the matrix A is said to be *singular*; otherwise, *nonsingular*. Thus, the homogeneous system has nontrivial solutions if and only if the matrix A is singular. The importance of this concept lies in the fact that a nonsingular matrix A possesses a unique inverse denoted by A^{-1}. This matrix has the property that $AA^{-1} = A^{-1}A = I$. Moreover, $A^{-1} = (\text{Adj}\,A)/(\det A)$, where $\text{Adj}\,A$ is an $n \times n$ matrix with elements α_{ji}.

A real *vector space* (*linear space*) V is a collection of objects called vectors, together with an addition and a multiplication by real numbers which satisfy the following axioms:

1. Given any pair of vectors u and v in V there exists a unique vector $u + v$ in V called the sum of u and v. It is required that

(i)   addition be commutative, i.e., $u + v = v + u$;

(ii) addition be associative, i.e., $u + (v + w) = (u + v) + w$;

(iii) there exists a vector 0 in V (called the *zero vector*) such that $u + 0 = 0 + u = u$ for all u in V;

(iv)   for each u in V, there exists a vector $-u$ in V such that $u + (-u) = 0$.

2.   Given any vector u in V and any real number α, there exists a unique vector $\alpha\, u$ in V called the product or scalar product of α and u. Given any two real numbers α and β it is required that

(i) $\alpha(u + v) = \alpha u + \alpha v$,

(ii) $(\alpha + \beta)u = \alpha u + \beta u$,

(iii) $(\alpha\beta)u = \alpha(\beta u)$,

(iv) $1\, u = u$.

A complex vector space is defined analogously.

There are numerous examples of real (complex) vector spaces which are of interest in analysis, some of these are the following:

(a)   The space $\mathbb{R}^n (C^n)$ is a real (complex) vector space if for all $u = (u_1, \ldots, u_n)$, $v = (v_1, \ldots, v_n)$ in $\mathbb{R}^n (C^n)$, and $\alpha \in \mathbb{R}(C)$,

$$u + v = (u_1 + v_1, \ldots, u_n + v_n)$$
$$\alpha u = (\alpha u_1, \ldots, \alpha u_n).$$

(b)   The space of all continuous functions in an interval J denoted by $C(J)$ is a real vector space if for all $y(x)$, $z(x) \in C(J)$, and $\alpha \in \mathbb{R}$,

$$(y + z)(x) = y(x) + z(x)$$
$$(\alpha y)(x) = \alpha y(x).$$

(c)   The function $u(x) = (u_1(x), \ldots, u_n(x))$, or $(u_1(x), \ldots, u_n(x))^T$, where $u_i(x)$, $1 \leq i \leq n$, are continuous in an interval J, is called a vector-valued function. The space of all continuous vector-valued functions in J denoted by $C_n(J)$ is a real vector space if for all $u(x)$, $v(x) \in C_n(J)$ and $\alpha \in \mathbb{R}$

$$(u + v)(x) = (u_1(x) + v_1(x), \ldots, u_n(x) + v_n(x))$$
$$(\alpha u)(x) = (\alpha u_1(x), \ldots, \alpha u_n(x)).$$

(d)   The matrix $A(x) = (a_{ij}(x))$, where $a_{ij}(x)$, $1 \leq i, j \leq n$, are continuous in an interval J, is called a matrix-valued function. The space of all continuous matrix-valued functions in J denoted by $C_{n \times n}(J)$ is a real vector space if for all $A(x), B(x) \in C_{n \times n}(J)$ and $\alpha \in \mathbb{R}$

$$(A + B)(x) = (a_{ij}(x) + b_{ij}(x))$$
$$(\alpha A)(x) = (\alpha a_{ij}(x)).$$

The space of all m (nonnegative integer) times continuously differentiable functions in J denoted by $C^{(m)}(J)$ is also a vector space. Similarly, $C_n^{(m)}(J)$ and $C_{n \times n}^{(m)}(J)$, where the derivatives of the function $u(x)$ and of the matrix $A(x)$ are defined as $u'(x) = (u_i'(x))$ and $A'(x) = (a_{ij}'(x))$, respectively, are vector spaces.

From the definition of determinants it is clear that for a given matrix $A(x) \in C_{n \times n}^{(m)}(J)$ the function $\det A(x) \in C^{(m)}(J)$. Later we shall need the differentiation of the function $\det A(x)$ which we shall now compute by using the expansion of $\det A$ given in (13.1). Since

$$\det A(x) = \sum_{j=1}^{n} a_{ij}(x)\alpha_{ij}(x),$$

it follows that

$$\frac{\partial \det A(x)}{\partial a_{ij}(x)} = \alpha_{ij}(x),$$

and hence

$$(\det A(x))' = \sum_{j=1}^{n}\sum_{i=1}^{n} \frac{\partial \det A(x)}{\partial a_{ij}(x)} \frac{da_{ij}(x)}{dx} = \sum_{j=1}^{n}\sum_{i=1}^{n} \alpha_{ij}(x)a_{ij}'(x),$$

which is equivalent to

$$(\det A(x))' = \begin{vmatrix} a_{11}'(x) & \cdots & a_{1n}'(x) \\ a_{21}(x) & \cdots & a_{2n}(x) \\ \cdots \\ a_{n1}(x) & \cdots & a_{nn}(x) \end{vmatrix} + \begin{vmatrix} a_{11}(x) & \cdots & a_{1n}(x) \\ a_{21}'(x) & \cdots & a_{2n}'(x) \\ \cdots \\ a_{n1}(x) & \cdots & a_{nn}(x) \end{vmatrix} + \cdots$$

$$+ \begin{vmatrix} a_{11}(x) & \cdots & a_{1n}(x) \\ a_{21}(x) & \cdots & a_{2n}(x) \\ \cdots \\ a_{n1}'(x) & \cdots & a_{nn}'(x) \end{vmatrix}.$$

$$(13.5)$$

Let V be a vector space and let $v^1, \ldots, v^m \in V$ be fixed vectors. These vectors are said to be *linearly independent* if the choice of constants $\alpha_1, \ldots, \alpha_m$ for which $\alpha_1 v^1 + \cdots + \alpha_m v^m = 0$ is the trivial choice, i.e., $\alpha_1 = \cdots = \alpha_m = 0$. Conversely, these vectors are said to be *linearly dependent* if there exist constants $\alpha_1, \ldots, \alpha_m$ not all zero such that $\alpha_1 v^1 + \cdots + \alpha_m v^m = 0$. The set $\{v^1, \ldots, v^m\}$ is a *basis* for V if for every $v \in V$, there is a unique choice of constants $\alpha_1, \ldots, \alpha_m$ for which $v = \alpha_1 v^1 + \cdots + \alpha_m v^m$. Note that this implies $v^1, \ldots, v^m$ is independent. If such a finite basis exists, we say that V is finite dimensional. Otherwise, it is called infinite dimensional. If V is a vector space with a basis $\{v^1, \ldots, v^m\}$, then every basis for V will contain exactly m vectors. The number m is called the *dimension* of V. A nonempty subset $W \subset V$ is called a *subspace* if W is closed under the operations of addition and multiplication in V.

For a given $n \times n$ matrix A, its columns (rows) generate a subspace whose dimension is called the column (row) rank of A. It is well known that the row rank and the column rank of A are equal to the same number r. This number r is called the *rank* of A. If $r = n$, i.e., $\det A \neq 0$, then Theorem 13.2 implies that the system (13.4) has a unique solution. However, if $r < n$, then (13.4) may not have any solution. In this case, the following result provides necessary and sufficient conditions for the system (13.4) to have at least one solution.

Theorem 13.3.   If the rank of A is $n - m$ $(1 \leq m \leq n)$, then the system (13.4) possesses a solution if and only if

$$Bb = 0, \tag{13.6}$$

where B is an $m \times n$ matrix whose row vectors are linearly independent vectors b^i, $1 \leq i \leq m$, satisfying $b^i A = 0$.

Further, in the case when (13.6) holds, any solution of (13.4) can be expressed as

$$u = \sum_{i=1}^{m} c_i u^i + Sb,$$

where c_i, $1 \leq i \leq m$, are arbitrary constants, u^i, $1 \leq i \leq m$, are m linearly independent column vectors satisfying $Au^i = 0$, and S is an $n \times n$ matrix independent of b such that $ASv = v$ for any column vector v satisfying $Bv = 0$.

The matrix S in the above result is not unique.

Lecture 14
Preliminary Results from Algebra and Analysis (Contd.)

The number λ, real or complex, is called an *eigenvalue* of the matrix A if there exists a nonzero real or complex vector v such that $Av = \lambda v$. The vector v is called an *eigenvector* corresponding to the eigenvalue λ. From Theorem 13.2, λ is an eigenvalue of A if and only if it is a solution of the *characteristic equation* $p(\lambda) = \det(A - \lambda I) = 0$. Since the matrix A is $n \times n$, $p(\lambda)$ is a polynomial of degree exactly n, and it is called the *characteristic polynomial* of A. Hence, from Theorem 13.1 it follows that A has exactly n eigenvalues counting their multiplicities.

In the case when the eigenvalues $\lambda_1, \ldots, \lambda_n$ of A are distinct it is easy to find the corresponding eigenvectors $v^1, \ldots, v^n$. For this, first we note that for the fixed eigenvalue λ_j of A at least one of the cofactors of $(a_{ii} - \lambda_j)$ in the matrix $(A - \lambda_j I)$ is nonzero. If not, then from (13.5) it follows that $p'(\lambda) = -[\text{cofactor of } (a_{11} - \lambda)] - \cdots - [\text{cofactor of } (a_{nn} - \lambda)]$, and hence $p'(\lambda_j) = 0$; i.e., λ_j was a multiple root, which is a contradiction to our assumption that λ_j is simple. Now let the cofactor of $(a_{kk} - \lambda_j)$ be different from zero, then one of the possible nonzero solution of the system $(A - \lambda_j I)v^j = 0$ is $v_i^j = $ cofactor of a_{ki} in $(A - \lambda_j I)$, $1 \leq i \leq n$, $i \neq k$, $v_k^j = $ cofactor of $(a_{kk} - \lambda_j)$ in $(A - \lambda_j I)$. Since for this choice of v^j, it follows from (13.2) that every equation, except the kth one, of the system $(A - \lambda_j I)v^j = 0$ is satisfied, and for the kth equation from (13.1), we have

$$\sum_{\substack{i=1 \\ i \neq k}}^{n} a_{ki}[\text{cofactor of } a_{ki}] + (a_{kk} - \lambda_j)[\text{cofactor of } (a_{kk} - \lambda_j)] = \det(A - \lambda_j I),$$

which is also zero. In conclusion this v^j is the eigenvector corresponding to the eigenvalue λ_j.

Example 14.1. The characteristic polynomial for the matrix

$$A = \begin{bmatrix} 2 & 1 & 0 \\ 1 & 3 & 1 \\ 0 & 1 & 2 \end{bmatrix}$$

is $p(\lambda) = -\lambda^3 + 7\lambda^2 - 14\lambda + 8 = -(\lambda - 1)(\lambda - 2)(\lambda - 4) = 0$. Thus, the eigenvalues are $\lambda_1 = 1$, $\lambda_2 = 2$ and $\lambda_3 = 4$. To find the corresponding

R.P. Agarwal and D. O'Regan, *An Introduction to Ordinary Differential Equations*,
doi: 10.1007/978-0-387-71276-5_14, © Springer Science + Business Media, LLC 2008

eigenvectors we have to consider the systems $(A - \lambda_i I)v^i = 0$, $i = 1, 2, 3$. For $\lambda_1 = 1$, we find

$$(A - \lambda_1 I) = \begin{bmatrix} 1 & 1 & 0 \\ 1 & 2 & 1 \\ 0 & 1 & 1 \end{bmatrix}.$$

Since the cofactor of $(a_{11} - \lambda_1) = 1 \neq 0$, we can take $v_1^1 = 1$, and then $v_2^1 = $ cofactor of $a_{12} = -1$, $v_3^1 = $ cofactor of $a_{13} = 1$, i.e., $v^1 = [1 \ -1 \ 1]^T$.

Next for $\lambda_2 = 2$ we have

$$(A - \lambda_2 I) = \begin{bmatrix} 0 & 1 & 0 \\ 1 & 1 & 1 \\ 0 & 1 & 0 \end{bmatrix}.$$

Since the cofactor of $(a_{22} - \lambda_2) = 0$, the choice $v_2^2 = $ cofactor of $(a_{22} - \lambda_2)$ is not correct. However, cofactor of $(a_{11} - \lambda_2) = $ cofactor of $(a_{33} - \lambda_2) = -1 \neq 0$ and we can take $v_1^2 = -1$ $(v_3^2 = -1)$, then $v_2^2 = $ cofactor of $a_{12} = 0$, $v_3^2 = $ cofactor of $a_{13} = 1$ $(v_1^2 = $ cofactor of $a_{31} = 1$, $v_2^2 = $ cofactor of $a_{32} = 0)$, i.e., $v^2 = [-1 \ 0 \ 1]^T$ $([1 \ 0 \ -1]^T)$.

Similarly, we can find $v^3 = [1 \ 2 \ 1]^T$.

For the eigenvalues and eigenvectors of an $n \times n$ matrix A, we have the following basic result.

Theorem 14.1. Let $\lambda_1, \ldots, \lambda_m$ be distinct eigenvalues of an $n \times n$ matrix A and $v^1, \ldots, v^m$ be corresponding eigenvectors. Then $v^1, \ldots, v^m$ are linearly independent.

Since $p(\lambda)$ is a polynomial of degree n, and A^m for all nonnegative integers m is defined, $p(A)$ is a well-defined matrix. For this matrix $p(A)$ we state the following well-known theorem.

Theorem 14.2 (Cayley–Hamilton Theorem). Let A be an $n \times n$ matrix and let $p(\lambda) = \det(A - \lambda I)$. Then $p(A) = 0$.

If A is a nonsingular matrix, then $\ln A$ is a well-defined matrix. This important result is stated in the following theorem.

Theorem 14.3. Let A be a nonsingular $n \times n$ matrix. Then there exists an $n \times n$ matrix B (called logarithm of A) such that $A = e^B$.

A real *normed vector space* is a real vector space V in which to each vector u there corresponds a real number $\|u\|$, called the *norm* of u, which satisfies the following conditions:

(i) $\|u\| \geq 0$, and $\|u\| = 0$ if and only if $u = 0$;

(ii)   for each $c \in \mathbb{R}$, $\|cu\| = |c|\|u\|$;

(iii) the triangle inequality $\|u + v\| \leq \|u\| + \|v\|$.

From the triangle inequality it immediately follows that

$$\left| \|u\| - \|v\| \right| \leq \|u - v\|. \tag{14.1}$$

The main conclusion we draw from this inequality is that the norm is a Lipschitz function and, therefore, in particular, a continuous real-valued function.

In the vector space $\mathbb{R}^n$ the following three norms are in common use

$$\text{absolute norm } \|u\|_1 = \sum_{i=1}^{n} |u_i|,$$

$$\text{Euclidean norm } \|u\|_2 = \left(\sum_{i=1}^{n} |u_i|^2 \right)^{1/2},$$

and

$$\text{maximum norm } \|u\|_\infty = \max_{1 \leq i \leq n} |u_i|.$$

The notations $\|\cdot\|_1$, $\|\cdot\|_2$, and $\|\cdot\|_\infty$ are justified because of the fact that all these norms are special cases of a more general norm

$$\|u\|_p = \left(\sum_{i=1}^{n} |u_i|^p \right)^{1/p}, \quad p \geq 1.$$

The set of all $n \times n$ matrices with real elements can be considered as equivalent to the vector space $\mathbb{R}^{n^2}$, with a special multiplicative operation added into the vector space. Thus, a matrix norm should satisfy the usual three requirements of a vector norm and, in addition, we require

(iv) $\|AB\| \leq \|A\|\|B\|$ for all $n \times n$ matrices A, B;

(v)   compatibility with the vector norm; i.e., if $\|\cdot\|_*$ is the norm in $\mathbb{R}^n$, then $\|Au\|_* \leq \|A\|\|u\|_*$ for all $u \in \mathbb{R}^n$ and any $n \times n$ matrix A.

Once in $\mathbb{R}^n$ a norm $\|\cdot\|_*$ is fixed, then an associated matrix norm is usually defined by

$$\|A\| = \sup_{u \neq 0} \frac{\|Au\|_*}{\|u\|_*}. \tag{14.2}$$

From (14.2) condition (v) is immediately satisfied. To show (iv) we use (v) twice, to obtain

$$\|ABu\|_* = \|A(Bu)\|_* \leq \|A\|\|Bu\|_* \leq \|A\|\|B\|\|u\|_*$$

and hence for all $u \neq 0$, we have

$$\frac{\|ABu\|_*}{\|u\|_*} \leq \|A\|\|B\|,$$

or

$$\|AB\| = \sup_{u \neq 0} \frac{\|ABu\|_*}{\|u\|_*} \leq \|A\|\|B\|.$$

The norm of the matrix A induced by the vector norm $\|u\|_*$ will be denoted by $\|A\|_*$. For the three norms $\|u\|_1$, $\|u\|_2$ and $\|u\|_\infty$ the corresponding matrix norms are

$$\|A\|_1 = \max_{1 \leq j \leq n} \sum_{i=1}^{n} |a_{ij}|, \quad \|A\|_2 = \sqrt{\rho(A^T A)} \quad \text{and} \quad \|A\|_\infty = \max_{1 \leq i \leq n} \sum_{j=1}^{n} |a_{ij}|,$$

where for a given $n \times n$ matrix B with eigenvalues $\lambda_1, \ldots, \lambda_n$ not necessarily distinct $\rho(B)$ is called the *spectral radius* of B and is defined as

$$\rho(B) = \max\{|\lambda_i|, \ 1 \leq i \leq n\}.$$

A sequence $\{u^m\}$ in a normed linear space V is said to converge to $u \in V$ if and only if $\|u - u^m\| \to 0$ as $m \to \infty$. In particular, a sequence of $n \times n$ matrices $\{A_m\}$ is said to converge to a matrix A if $\|A - A_m\| \to 0$ as $m \to \infty$. Further, if $A_m = (a_{ij}^{(m)})$ and $A = (a_{ij})$, then it is same as $a_{ij}^{(m)} \to a_{ij}$ for all $1 \leq i, \ j \leq n$. Combining this definition with the Cauchy criterion for sequences of real numbers, we have the following: the sequence $\{A_m\}$ converges to a limit if and only if $\|A_k - A_\ell\| \to 0$ as $k, \ \ell \to \infty$. The series $\sum_{k=0}^{\infty} A_k$ is said to converge if and only if the sequence of its partial sums $\{\sum_{k=0}^{m} A_k\}$ is convergent. For example, the exponential series

$$e^A = I + \sum_{k=1}^{\infty} \frac{A^k}{k!}$$

converges for any matrix A. Indeed, it follows from

$$\left\| \sum_{k=0}^{m+p} A_k - \sum_{k=0}^{m} A_k \right\| = \left\| \sum_{k=m+1}^{m+p} \frac{A^k}{k!} \right\| \leq \sum_{k=m+1}^{m+p} \frac{\|A^k\|}{k!} \leq e^{\|A\|}.$$

Hence, for any $n \times n$ matrix A, e^A is a well-defined $n \times n$ matrix. Further, from Problem 14.3 we have $e^A e^{-A} = e^{(A-A)} = I$, and hence $(\det e^A)(\det e^{-A}) = 1$; i.e., the matrix e^A is always nonsingular.

Similarly, for a real number x, e^{Ax} is defined as

$$e^{Ax} = I + \sum_{k=1}^{\infty} \frac{(Ax)^k}{k!}.$$

Since each element of e^{Ax} is defined as a convergent power series, e^{Ax} is differentiable and it follows that

$$\left(e^{Ax}\right)' = \sum_{k=1}^{\infty} \frac{A^k x^{k-1}}{(k-1)!} = \sum_{k=1}^{\infty} A \frac{(Ax)^{k-1}}{(k-1)!} = Ae^{Ax} = e^{Ax}A.$$

In a normed linear space V norms $\|\cdot\|$ and $\|\cdot\|_*$ are said to be *equivalent* if there exist positive constants m and M such that for all $u \in V$, $m\|u\| \leq \|u\|_* \leq M\|u\|$. It is well known that in $\mathbb{R}^n$ all the norms are equivalent. Hence, unless otherwise stated, in $\mathbb{R}^n$ we shall always consider $\|\cdot\|_1$ norm and the subscript 1 will be dropped.

Problems

14.1. Let $\lambda_1, \ldots, \lambda_n$ be the (not necessarily distinct) eigenvalues of an $n \times n$ matrix A. Show the following:

(i)   The eigenvalues of A^T are $\lambda_1, \ldots, \lambda_n$.

(ii) For any constant α the eigenvalues of αA are $\alpha\lambda_1, \ldots, \alpha\lambda_n$.

(iii) $\sum_{i=1}^{n} \lambda_i = \operatorname{Tr} A$.

(iv) $\prod_{i=1}^{n} \lambda_i = \det A$.

(v)   If A^{-1} exists, then the eigenvalues of A^{-1} are $1/\lambda_1, \ldots, 1/\lambda_n$.

(vi) For any polynomial $P_n(x)$ the eigenvalues of $P_n(A)$ are $P_n(\lambda_1), \ldots, P_n(\lambda_n)$.

(vii) If A is upper (lower) triangular, i.e., $a_{ij} = 0$, $i > j$ ($i < j$), then the eigenvalues of A are the diagonal elements of A.

(viii) If A is real and λ_1 is complex with the corresponding eigenvector v^1, then there exists at least one i, $2 \leq i \leq n$, such that $\lambda_i = \bar{\lambda}_1$ and for such an i, $\bar{v}^1$ is the corresponding eigenvector.

14.2. Let $n \times n$ matrices $A(x)$, $B(x) \in C^{(1)}_{n \times n}(J)$, and the function $u(x) \in C^{(1)}_n(J)$. Show the following:

(i) $\dfrac{d}{dx}(A(x)B(x)) = \dfrac{dA}{dx}B(x) + A(x)\dfrac{dB}{dx}$.

(ii) $\dfrac{d}{dx}(A(x)u(x)) = \dfrac{dA}{dx}u(x) + A(x)\dfrac{du}{dx}$.

(iii) $\dfrac{d}{dx}(A^{-1}(x)) = -A^{-1}(x)\dfrac{dA}{dx}A^{-1}(x)$, provided $A^{-1}(x)$ exists.

14.3. Prove the following:

(i) $\|A^k\| \leq \|A\|^k$, $k = 0, 1, \ldots$.

(ii)   For any nonsingular matrix C, $C^{-1}e^A C = \exp(C^{-1}AC)$.

(iii) $e^{Ax}e^{Bx} = e^{(A+B)x}$ and $Be^{Ax} = e^{Ax}B$ for all real x if and only if $BA = AB$.

(iv)   For any function $u(x) \in C_n[\alpha, \beta]$,

$$\left\| \int_\alpha^\beta u(t)dt \right\| \leq \int_\alpha^\beta \|u(t)\| dt.$$

Answers or Hints

14.1. For each part use the properties of determinants. In particular, for (iii), since

$$\det (A - \lambda I) = (-1)^n (\lambda - \lambda_1) \cdots (\lambda - \lambda_n)$$
$$= (a_{11} - \lambda) \cdot \text{cofactor } (a_{11} - \lambda) + \sum_{j=2}^n a_{1j} \cdot \text{cofactor } a_{1j},$$

and since each term $a_{1j} \cdot \text{cofactor } a_{1j}$ is a polynomial of degree at most $n - 2$, on comparing the coefficients of λ^{n-1}, we get

$$(-1)^{n+1} \sum_{i=1}^n \lambda_i = \text{coefficient of } \lambda^{n-1} \text{ in } (a_{11} - \lambda) \cdot \text{cofactor } (a_{11} - \lambda).$$

Therefore, an easy induction implies

$$(-1)^{n+1} \sum_{i=1}^n \lambda_i = \text{coefficient of } \lambda^{n-1} \text{ in } (a_{11} - \lambda) \cdots (a_{nn} - \lambda)$$
$$= (-1)^{n-1} \sum_{i=1}^n a_{ii}.$$

14.2. For part (iii) use $A(x)A^{-1}(x) = I$.

14.3. (i) Use induction. (ii) $C^{-1} \left(\sum_{i=0}^\infty A^i/i! \right) C = \sum_{i=0}^\infty (C^{-1}AC)^i/i!$. (iii) $(A + B)^2 = A^2 + AB + BA + B^2 = A^2 + 2AB + B^2$ if and only if $AB = BA$. (iv) Use the fact that $\| \int_\alpha^\beta u(t)dt \| = \sum_{i=1}^n |\int_\alpha^\beta u_i(t)dt|$.

Lecture 15

Existence and Uniqueness of Solutions of Systems

So far we have concentrated on the existence and uniqueness of solutions of scalar initial value problems. It is natural to extend these results to a system of first-order DEs and higher-order DEs. We consider a system of first-order DEs of the form

$$
\begin{aligned}
u_1' &= g_1(x, u_1, \ldots, u_n) \\
u_2' &= g_2(x, u_1, \ldots, u_n) \\
&\quad \cdots \\
u_n' &= g_n(x, u_1, \ldots, u_n).
\end{aligned}
\tag{15.1}
$$

Such systems arise frequently in many branches of applied sciences, especially in the analysis of vibrating mechanical systems with several degrees of freedom. Furthermore, these systems have mathematical importance in themselves, e.g., each nth-order DE (1.6) is equivalent to a system of n first-order equations. Indeed, if we take $y^{(i)} = u_{i+1}$, $0 \le i \le n-1$, then the equation (1.6) can be written as

$$
\begin{aligned}
u_i' &= u_{i+1}, \quad 1 \le i \le n-1 \\
u_n' &= f(x, u_1, \ldots, u_n),
\end{aligned}
\tag{15.2}
$$

which is of the type (15.1).

Throughout, we shall assume that the functions $g_1, \ldots, g_n$ are continuous in some domain E of $(n+1)$-dimensional space $\mathbb{R}^{n+1}$. By a solution of (15.1) in an interval J we mean a set of n functions $u_1(x), \ldots, u_n(x)$ such that (i) $u_1'(x), \ldots, u_n'(x)$ exist for all $x \in J$, (ii) for all $x \in J$ the points $(x, u_1(x), \ldots, u_n(x)) \in E$, and (iii) $u_i'(x) = g_i(x, u_1(x), \ldots, u_n(x))$ for all $x \in J$. In addition to the differential system (15.1) there may also be given initial conditions of the form

$$
u_1(x_0) = u_1^0, \quad u_2(x_0) = u_2^0, \ldots, u_n(x_0) = u_n^0,
\tag{15.3}
$$

where x_0 is a specified value of x in J and $u_1^0, \ldots, u_n^0$ are prescribed numbers such that $(x_0, u_1^0, \ldots, u_n^0) \in E$. The differential system (15.1) together with the initial conditions (15.3) forms an initial value problem.

To study the existence and uniqueness of the solutions of (15.1), (15.3), there are two possible approaches, either directly imposing sufficient conditions on the functions $g_1, \ldots, g_n$ and proving the results, or alternatively

R.P. Agarwal and D. O'Regan, *An Introduction to Ordinary Differential Equations*,
doi: 10.1007/978-0-387-71276-5_15, © Springer Science + Business Media, LLC 2008

using vector notations to write (15.1), (15.3) in a compact form and then proving the results. We shall prefer to use the second approach since then the proofs are very similar to the scalar case.

By setting

$$u(x) = (u_1(x), \ldots, u_n(x)) \quad \text{and} \quad g(x, u) = (g_1(x, u), \ldots, g_n(x, u))$$

and agreeing that differentiation and integration are to be performed component-wise, i.e., $u'(x) = (u_1'(x), \ldots, u_n'(x))$ and

$$\int_\alpha^\beta u(x)dx = \left(\int_\alpha^\beta u_1(x)dx, \ldots, \int_\alpha^\beta u_n(x)dx \right),$$

the problem (15.1), (15.3) can be written as

$$u' = g(x, u), \quad u(x_0) = u^0, \tag{15.4}$$

which is exactly the same as (7.1) except now u and u' are the functions defined in J; and taking the values in $\mathbb{R}^n$, $g(x, u)$ is a function from $E \subseteq \mathbb{R}^{n+1}$ to $\mathbb{R}^n$ and $u^0 = (u_1^0, \ldots, u_n^0)$.

The function $g(x, u)$ is said to be continuous in E if each of its components is continuous in E. The function $g(x, u)$ is defined to be uniformly Lipschitz continuous in E if there exists a nonnegative constant L (Lipschitz constant) such that

$$\|g(x, u) - g(x, v)\| \leq L\|u - v\| \tag{15.5}$$

for all (x, u), (x, v) in the domain E. For example, let $g(x, u) = (a_{11}u_1 + a_{12}u_2, a_{21}u_1 + a_{22}u_2)$ and $E = \mathbb{R}^3$, then

$$
\begin{aligned}
&\|g(x, u) - g(x, v)\| \\
&= \|(a_{11}(u_1 - v_1) + a_{12}(u_2 - v_2), a_{21}(u_1 - v_1) + a_{22}(u_2 - v_2))\| \\
&= |a_{11}(u_1 - v_1) + a_{12}(u_2 - v_2)| + |a_{21}(u_1 - v_1) + a_{22}(u_2 - v_2)| \\
&\leq |a_{11}||u_1 - v_1| + |a_{12}||u_2 - v_2| + |a_{21}||u_1 - v_1| + |a_{22}||u_2 - v_2| \\
&= [|a_{11}| + |a_{21}|]|u_1 - v_1| + [|a_{12}| + |a_{22}|]|u_2 - v_2| \\
&\leq \max\{|a_{11}| + |a_{21}|, |a_{12}| + |a_{22}|\}[|u_1 - v_1| + |u_2 - v_2|] \\
&= \max\{|a_{11}| + |a_{21}|, |a_{12}| + |a_{22}|\}\|u - v\|.
\end{aligned}
$$

Hence, the Lipschitz constant is

$$L = \max\{|a_{11}| + |a_{21}|, |a_{12}| + |a_{22}|\}.$$

The following result provides sufficient conditions for the function $g(x, u)$ to satisfy the Lipschitz condition (15.5).

Theorem 15.1. Let the domain E be convex and for all (x, u) in E the partial derivatives $\partial g/\partial u_k,\ k = 1, \ldots, n$ exist and $\|\partial g/\partial u\| \leq L$. Then the function $g(x, u)$ satisfies the Lipschitz condition (15.5) in E with Lipschitz constant L.

Proof. Let (x, u) and (x, v) be fixed points in E. Then since E is convex, for all $0 \leq t \leq 1$ the points $(x, v + t(u - v))$ are in E. Thus, the vector-valued function $G(t) = g(x, v + t(u - v)),\ 0 \leq t \leq 1$ is well defined, also

$$G'(t) = (u_1 - v_1)\frac{\partial g}{\partial u_1}(x, v + t(u - v)) + \cdots$$
$$+ (u_n - v_n)\frac{\partial g}{\partial u_n}(x, v + t(u - v))$$

and hence

$$\|G'(t)\| \leq \sum_{i=1}^{n}\left|\frac{\partial g_i}{\partial u_1}(x, v + t(u - v))\right||u_1 - v_1| + \cdots$$
$$+ \sum_{i=1}^{n}\left|\frac{\partial g_i}{\partial u_n}(x, v + t(u - v))\right||u_n - v_n|$$
$$\leq L[|u_1 - v_1| + \cdots + |u_n - v_n|] = L\|u - v\|.$$

Now from the relation

$$g(x, u) - g(x, v) = G(1) - G(0) = \int_0^1 G'(t)dt$$

we find that

$$\|g(x, u) - g(x, v)\| \leq \int_0^1 \|G'(t)\|dt \leq L\|u - v\|. \quad \blacksquare$$

As an example once again we consider

$$g(x, u) = (a_{11}u_1 + a_{12}u_2,\ a_{21}u_1 + a_{22}u_2).$$

Since

$$\frac{\partial g}{\partial u_1} = (a_{11},\ a_{21}), \quad \frac{\partial g}{\partial u_2} = (a_{12},\ a_{22}),$$
$$\left\|\frac{\partial g}{\partial u}\right\| = \max\{|a_{11}| + |a_{21}|,\ |a_{12}| + |a_{22}|\} = L,$$

as it should be.

Next arguing as in Theorem 7.1, we see that if $g(x, u)$ is continuous in the domain E, then any solution of (15.4) is also a solution of the integral equation

$$u(x) = u^0 + \int_{x_0}^{x} g(t, u(t))dt \tag{15.6}$$

and conversely.

To find a solution of the integral equation (15.6) the Picard method of successive approximations is equally useful. Let $u^0(x)$ be any continuous function which we assume to be an initial approximation of the solution, then we define approximations successively by

$$u^{m+1}(x) = u^0 + \int_{x_0}^x g(t, u^m(t))dt, \quad m = 0, 1, \ldots \qquad (15.7)$$

and, as before, if the sequence of functions $\{u^m(x)\}$ converges uniformly to a continuous function $u(x)$ in some interval J containing x_0 and for all $x \in J$, the points $(x, u(x)) \in E$, then this function $u(x)$ will be a solution of the integral equation (15.6).

Example 15.1. For the initial value problem

$$\begin{aligned}
u_1' &= x + u_2 \\
u_2' &= x + u_1 \\
u_1(0) &= 1, \quad u_2(0) = -1
\end{aligned} \qquad (15.8)$$

we take $u^0(x) = (1, -1)$, to obtain

$$\begin{aligned}
u^1(x) &= (1, -1) + \int_0^x (t - 1, t + 1)dt = \left(1 - x + \frac{x^2}{2}, -1 + x + \frac{x^2}{2}\right) \\
u^2(x) &= (1, -1) + \int_0^x \left(t - 1 + t + \frac{t^2}{2}, \ t + 1 - t + \frac{t^2}{2}\right) dt \\
&= \left(1 - x + \frac{2x^2}{2} + \frac{x^3}{3!}, \ -1 + x + \frac{x^3}{3!}\right) \\
u^3(x) &= \left(1 - x + \frac{2x^2}{2} + \frac{x^4}{4!}, \ -1 + x + \frac{2x^3}{3!} + \frac{x^4}{4!}\right) \\
u^4(x) &= \left(1 - x + \frac{2x^2}{2} + \frac{2x^4}{4!} + \frac{x^5}{5!}, \ -1 + x + \frac{2x^3}{3!} + \frac{x^5}{5!}\right) \\
&= \left(-(1 + x) + \left(2 + \frac{2x^2}{2!} + \frac{2x^4}{4!} + \frac{x^5}{5!}\right), \right. \\
&\qquad \left. -(1 + x) + \left(2x + \frac{2x^3}{3!} + \frac{x^5}{5!}\right)\right)
\end{aligned}$$

$\cdots$.

Hence, the sequence $\{u^m(x)\}$ exists for all real x and converges to $u(x) = (-(1+x) + e^x + e^{-x}, \ -(1+x) + e^x - e^{-x})$, which is the solution of the initial value problem (15.8).

Now we shall state several results for the initial value problem (15.4) which are analogous to those proved in earlier lectures for the problem (7.1).

Theorem 15.2 (Local Existence Theorem). Let the following conditions hold:

(i) $g(x, u)$ is continuous in $\Omega : |x - x_0| \leq a$, $\|u - u^0\| \leq b$ and hence there exists a $M > 0$ such that $\|g(x, u)\| \leq M$ for all $(x, u) \in \Omega$.

(ii) $g(x, u)$ satisfies a uniform Lipschitz condition (15.5) in Ω.

(iii) $u^0(x)$ is continuous in $|x - x_0| \leq a$ and $\|u^0(x) - u^0\| \leq b$.

Then the sequence $\{u^m(x)\}$ generated by the Picard iterative scheme (15.7) converges to the unique solution $u(x)$ of the problem (15.4). This solution is valid in the interval $J_h : |x - x_0| \leq h = \min\{a, b/M\}$. Further, for all $x \in J_h$, the following error estimate holds

$$\|u(x) - u^m(x)\| \leq N e^{Lh} \min\left\{1, \frac{(Lh)^m}{m!}\right\}, \quad m = 0, 1, \ldots$$

where $\|u^1(x) - u^0(x)\| \leq N$.

Theorem 15.3 (Global Existence Theorem). Let the following conditions hold:

(i) $g(x, u)$ is continuous in $\Delta : |x - x_0| \leq a$, $\|u\| < \infty$.

(ii) $g(x, u)$ satisfies a uniform Lipschitz condition (15.5) in Δ.

(iii) $u^0(x)$ is continuous in $|x - x_0| \leq a$.

Then the sequence $\{u^m(x)\}$ generated by the Picard iterative scheme (15.7) exists in the entire interval $|x - x_0| < a$, and converges to the unique solution $u(x)$ of the problem (15.4).

Corollary 15.4. Let $g(x, u)$ be continuous in $\mathbb{R}^{n+1}$ and satisfy a uniform Lipschitz condition (15.5) in each $\Delta_a : |x| \leq a$, $\|u\| < \infty$ with the Lipschitz constant L_a. Then the problem (15.4) has a unique solution which exists for all x.

Theorem 15.5 (Peano's Existence Theorem). Let $g(x, u)$ be continuous and bounded in Δ. Then the problem (15.4) has at least one solution in $|x - x_0| \leq a$.

Definition 15.1. Let $g(x, u)$ be continuous in a domain E. A function $u(x)$ defined in J is said to be an ϵ-approximate solution of the differential system $u' = g(x, u)$ if (i) $u(x)$ is continuous for all x in J, (ii) for all $x \in J$ the points $(x, u(x)) \in E$, (iii) $u(x)$ has a piecewise continuous derivative in J which may fail to be defined only for a finite number of points, say, $x_1, x_2, \ldots, x_k$, and (iv) $\|u'(x) - g(x, u(x))\| \leq \epsilon$ for all $x \in J$, $x \neq x_i$, $i = 1, 2, \ldots, k$.

Theorem 15.6. Let $g(x, u)$ be continuous in Ω, and hence there exists a $M > 0$ such that $\|g(x, u)\| \leq M$ for all $(x, u) \in \Omega$. Then for any $\epsilon > 0$, there

exists an ϵ-approximate solution $u(x)$ of the differential system $u' = g(x, u)$ in the interval J_h such that $u(x_0) = u^0$.

Theorem 15.7 (Cauchy–Peano's Existence Theorem).
Let the conditions of Theorem 15.7 be satisfied. Then the problem (15.4) has at least one solution in J_h.

Lecture 16
Existence and Uniqueness of Solutions of Systems (Contd.)

In this lecture we shall continue extending the results for the initial value problem (15.4) some of which are analogous to those proved in earlier lectures for the problem (7.1).

Theorem 16.1 (Continuation of Solutions). Assume that $g(x, u)$ is continuous in E and $u(x)$ is a solution of the problem (15.4) in an interval J. Then $u(x)$ can be extended as a solution of (15.4) to the boundary of E.

Corollary 16.2. Assume that $g(x, u)$ is continuous in

$$E_1 = \{(x, u) \in E : \ x_0 \leq x < x_0 + a, \ a < \infty, \ \|u\| < \infty\}.$$

If $u(x)$ is any solution of (15.4), then the largest interval of existence of $u(x)$ is either $[x_0, x_0 + a]$ or $[x_0, x_0 + \alpha)$, $\alpha < a$ and $\|u(x)\| \to \infty$ as $x \to x_0 + \alpha$.

Theorem 16.3 (Perron's Uniqueness Theorem). Let $f(x, y)$, $f(x, 0) \equiv 0$, be a nonnegative continuous function defined in the rectangle $x_0 \leq x \leq x_0 + a$, $0 \leq y \leq 2b$. For every $x_1 \in (x_0, x_0 + a)$, let $y(x) \equiv 0$ be the only differentiable function satisfying the initial value problem

$$y' = f(x, y), \quad y(x_0) = 0 \tag{16.1}$$

in the interval $[x_0, x_1)$. Further, let $g(x, u)$ be continuous in $\Omega_+ : x_0 \leq x \leq x_0 + a$, $\|u - u^0\| \leq b$ and

$$\|g(x, u) - g(x, v)\| \ \leq \ f(x, \|u - v\|) \tag{16.2}$$

for all (x, u), $(x, v) \in \Omega_+$. Then the problem (15.4) has at most one solution in $[x_0, x_0 + a]$.

Proof. Suppose $u(x)$ and $v(x)$ are any two solutions of (15.4) in $[x_0, x_0 + a]$. Let $y(x) = \|u(x) - v(x)\|$, then clearly $y(x_0) = 0$, and from Problem 11.5 it follows that

$$D^+ y(x) \ \leq \ \|u'(x) - v'(x)\| \ = \ \|g(x, u(x)) - g(x, v(x))\|. \tag{16.3}$$

R.P. Agarwal and D. O'Regan, *An Introduction to Ordinary Differential Equations*, doi: 10.1007/978-0-387-71276-5_16, © Springer Science + Business Media, LLC 2008

Using inequality (16.2) in (16.3), we obtain $D^+y(x) \leq f(x, y(x))$. Therefore, from Theorem 11.6 it follows that $y(x) \leq r(x)$, $x \in [x_0, x_1)$ for any $x_1 \in (x_0, x_0 + a)$, where $r(x)$ is the maximal solution of (16.1). However, from the hypothesis $r(x) \equiv 0$, and hence $y(x) \equiv 0$ in $[x_0, x_1)$. This proves the theorem. ∎

In Theorem 16.3 the function $f(x, y) = h(x)y$, where $h(x) \geq 0$ is continuous in $[x_0, x_0 + a]$ is admissible, i.e., it includes the Lipschitz uniqueness criterion.

For our next result we need the following lemma.

Lemma 16.4. Let $f(x, y)$ be a nonnegative continuous function for $x_0 < x \leq x_0 + a$, $0 \leq y \leq 2b$ with the property that the only solution $y(x)$ of the DE $y' = f(x, y)$ in any interval (x_0, x_1) where $x_1 \in (x_0, x_0 + a)$ for which $y'_+(x_0)$ exists, and

$$y(x_0) = y'_+(x_0) = 0 \tag{16.4}$$

is $y(x) \equiv 0$. Further, let $f_1(x, y)$ be a nonnegative continuous function for $x_0 \leq x \leq x_0 + a$, $0 \leq y \leq 2b$, $f_1(x, 0) \equiv 0$ and

$$f_1(x, y) \leq f(x, y), \quad x \neq x_0. \tag{16.5}$$

Then for every $x_1 \in (x_0, x_0 + a)$, $y_1(x) \equiv 0$ is the only differentiable function in $[x_0, x_1)$, which satisfies

$$y_1' = f_1(x, y_1), \quad y_1(x_0) = 0. \tag{16.6}$$

Proof. Let $r(x)$ be the maximal solution of (16.6) in $[x_0, x_1)$. Since $f_1(x, 0) \equiv 0$, $y_1(x) \equiv 0$ is a solution of the problem (16.6). Thus, $r(x) \geq 0$ in $[x_0, x_1)$. Hence, it suffices to show that $r(x) = 0$ in $[x_0, x_1)$. Suppose, on the contrary, that there exists a x_2, $x_0 < x_2 < x_1$ such that $r(x_2) > 0$. Then because of the inequality (16.5), we have

$$r'(x) \leq f(x, r(x)), \quad x_0 < x \leq x_2.$$

If $\rho(x)$ is the minimal solution of

$$y' = f(x, y), \quad y(x_2) = r(x_2),$$

then an application of Problem 11.6 implies that

$$\rho(x) \leq r(x) \tag{16.7}$$

as long as $\rho(x)$ exists to the left of x_2. The solution $\rho(x)$ can be continued to $x = x_0$. If $\rho(x_3) = 0$, for some x_3, $x_0 < x_3 < x_2$, we can affect the continuation by defining $\rho(x) = 0$ for $x_0 < x < x_3$. Otherwise, (16.7) ensures the possibility of continuation. Since $r(x_0) = 0$, $\lim_{x \to x_0^+} \rho(x) = 0$, and we define $\rho(x_0) = 0$. Furthermore, since $f_1(x, y)$ is continuous at $(x_0, 0)$ and $f_1(x_0, 0) = 0$, $r'_+(x_0)$ exists and is equal to zero. This, because of (16.7), implies that $\rho'_+(x_0)$ exists and $\rho'_+(x_0) = 0$. Thus, $\rho'(x) = f(x, \rho(x))$, $\rho(x_0) = 0$, $\rho'_+(x_0) = 0$, and hence from the hypothesis on $f(x, y)$ it follows that $\rho(x) \equiv 0$. This contradicts the assumption that $\rho(x_2) = r(x_2) > 0$. Therefore, $r(x) \equiv 0$.   ∎

Theorem 16.5 (Kamke's Uniqueness Theorem).   Let $f(x, y)$ be as in Lemma 16.4, and $g(x, u)$ as in Theorem 16.3, except that the condition (16.2) holds for all (x, u), $(x, v) \in \Omega_+$, $x \neq x_0$. Then the problem (15.4) has at most one solution in $[x_0, x_0 + a]$.

Proof. Define the function

$$f_g(x, y) = \sup_{\|u-v\|=y} \|g(x, u) - g(x, v)\| \qquad (16.8)$$

for $x_0 \leq x \leq x_0 + a$, $0 \leq y \leq 2b$. Since $g(x, u)$ is continuous in Ω_+, the function $f_g(x, y)$ is continuous for $x_0 \leq x \leq x_0 + a$, $0 \leq y \leq 2b$. From (16.8) it is clear that the condition (16.2) holds for the function $f_g(x, y)$. Moreover, $f_g(x, y) \leq f(x, y)$ for $x_0 < x \leq x_0 + a$, $0 \leq y \leq 2b$. Lemma 16.4 is now applicable with $f_1(x, y) = f_g(x, y)$ and therefore $f_g(x, y)$ satisfies the assumptions of Theorem 16.3. This completes the proof.   ∎

Kamke's uniqueness theorem is evidently more general than that of Perron and it includes as special cases many known criteria, e.g., the following:

1.   Osgood's criterion in the interval $[x_0, x_0 + a]$: $f(x, y) = w(y)$, where the function $w(y)$ is as in Lemma 10.3.

2.   Nagumo's criterion in the interval $[x_0, x_0 + a]$: $f(x, y) = ky/(x - x_0)$, $k \leq 1$.

3.   Krasnoselski–Krein criterion in the interval $[x_0, x_0 + a]$:

$$f(x, y) = \min\left\{\frac{ky}{x - x_0}, Cy^\alpha\right\}, \quad C > 0, \quad 0 < \alpha < 1, \quad k(1 - \alpha) < 1.$$

Theorem 16.6 (Continuous Dependence on Initial Conditions).   Let the following conditions hold:

(i) $g(x, u)$ is continuous and bounded by M in a domain E containing the points (x_0, u^0) and (x_1, u^1).

(ii) $g(x, u)$ satisfies a uniform Lipschitz condition (15.5) in E.

(iii) $h(x, u)$ is continuous and bounded by M_1 in E.

(iv) $u(x)$ and $v(x)$ are the solutions of the initial value problems (15.4) and

$$v' = g(x, v) + h(x, v), \quad v(x_1) = u^1,$$

respectively, which exist in an interval J containing x_0 and x_1.

Then for all $x \in J$, the following inequality holds:

$$\|u(x) - v(x)\| \leq \left(\|u^0 - u^1\| + (M + M_1)|x_1 - x_0| + \frac{1}{L}M_1 \right)$$

$$\times \exp\left(L|x - x_0|\right) - \frac{1}{L}M_1.$$

Theorem 16.7 (Differentiation with Respect to Initial Conditions). Let the following conditions be satisfied:

(i) $g(x, u)$ is continuous and bounded by M in a domain E containing the point (x_0, u^0).

(ii)   The matrix $\partial g(x, u)/\partial u$ exists and is continuous and bounded by L in E.

(iii) The solution $u(x, x_0, u^0)$ of the initial value problem (15.4) exists in an interval J containing x_0.

Then the following hold:

1. The solution $u(x, x_0, u^0)$ is differentiable with respect to u^0, and for each j $(1 \leq j \leq n)$, $v^j(x) = \partial u(x, x_0, u^0)/\partial u_j^0$ is the solution of the initial value problem

$$v' = \frac{\partial g}{\partial u}(x, u(x, x_0, u^0))v \tag{16.9}$$

$$v(x_0) = e^j = (0, \ldots, 0, 1, 0, \ldots, 0). \tag{16.10}$$

2. The solution $u(x, x_0, u^0)$ is differentiable with respect to x_0 and $v(x) = \partial u(x, x_0, u^0)/\partial x_0$ is the solution of the differential system (16.9), satisfying the initial condition

$$v(x_0) = -g(x_0, u^0). \tag{16.11}$$

Finally, in this lecture we shall consider the differential system

$$u' = g(x, u, \lambda), \tag{16.12}$$

where $\lambda = (\lambda_1, \ldots, \lambda_m) \in \mathbb{R}^m$ is a parameter.

If in (16.12) we treat $\lambda_1, \ldots, \lambda_m$ as new variables, then

$$\frac{d\lambda_i}{dx} = 0, \quad 1 \leq i \leq m. \tag{16.13}$$

Thus, the new system consisting of (16.12) and (16.13) is exactly of the form (15.1), but instead of n, now it is $(n+m)$-dimensional. Hence, for the initial value problem

$$u' = g(x, u, \lambda), \quad u(x_0) = u^0, \tag{16.14}$$

the result analogous to Theorem 12.4 can be stated as follows.

Theorem 16.8. Let the following conditions be satisfied:

(i) $g(x, u, \lambda)$ is continuous and bounded by M in a domain $E \subset \mathbb{R}^{n+m+1}$ containing the point (x_0, u^0, λ^0).

(ii) The matrix $\partial g(x, u, \lambda)/\partial u$ exists and is continuous and bounded by L in E.

(iii) The $n \times m$ matrix $\partial g(x, u, \lambda)/\partial \lambda$ exists and is continuous and bounded by L_1 in E.

Then the following hold:

1. There exist positive numbers h and ϵ such that for any λ satisfying $\|\lambda - \lambda^0\| \leq \epsilon$, there exists a unique solution $u(x, \lambda)$ of the problem (16.14) in the interval $|x - x_0| \leq h$.

2. For all λ^i such that $\|\lambda^i - \lambda^0\| \leq \epsilon$, $i = 1, 2$, and x in $|x - x_0| \leq h$ the following inequality holds:

$$\|u(x, \lambda^1) - u(x, \lambda^2)\| \leq \frac{L_1}{L}\|\lambda^1 - \lambda^2\|(\exp(L|x - x_0|) - 1).$$

3. The solution $u(x, \lambda)$ is differentiable with respect to λ and for each j $(1 < j < m)$, $v^j(x, \lambda) = \partial u(x, \lambda)/\partial \lambda_j$ is the solution of the initial value problem

$$v'(x, \lambda) = \frac{\partial g}{\partial u}(x, u(x, \lambda), \lambda)v(x, \lambda) + \frac{\partial g}{\partial \lambda_j}(x, u(x, \lambda), \lambda) \quad (16.15)$$

$$v(x_0, \lambda) = 0. \quad (16.16)$$

Problems

16.1. Solve the following problems by using Picard's method of successive approximations:

(i) $u' = \begin{bmatrix} 0 & 1 \\ -1 & 0 \end{bmatrix} u$, $u(0) = \begin{bmatrix} 0 \\ 1 \end{bmatrix}$.

(ii) $u' = \begin{bmatrix} 0 & 1 \\ 1 & 0 \end{bmatrix} u + \begin{bmatrix} x \\ x \end{bmatrix}$, $u(0) = \begin{bmatrix} 2 \\ -2 \end{bmatrix}$.

16.2. Show that the problem (1.6), (1.8) is equivalent to the integral equation

$$y(x) = \sum_{i=0}^{n-1} \frac{(x-x_0)^i}{i!} y_i + \frac{1}{(n-1)!} \int_{x_0}^{x} (x-t)^{n-1} f(t, y(t), y'(t), \ldots, y^{(n-1)}(t))dt.$$

16.3. Let the following conditions hold:

(i) $f(x, \phi_0, \ldots, \phi_{n-1})$ is continuous in

$$\Omega_1 : |x - x_0| \le a, \quad \sum_{i=0}^{n-1} |\phi_i - y_i| \le b$$

and hence there exists a $M > 0$ such that $\sup_{\Omega_1} |f(x, \phi_0, \ldots, \phi_{n-1})| \le M$.
(ii) $f(x, \phi_0, \ldots, \phi_{n-1})$ satisfies a uniform Lipschitz condition in Ω_1, i.e., for all $(x, \phi_0, \ldots, \phi_{n-1}), (x, \psi_0, \ldots, \psi_{n-1}) \in \Omega_1$ there exists a constant L such that

$$|f(x, \phi_0, \ldots, \phi_{n-1}) - f(x, \psi_0, \ldots, \psi_{n-1})| \le L \sum_{i=0}^{n-1} |\phi_i - \psi_i|.$$

Show that the problem (1.6), (1.8) has a unique solution in the interval $J_h : |x - x_0| \le h = \min\{a, b/M_1\}$, where $M_1 = M + b + \sum_{i=0}^{n-1} |y_i|$.

16.4. Let $y(x)$ and $z(x)$ be two solutions of the DE

$$y^{(n)} + p_1(x)y^{(n-1)} + \cdots + p_n(x)y = r(x) \qquad (16.17)$$

in the interval J containing the point x_0. Show that for all x in J

$$u(x_0) \exp(-2K|x - x_0|) \le u(x) \le u(x_0) \exp(2K|x - x_0|),$$

where

$$K = 1 + \sum_{i=1}^{n} \sup_{x \in J} |p_i(x)| \quad \text{and} \quad u(x) = \sum_{i=0}^{n-1} (y^{(i)}(x) - z^{(i)}(x))^2.$$

*16.5. Consider the initial value problem

$$y'' + \alpha(y, y')y' + \beta(y) = f(x), \quad y(0) = y_0, \quad y'(0) = y_1$$

where $\alpha(y, y'), \beta(y)$ are continuous together with their first-order partial derivatives, and $f(x)$ is continuous and bounded on $\mathbb{R}$, $\alpha \ge 0$, $y\beta(y) \ge 0$. Show that this problem has a unique solution and it can be extended to $[0, \infty)$.

16.6. Using an example of the form

$$u_1' = u_2$$
$$u_2' = -u_1$$

observe that a generalization of Theorem 11.1 to systems of first-order DEs with inequalities interpreted component-wise is in general not true.

Answers or Hints

16.1. (i) $(\sin x, \cos x)^T$. (ii) $(e^x + 2e^{-x} - x - 1, e^x - 2e^{-x} - x - 1)^T$.

16.2. Use Taylor's formula.

16.3. Write (1.6), (1.8) in system form and then apply Theorem 15.2.

16.4. Use the inequality $2|a||b| \leq a^2 + b^2$ to get $-2Ku(x) \leq u'(x) \leq 2Ku(x)$.

16.6. Let $J = [0, \pi)$, $u(x) = (\sin x, \cos x)^T$ and $v(x) = (-\epsilon, 1 - \epsilon)^T$, $0 < \epsilon < 1$.

Lecture 17
General Properties
of Linear Systems

If the system (15.1) is linear, i.e.,

$$g_i(x, u) = a_{i1}(x)u_1 + a_{i2}(x)u_2 + \cdots + a_{in}(x)u_n + b_i(x), \quad 1 \le i \le n,$$

then it can be written as

$$u' = A(x)u + b(x), \tag{17.1}$$

where $A(x)$ is an $n \times n$ matrix with elements $a_{ij}(x)$, $b(x)$ is an $n \times 1$ vector with components $b_i(x)$, and $u(x)$ is an $n \times 1$ unknown vector with components $u_i(x)$.

The existence and uniqueness of solutions of the differential system (17.1) together with the initial condition

$$u(x_0) = u^0 \tag{17.2}$$

in an interval J containing x_0 follows from Corollary 15.4 provided the functions $a_{ij}(x)$, $b_i(x)$, $1 \le i$, $j \le n$ are continuous in J which we shall assume throughout.

The principle of superposition for the first-order linear DEs given in Problem 5.2 holds for the differential system (17.1) also, and it is stated as follows: If $u(x)$ is a solution of the differential system $u' = A(x)u + b^1(x)$ and $v(x)$ is a solution of $v' = A(x)v + b^2(x)$, then $\phi(x) = c_1 u(x) + c_2 v(x)$ is a solution of the differential system $\phi' = A(x)\phi + c_1 b^1(x) + c_2 b^2(x)$. For this, we have

$$
\begin{aligned}
\phi'(x) &= c_1 u'(x) + c_2 v'(x) \\
&= c_1(A(x)u(x) + b^1(x)) + c_2(A(x)v(x) + b^2(x)) \\
&= A(x)(c_1 u(x) + c_2 v(x)) + c_1 b^1(x) + c_2 b^2(x) \\
&= A(x)\phi(x) + c_1 b^1(x) + c_2 b^2(x).
\end{aligned}
$$

In particular, if $b^1(x) = b^2(x) \equiv 0$, i.e., $u(x)$ and $v(x)$ are solutions of the homogeneous differential system

$$u' = A(x)u, \tag{17.3}$$

R.P. Agarwal and D. O'Regan, *An Introduction to Ordinary Differential Equations*, doi: 10.1007/978-0-387-71276-5_17, © Springer Science + Business Media, LLC 2008

then $c_1 u(x) + c_2 v(x)$ is also a solution. Thus, solutions of the homogeneous differential system (17.3) form a vector space. Further, if $b^1(x) = b^2(x)$, $c_1 = 1$, $c_2 = -1$ and $u(x)$ is a solution of (17.1), then $v(x)$ is also a solution of (17.1) if and only if $u(x) - v(x)$ is a solution of (17.3). Thus, the general solution of (17.1) is obtained by adding to a particular solution of (17.1) the general solution of the corresponding homogeneous system (17.3).

To find the dimension of the vector space of the solutions of (17.3) we need to define the concept of linear independence and dependence of vector-valued functions. The vector-valued functions $u^1(x), \ldots, u^m(x)$ defined in an interval J are said to be *linearly independent* in J, if the relation $c_1 u^1(x) + \cdots + c_m u^m(x) = 0$ for all x in J implies that $c_1 = \cdots = c_m = 0$. Conversely, these functions are said to be *linearly dependent* if there exist constants $c_1, \ldots, c_m$ not all zero such that $c_1 u^1(x) + \cdots + c_m u^m(x) = 0$ for all $x \in J$.

Let m functions $u^1(x), \ldots, u^m(x)$ be linearly dependent in J and $c_k \neq 0$; then we have

$$u^k(x) = -\frac{c_1}{c_k} u^1(x) - \cdots - \frac{c_{k-1}}{c_k} u^{k-1}(x) - \frac{c_{k+1}}{c_k} u^{k+1}(x) - \cdots - \frac{c_m}{c_k} u^m(x),$$

i.e., $u^k(x)$ (and hence at least one of these functions) can be expressed as a linear combination of the remaining $m-1$ functions. On the other hand, if one of these functions, say, $u^k(x)$, is a linear combination of the remaining $m-1$ functions, so that

$$u^k(x) = c_1 u^1(x) + \cdots + c_{k-1} u^{k-1}(x) + c_{k+1} u^{k+1}(x) + \cdots + c_m u^m(x),$$

then obviously these functions are linearly dependent. Hence, if two functions are linearly dependent in J, then each one of these functions is identically equal to a constant times the other function, while if two functions are linearly independent, then it is impossible to express either function as a constant times the other. The concept of linear independence allows us to distinguish when the given functions are "essentially" different.

Example 17.1. (i) The functions $1, x, \ldots, x^{m-1}$ are linearly independent in every interval J. For this, $c_1 + c_2 x + \cdots + c_m x^{m-1} \equiv 0$ in J implies that $c_1 = \cdots = c_m = 0$. If any c_k were not zero, then the equation $c_1 + c_2 x + \cdots + c_m x^{m-1} = 0$ could hold for at most $m-1$ values of x, whereas it must hold for all x in J.

(ii) The functions

$$u^1(x) = \begin{bmatrix} e^x \\ e^x \end{bmatrix}, \quad u^2(x) = \begin{bmatrix} e^{2x} \\ 3e^{2x} \end{bmatrix}$$

are linearly independent in every interval J. Indeed,

$$c_1 \begin{bmatrix} e^x \\ e^x \end{bmatrix} + c_2 \begin{bmatrix} e^{2x} \\ 3e^{2x} \end{bmatrix} = 0$$

implies that $c_1 e^x + c_2 e^{2x} = 0$ and $c_1 e^x + 3c_2 e^{2x} = 0$, which is possible only for $c_1 = c_2 = 0$.

(iii) The functions

$$u^1(x) = \begin{bmatrix} \sin x \\ \cos x \end{bmatrix}, \quad u^2(x) = \begin{bmatrix} 0 \\ 0 \end{bmatrix}$$

are linearly dependent.

For the given n vector-valued functions $u^1(x), \ldots, u^n(x)$ the determinant $W(u^1, \ldots, u^n)(x)$ or $W(x)$, when there is no ambiguity, defined by

$$\begin{vmatrix} u_1^1(x) & \cdots & u_1^n(x) \\ u_2^1(x) & \cdots & u_2^n(x) \\ \cdots \\ u_n^1(x) & \cdots & u_n^n(x) \end{vmatrix}$$

is called the *Wronskian* of these functions. This determinant is closely related to the question of whether $u^1(x), \ldots, u^n(x)$ are linearly independent. In fact, we have the following result.

Theorem 17.1. If the Wronskian $W(x)$ of n vector-valued functions $u^1(x), \ldots, u^n(x)$ is different from zero for at least one point in an interval J; then these functions are linearly independent in J.

Proof. Let $u^1(x), \ldots, u^n(x)$ be linearly dependent in J, then there exist n constants $c_1, \ldots, c_n$ not all zero such that $\sum_{i=1}^{n} c_i u^i(x) = 0$ in J. This is the same as saying the homogeneous system of equations $\sum_{i=1}^{n} u_k^i(x)c_i = 0$, $1 \leq k \leq n$, $x \in J$, has a nontrivial solution. However, from Theorem 13.2 this homogeneous system for each $x \in J$ has a nontrivial solution if and only if $W(x) = 0$. But $W(x) \neq 0$ for at least one x in J, and, therefore $u^1(x), \ldots, u^n(x)$ cannot be linearly dependent.

In general the converse of this theorem is not true. For instance, for

$$u^1(x) = \begin{bmatrix} x \\ 1 \end{bmatrix}, \quad u^2(x) = \begin{bmatrix} x^2 \\ x \end{bmatrix},$$

which are linearly independent in any interval J, $W(u^1, u^2)(x) = 0$ in J. This example also shows that $W(u^1, u^2)(x) \neq 0$ in J is not necessary for the linear independence of $u^1(x)$ and $u^2(x)$ in J, and $W(u^1, u^2)(x) = 0$ in J may not imply that $u^1(x)$ and $u^2(x)$ are linearly dependent in J. Thus, the only conclusion we have is $W(x) \neq 0$ in J implies that $u^1(x), \ldots, u^n(x)$ are linearly independent in J, and linear dependence of these functions in J implies that $W(x) = 0$ in J.

The converse of Theorem 17.1 holds if $u^1(x), \ldots, u^n(x)$ are the solutions of the homogeneous differential system (17.3) in J. This we shall prove in the following theorem.

Theorem 17.2. Let $u^1(x), \ldots, u^n(x)$ be linearly independent solutions of the differential system (17.3) in J. Then $W(x) \neq 0$ for all $x \in J$.

Proof. Let x_0 be a point in J where $W(x_0) = 0$, then from Theorem 13.2 there exist constants $c_1, \ldots, c_n$ not all zero such that $\sum_{i=1}^{n} c_i u^i(x_0) = 0$. Since $u(x) = \sum_{i=1}^{n} c_i u^i(x)$ is a solution of (17.3), and $u(x_0) = 0$, from the uniqueness of the solutions it follows that $u(x) = \sum_{i=1}^{n} c_i u^i(x) = 0$ in J. However, the functions $u^1(x), \ldots, u^n(x)$ are linearly independent in J so we must have $c_1 = \cdots = c_n = 0$. This contradiction completes the proof. ∎

Thus, on combining the results of Theorems 17.1 and 17.2 we find that the solutions $u^1(x), \ldots, u^n(x)$ of the differential system (17.3) are linearly independent in J if and only if there exists at least one point $x_0 \in J$ such that $W(x_0) \neq 0$, i.e, the vectors $u^1(x_0), \ldots, u^n(x_0)$ are linearly independent. Hence, the solutions $u^1(x), \ldots, u^n(x)$ of the system (17.3) satisfying the initial conditions

$$u^i(x_0) = e^i, \quad i = 1, \ldots, n \tag{17.4}$$

are linearly independent in J. This proves the existence of n linearly independent solutions of the differential system (17.3) in J.

Now let $u(x)$ be any solution of the differential system (17.3) in J such that $u(x_0) = u^0$. Then from the existence and uniqueness of the solutions of the initial value problem (17.3), (17.2) it is immediate that $u(x) = \sum_{i=1}^{n} u_i^0 u^i(x)$, where $u^i(x)$ is the solution of the problem (17.3), (17.4). Thus, every solution of (17.3) can be expressed as a linear combination of the n linearly independent solutions of (17.3), (17.4). In conclusion, we find that the vector space of all solutions of (17.3) is of dimension n.

Finally, in this lecture we shall prove the following result, which gives a relation between the Wronskian $W(x)$ and the matrix $A(x)$.

Theorem 17.3 (Abel's Formula). Let $u^1(x), \ldots, u^n(x)$ be the solutions of the differential system (17.3) in J and $x_0 \in J$. Then for all $x \in J$,

$$W(x) = W(x_0) \exp \left(\int_{x_0}^{x} \operatorname{Tr} A(t) dt \right). \tag{17.5}$$

Proof. In view of (13.5) the derivative of the Wronskian $W(x)$ can be written as

$$W'(x) \; = \; \sum_{i=1}^{n} \begin{vmatrix} u_1^1(x) & \cdots & u_1^n(x) \\ & \cdots & \\ u_{i-1}^1(x) & \cdots & u_{i-1}^n(x) \\ u_i^1{}'(x) & \cdots & u_i^n{}'(x) \\ u_{i+1}^1(x) & \cdots & u_{i+1}^n(x) \\ & \cdots & \\ u_n^1(x) & \cdots & u_n^n(x) \end{vmatrix} . \qquad (17.6)$$

In the ith determinant of the right side of (17.6) we use the differential system (17.3) to replace $u_i^j{}'(x)$ by $\sum_{k=1}^{n} a_{ik}(x)u_k^j(x)$, and multiply the first row by $a_{i1}(x)$, the second row by $a_{i2}(x)$, and so on—except the ith row—and subtract their sum from the ith row, to get

$$W'(x) \; = \; \sum_{i=1}^{n} a_{ii}(x)W(x) \; = \; (\mathrm{Tr}\, A(x))W(x). \qquad (17.7)$$

Integration of the first-order DE (17.7) from x_0 to x gives the required relation (17.5). ■

Example 17.2. For the differential system

$$u' \; = \; \begin{bmatrix} 0 & 1 \\ -\dfrac{2}{(x^2+2x-1)} & \dfrac{2(x+1)}{(x^2+2x-1)} \end{bmatrix} u, \quad x \neq -1 \pm \sqrt{2}$$

it is easy to verify that

$$u^1(x) \; = \; \begin{bmatrix} x+1 \\ 1 \end{bmatrix} \quad \text{and} \quad u^2(x) \; = \; \begin{bmatrix} x^2+1 \\ 2x \end{bmatrix}$$

are two linearly independent solutions. Also,

$$W(u^1, u^2)(x) \; = \; \begin{vmatrix} x+1 & x^2+1 \\ 1 & 2x \end{vmatrix} \; = \; x^2+2x-1$$

and

$$\exp\left(\int_{x_0}^{x} \mathrm{Tr}\, A(t)dt \right) \; = \; \exp\left(\int_{x_0}^{x} \frac{2(t+1)}{(t^2+2t-1)} dt \right) \; = \; \frac{x^2+2x+1}{x_0^2+2x_0-1},$$
$$x_0 \neq -1 \pm \sqrt{2}.$$

Substituting these expressions in (17.5) we see that it holds for all x.

We finish this lecture with the remark that the relation (17.5) among other things says that $W(x)$ is either identically zero or never zero in J.

Problems

17.1. Show that the linear differential system (17.1) remains linear after the change of the independent variable $x = p(t)$.

17.2. Matrices of the form

$$
\begin{bmatrix}
0 & 1 & 0 & \cdots & 0 \\
0 & 0 & 1 & \cdots & 0 \\
\cdots & & & & \\
0 & 0 & 0 & \cdots & 1 \\
-p_n & -p_{n-1} & -p_{n-2} & \cdots & -p_1
\end{bmatrix}
$$

are called *companion matrices*. Show the following:

(i)   If $y(x)$ satisfies the nth-order linear homogeneous DE

$$
y^{(n)} + p_1(x)y^{(n-1)} + \cdots + p_n(x)y = 0 \tag{17.8}
$$

and if the vector-valued function $u(x)$ is defined by $u_i(x) = y^{(i-1)}(x)$, $i = 1, 2, \ldots, n$, then $u' = A(x)u$ with $A(x)$ in the companion form.

(ii)   If $y_k(x)$, $1 \le k \le n$ are n solutions of (17.8), then

$$
u^k(x) = (y_k(x), y_k'(x), \ldots, y_k^{(n-1)}(x))^T, \quad 1 \le k \le n
$$

satisfy the system $u' = A(x)u$.

(iii) $W(u^1, \ldots, u^n)(x) = W(u^1, \ldots, u^n)(x_0) \exp\left(-\int_{x_0}^{x} p_1(t)dt\right).$

17.3. The Wronskian of n functions $y_1(x), \ldots, y_n(x)$ which are $(n-1)$ times differentiable in an interval J is defined by the determinant

$$
W(y_1, \ldots, y_n)(x) = \begin{vmatrix}
y_1(x) & \cdots & y_n(x) \\
y_1'(x) & \cdots & y_n'(x) \\
\cdots & & \\
y_1^{(n-1)}(x) & \cdots & y_n^{(n-1)}(x)
\end{vmatrix}.
$$

Show the following:

(i)   If $W(y_1, \ldots, y_n)(x)$ is different from zero for at least one point in J, then the functions $y_1(x), \ldots, y_n(x)$ are linearly independent in J.

(ii) If the functions $y_1(x), \ldots, y_n(x)$ are linearly dependent in J, then the Wronskian $W(y_1, \ldots, y_n)(x) = 0$ in J.

(iii) The converse of (i) as well as of (ii) is not necessarily true.

(iv)   If $W(y_1, \ldots, y_n)(x) = 0$ in J, but for some set of $(n-1)$, y's (say, without loss of generality, all but $y_n(x)$) $W(y_1, \ldots, y_{n-1})(x) \neq 0$ for all $x \in J$, then the functions $y_1(x), \ldots, y_n(x)$ are linearly dependent in J.

17.4. Let $y_1(x), \ldots, y_n(x)$ be n times continuously differentiable functions in an interval J and $W(y_1, \ldots, y_n)(x) \neq 0$. Show that

$$\frac{W(y_1, \ldots, y_n, y)(x)}{W(y_1, \ldots, y_n)(x)} = 0$$

is an nth-order linear homogeneous DE for which $y_1(x), \ldots, y_n(x)$ are solutions.

17.5. Show that the DE (17.8) can be transformed to its normal form

$$z^{(n)} + q_2(x)z^{(n-2)} + q_3(x)z^{(n-3)} + \cdots + q_n(x)z = 0,$$

where

$$z(x) = y(x) \exp\left(\frac{1}{n}\int^x p_1(t)dt\right),$$

provided $p_1(x)$ is $(n-1)$ times differentiable in J. In particular for $n = 2$, the normal form of (17.8) is

$$z'' + \left(p_2(x) - \frac{1}{2}p_1'(x) - \frac{1}{4}p_1^2(x)\right)z = 0.$$

17.6. Let $\phi_1(x) \neq 0$ in J be a solution of the DE (17.8). If $v_2, \ldots, v_n$ are linearly independent solutions of the DE

$$\phi_1 v^{(n-1)} + \cdots + [n\phi_1^{(n-1)} + p_1(x)(n-1)\phi_1^{(n-2)} + \cdots + p_{n-1}(x)\phi_1]v = 0$$

and if $v_k = u_k'$, $k = 2, \ldots, n$, then $\phi_1, u_2\phi_1, \ldots, u_n\phi_1$ are linearly independent solutions of (17.8) in J. In particular, for $n = 2$ the second linearly independent solution of (17.8) is given by

$$\phi_2(x) = \phi_1(x) \int_{x_0}^x \frac{1}{(\phi_1(t))^2} \exp\left(-\int_{x_0}^t p_1(s)ds\right) dt$$

(see (6.5)).

17.7. Let $u(x), v(x)$ and $w(x)$ be the solutions of the DE $y''' + y = 0$ satisfying $u(0) = 1$, $u'(0) = 0$, $u''(0) = 0$; $v(0) = 0$, $v'(0) = 1$, $v''(0) = 0$; $w(0) = 0$, $w'(0) = 0$, $w''(0) = 1$. Without solving the DE, show the following:

(i) $u'(x) = -w(x)$.

(ii) $v'(x) = u(x)$.

(iii) $w'(x) = v(x)$.

(iv) $W(u, v, w) = u^3 - v^3 + w^3 + 3uvw = 1$.

Answers or Hints

17.1. For each k, $0 \leq k \leq n$ it is easy to find

$$\frac{d^k y}{dx^k} = \sum_{i=0}^{k} p_{ki}(t) \frac{d^i y}{dt^i},$$

where $p_{ki}(t)$ are some suitable functions.

17.2. To show (i) and (ii) write the equation (17.8) in a system form. part (iii) follows from Abel's formula (17.5).

17.3. (i) If $y_i(x)$, $1 \leq i \leq n$ are linearly dependent, then there exist non-trivial c_i, $1 \leq i \leq n$ such that $\sum_{i=1}^{n} c_i y_i(x) = 0$ for all $x \in J$. Differentiating this, we obtain $\sum_{i=1}^{n} c_i y_i^{(k)}(x) = 0$, $k = 0, 1, \ldots, n-1$ for all $x \in J$. But, this implies $W(x) = 0$. (ii) Clear from part (i). (iii) Consider the functions $y_1(x) = x^3$, $y_2(x) = x^2|x|$. (iv) $W(y_1, \ldots, y_{n-1}, y)(x) = 0$ is a linear homogeneous DE such that the coefficient of $y^{(n-1)}(x)$ is $W(y_1, \ldots, y_{n-1})(x) \neq 0$. For this $(n-1)$th-order DE there are n solutions $y_i(x)$, $1 \leq i \leq n$.

17.4. See part (iv) of Problem 17.3.

17.5. Use Leibnitz's formula.

17.6. Since $\phi_1(x)$ is a solution of the DE (17.8), $y = u\phi_1$ will also be its solution provided $0 = (u\phi_1)^{(n)} + p_1(x)(u\phi_1)^{(n-1)} + \cdots + p_n(x)(u\phi_1) = \phi_1 v^{(n-1)} + (n\phi_1' + p_1(x)\phi_1)v^{(n-2)} + \cdots + (n\phi_1^{(n-1)} + (n-1)p_1(x)\phi_1^{(n-2)} + \cdots + p_{n-1}(x)\phi_1)v$, where $v = u'$. The coefficient of $v^{(n-1)}$ is ϕ_1, and hence if $\phi_1(x) \neq 0$ in J, then the above $(n-1)$th-order DE has $n-1$ linearly independent solutions $v_2, \ldots, v_n$ in J. If $x_0 \in J$, and $u_k(x) = \int_{x_0}^{x} v_k(t)dt$, $2 \leq k \leq n$ then the functions $\phi_1, u_2\phi_1, \ldots, u_n\phi_1$ are solutions of (17.8). To show that these solutions are linearly independent, let $c_1\phi_1 + c_2 u_2\phi_1 + \cdots + c_n u_n\phi_1 = 0$, where not all $c_1, \ldots, c_n$ are zero. However, since $\phi_1 \neq 0$ in J, this implies that $c_1 + c_2 u_2 + \cdots + c_n u_n = 0$, and on differentiation we obtain $c_2 u_2' + \cdots + c_n u_n' = 0$, i.e., $c_2 v_2 + \cdots + c_n v_n = 0$. Since $v_2, \ldots, v_n$ are linearly independent $c_2 = \cdots = c_n = 0$, which in turn also imply that $c_1 = 0$.

17.7. Since $W(u, v, w)(0) = 1$, u, v, w are linearly independent solutions of $y''' + y = 0$. Now since $(u')''' + (u') = 0$, u' is also a solution of $y''' + y = 0$. Hence, there exist nonzero constants c_1, c_2, c_3 such that $u'(x) = c_1 u(x) + c_2 v(x) + c_3 w(x)$. Part (iv) follows from parts (i)–(iii) and Abel's formula.

Lecture 18
Fundamental Matrix Solution

In our previous lecture we have seen that any solution $u(x)$ of the differential system (17.3) satisfying $u(x_0) = u^0$ can be written as $u(x) = \sum_{i=1}^{n} u_i^0 u^i(x)$, where $u^i(x)$ is the solution of the initial value problem (17.3), (17.4). In matrix notation this solution can be written as $u(x) = \Phi(x, x_0)u^0$, where $\Phi(x, x_0)$ is an $n \times n$ matrix whose ith column is $u^i(x)$. The matrix $\Phi(x, x_0)$ is called the *principal fundamental matrix*; however, some authors prefer to call it *evolution* or *transition matrix*. It is easy to verify that $\Phi(x, x_0)$ is a solution of the matrix initial value problem

$$\frac{d\Phi}{dx} = A(x)\Phi, \quad \Phi(x_0) = I. \tag{18.1}$$

The fact that the initial value problem (18.1) has a unique solution $\Phi(x, x_0)$ in J can be proved exactly as for the problem (17.1), (17.2). Moreover, the iterative scheme

$$\Phi^{m+1}(x) = I + \int_{x_0}^{x} A(t)\Phi^m(t)dt, \quad m = 0, 1, \ldots \tag{18.2}$$

$$\Phi^0(x) = I \tag{18.3}$$

converges to $\Phi(x, x_0)$, and

$$\Phi(x, x_0) = I + \int_{x_0}^{x} A(t)dt + \int_{x_0}^{x} \int_{x_0}^{t} A(t)A(t_1)dt_1 dt + \cdots. \tag{18.4}$$

The series (18.4) is called *Peano–Baker series* for the solution of (18.1). If A is an $n \times n$ constant matrix, then it can be taken out from the integrals and (18.4) becomes

$$\Phi(x, x_0) = I + A \int_{x_0}^{x} dt + A^2 \int_{x_0}^{x} \int_{x_0}^{t} dt_1 dt + \cdots$$

$$= I + \sum_{m=1}^{\infty} \frac{[A(x - x_0)]^m}{m!} = \exp(A(x - x_0)).$$

Summarizing this discussion, specifically we have proved the following theorem.

R.P. Agarwal and D. O'Regan, *An Introduction to Ordinary Differential Equations*, doi: 10.1007/978-0-387-71276-5_18, © Springer Science + Business Media, LLC 2008

Theorem 18.1. The matrix

$$\Phi(x, x_0) \;=\; \exp(A(x - x_0)) \tag{18.5}$$

is the principal fundamental matrix of the system

$$u' \;=\; Au, \tag{18.6}$$

where A is a constant matrix.

Example 18.1. For the matrix $A = \begin{bmatrix} 0 & 1 \\ -1 & 0 \end{bmatrix}$ it is easily seen that $A^{4m+1} = A$, $A^{4m+2} = -I$, $A^{4m+3} = -A$, $A^{4m+4} = I$ for $m = 0, 1, \ldots$ and hence the series (18.4) can be summed, to obtain

$$\begin{bmatrix} \cos(x - x_0) & \sin(x - x_0) \\ -\sin(x - x_0) & \cos(x - x_0) \end{bmatrix}.$$

In our previous lecture we have also proved that the solutions $u^1(x), \ldots,$ $u^n(x)$ of the differential system (17.3) are linearly independent in J if and only if there exists at least one point $x_0 \in J$ such that $W(x_0) \neq 0$. If these solutions are linearly independent, then the set $u^1(x), \ldots, u^n(x)$ is called a *fundamental system of solutions* of (17.3). Further, the $n \times n$ matrix $\Psi(x) - [u^1(x), \ldots, u^n(x)]$ is called the *fundamental matrix* of (17.3). For this matrix $\Psi(x)$, we shall prove the following result.

Theorem 18.2. If $\Psi(x)$ is a fundamental matrix of the differential system (17.3), then for any constant nonsingular $n \times n$ matrix C, $\Psi(x)C$ is also a fundamental matrix of (17.3), and every fundamental matrix of (17.3) is of the form $\Psi(x)C$ for some constant nonsingular $n \times n$ matrix C.

Proof. Obviously, $\Psi'(x) = A(x)\Psi(x)$, and hence $\Psi'(x)C = A(x)\Psi(x)C$, which is the same as $(\Psi(x)C)' = A(x)(\Psi(x)C)$, i.e., $\Psi(x)$ and $\Psi(x)C$ both are solutions of the same matrix differential system $\Phi' = A(x)\Phi$. Further, since $\det \Psi(x) \neq 0$ and $\det C \neq 0$ it follows that $\det(\Psi(x)C) \neq 0$, and hence $\Psi(x)C$ is also a fundamental matrix solution of (17.3). Conversely, let $\Psi_1(x)$ and $\Psi_2(x)$ be two fundamental matrix solutions of (17.3). If $\Psi_2^{-1}(x)\Psi_1(x) = C(x)$, i.e., $\Psi_1(x) = \Psi_2(x)C(x)$, then we find that $\Psi_1'(x) = \Psi_2'(x)C(x) + \Psi_2(x)C'(x)$ which is the same as

$$A(x)\Psi_1(x) \;=\; A(x)\Psi_2(x)C(x) + \Psi_2(x)C'(x) \;=\; A(x)\Psi_1(x) + \Psi_2(x)C'(x);$$

i.e., $\Psi_2(x)C'(x) = 0$, or $C'(x) = 0$, and hence $C(x)$ is a constant matrix. Further, since both $\Psi_1(x)$ and $\Psi_2(x)$ are nonsingular, this constant matrix is also nonsingular. ∎

As a consequence of this theorem, we find

$$\Phi(x, x_0) \;=\; \Psi(x)\Psi^{-1}(x_0) \tag{18.7}$$

and the solution of the initial value problem (17.3), (17.2) can be written as

$$u(x) \; = \; \Psi(x)\Psi^{-1}(x_0)u^0. \tag{18.8}$$

Since the product of matrices is not commutative, for a given constant nonsingular matrix C, $C\Psi(x)$ need not be a fundamental matrix solution of the differential system (17.3). Further, two different homogeneous systems cannot have the same fundamental matrix, i.e., $\Psi(x)$ determines the matrix $A(x)$ in (17.3) uniquely by the relation $A(x) = \Psi'(x)\Psi^{-1}(x)$. However, from Theorem 18.2 the converse is not true.

Now differentiating the relation $\Psi(x)\Psi^{-1}(x) = I$, we obtain

$$\Psi'(x)\Psi^{-1}(x) + \Psi(x)(\Psi^{-1}(x))' \; = \; 0$$

and hence

$$(\Psi^{-1}(x))' \; = \; -\Psi^{-1}(x)A(x),$$

which is the same as

$$\left((\Psi^{-1}(x))^T\right)' \; = \; -A^T(x)(\Psi^{-1}(x))^T. \tag{18.9}$$

Therefore, $(\Psi^{-1}(x))^T$ is a fundamental matrix of the differential system

$$u' \; = \; -A^T(x)u. \tag{18.10}$$

The system (18.10) is called the *adjoint system* to the differential system (17.3). This relationship is symmetric in the sense that (17.3) is the adjoint system to (18.10) and vice versa.

An important property of adjoint systems is given in the following result.

Theorem 18.3.   If $\Psi(x)$ is a fundamental matrix of the differential system (17.3), then $\chi(x)$ is a fundamental matrix of its adjoint system (18.10) if and only if

$$\chi^T(x)\Psi(x) \; = \; C, \tag{18.11}$$

where C is a constant nonsingular $n \times n$ matrix.

Proof.   If $\Psi(x)$ is a fundamental matrix of the differential system (17.3), then from (18.9) it follows that $(\Psi^{-1}(x))^T$ is a fundamental matrix of the differential system (18.10), and hence Theorem 18.2 gives $\chi(x) = (\Psi^{-1}(x))^T D$, where D is a constant nonsingular $n \times n$ matrix. Thus, on using the fact that $(\Psi^{-1}(x))^T$ is a fundamental matrix we have $\Psi^T(x)\chi(x) = D$, which is the same as $\chi^T(x)\Psi(x) = D^T$. Therefore, (18.11) holds with $C = D^T$. Conversely, if $\Psi(x)$ is a fundamental matrix of (17.3) satisfying (18.11), then we have $\Psi^T(x)\chi(x) = C^T$ and hence $\chi(x) = (\Psi^T(x))^{-1}C^T$. Thus, by Theorem 18.2, $\chi(x)$ is a fundamental matrix of the adjoint system (18.10).   ∎

As a consequence of this theorem, if $A(x) = -A^T(x)$, then $(\Psi^T(x))^{-1}$ being a fundamental matrix of (18.10) is also a fundamental matrix of the differential system (17.3). Thus, Theorem 18.2 gives $\Psi(x) = (\Psi^T(x))^{-1}C$, which is the same as $\Psi^T(x)\Psi(x) = C$. Hence, in this particular case the Euclidean length of any solution of the differential system (17.3) is a constant.

Now we shall show that the method of variation of parameters used in Lectures 5 and 6 to find the solutions of nonhomogeneous first and second-order DEs is equally applicable for the nonhomogeneous system (17.1). For this, we seek a vector-valued function $v(x)$ such that $\Phi(x, x_0)v(x)$ is a solution of the system (17.1). We have

$$\Phi'(x, x_0)v(x) + \Phi(x, x_0)v'(x) = A(x)\Phi(x, x_0)v(x) + b(x),$$

which reduces to give

$$\Phi(x, x_0)v'(x) = b(x);$$

and hence from Problem 18.2, it follows that

$$v'(x) = \Phi^{-1}(x, x_0)b(x) = \Phi(x_0, x)b(x).$$

Thus, the function $v(x)$ can be written as

$$v(x) = v(x_0) + \int_{x_0}^{x} \Phi(x_0, t)b(t)dt.$$

Finally, since $u(x_0) = \Phi(x_0, x_0)v(x_0) = v(x_0)$, the solution of the initial value problem (17.1), (17.2) takes the form

$$u(x) = \Phi(x, x_0)u^0 + \Phi(x, x_0)\int_{x_0}^{x} \Phi(x_0, t)b(t)dt, \qquad (18.12)$$

which from Problem 18.2 is the same as

$$u(x) = \Phi(x, x_0)u^0 + \int_{x_0}^{x} \Phi(x, t)b(t)dt. \qquad (18.13)$$

If we use the relation (18.7) in (18.12), then it is the same as

$$u(x) = \Psi(x)c + \int_{x_0}^{x} \Psi(x)\Psi^{-1}(t)b(t)dt, \qquad (18.14)$$

where the constant vector $c = \Psi^{-1}(x_0)u^0$. Hence, we have an explicit representation of the general solution of (17.1) in terms of any fundamental matrix $\Psi(x)$ of the differential system (17.3).

In the case when $A(x)$ is a constant matrix A, relation (18.5) can be used in (18.13), to obtain

$$u(x) = e^{A(x-x_0)}u^0 + \int_{x_0}^{x} e^{A(x-t)}b(t)dt. \qquad (18.15)$$

Example 18.2.   Consider the system

$$u' = \begin{bmatrix} 0 & 1 \\ -2 & 3 \end{bmatrix} u + \begin{bmatrix} 1 \\ 1 \end{bmatrix}. \tag{18.16}$$

For the corresponding homogeneous system

$$u' = \begin{bmatrix} 0 & 1 \\ -2 & 3 \end{bmatrix} u$$

it is easy to verify that the principal fundamental matrix is

$$\Phi(x,0) = \begin{bmatrix} 2e^x - e^{2x} & -e^x + e^{2x} \\ 2e^x - 2e^{2x} & -e^x + 2e^{2x} \end{bmatrix} = \begin{bmatrix} e^x & e^{2x} \\ e^x & 2e^{2x} \end{bmatrix} \begin{bmatrix} 2 & -1 \\ -1 & 1 \end{bmatrix}.$$

Thus, the solution of (18.16) satisfying $u(0) = u^0$ can be written as

$$\begin{aligned}
u(x) &= \begin{bmatrix} e^x & e^{2x} \\ e^x & 2e^{2x} \end{bmatrix} \begin{bmatrix} 2 & -1 \\ -1 & 1 \end{bmatrix} u^0 \\
&\quad + \begin{bmatrix} e^x & e^{2x} \\ e^x & 2e^{2x} \end{bmatrix} \int_0^x \begin{bmatrix} 2e^{-t} & -e^{-t} \\ -e^{-2t} & e^{-2t} \end{bmatrix} \begin{bmatrix} 1 \\ 1 \end{bmatrix} dt \\
&= \begin{bmatrix} 2e^x - e^{2x} & -e^x + e^{2x} \\ 2e^x - 2e^{2x} & -e^x + 2e^{2x} \end{bmatrix} u^0 + (e^x - 1) \begin{bmatrix} 1 \\ 1 \end{bmatrix}.
\end{aligned}$$

Problems

18.1.   Show that the vector-valued function

$$u(x) = \exp\left(\int_{x_0}^x A(t)dt \right) u^0$$

is not a solution of the differential system (17.3) unless $A(x)$ and $\int_{x_0}^x A(t)dt$ commute for all x.

18.2.   Let $\Phi(x, x_0)$ be the principal fundamental matrix of the system (17.3) in an interval J. Show that $\Phi(x, x_0) = \Phi(x, x_1)\Phi(x_1, x_0)$, where $x_1 \in J$, and hence, in particular, $\Phi^{-1}(x, x_0) = \Phi(x_0, x)$, and $\Phi(x, x) = I$ for all $x \in J$.

18.3.   For the $n \times n$ matrix $\Phi(x, t)$ appearing in (18.13), show the following:

(i) $\partial \Phi(x, t)/\partial x = A(x)\Phi(x, t)$.

(ii) $\partial \Phi(x,t)/\partial t = -\Phi(x,t)A(t)$.

(iii) $\Phi(x,t) = I + \int_t^x A(s)\Phi(s,t)ds$.

(iv) $\Phi(x,t) = I + \int_t^x \Phi(x,s)A(s)ds$.

18.4. Show that a square nonsingular matrix $\Phi(.,.)$ which depends on two arguments and is differentiable with respect to each argument in J is a principal fundamental matrix if $\Phi(x_0,x_0) = I$ for all x_0 in J and the matrix

$$\left[\frac{d}{dx}\Phi(x,x_0)\right]\Phi^{-1}(x,x_0)$$

depends on x alone.

18.5. Let $\Phi(x) = \Phi(x,0)$ be the principal fundamental matrix of the system (18.6). Show the following:

(i) For any real x_0, $\Phi_1(x) = \Phi(x - x_0)$ is a fundamental matrix.

(ii) $\Phi(x - x_0) = \Phi(x)\Phi(-x_0) = \Phi(x)\Phi^{-1}(x_0)$, and hence $\Phi(-x_0) = \Phi^{-1}(x_0)$.

18.6. The linear differential system (17.3) is said to be *self-adjoint* when $A(x) = -A^T(x)$ for all x in J. Let $\Phi(x,x_0)$ be the principal fundamental matrix of the system (17.3), and $\Psi(x,x_0)$ be the principal fundamental matrix of the adjoint system (18.10). Show that the differential system (17.3) is self-adjoint if and only if $\Phi(x,x_0) = \Psi^T(x_0,x)$.

18.7. Let the matrix $A(x)$ be such that $a_{ij}(x) \geq 0$ for all $i \neq j$ and all $x \geq x_0$. Show the following:

(i) Every element of the principal fundamental matrix $\Phi(x,x_0)$ of the system (17.3) is nonnegative for all $x \geq x_0$.

(ii) If $u(x)$ and $v(x)$ are two solutions of the differential system (17.3) satisfying $u_i(x_0) \geq v_i(x_0)$, $1 \leq i \leq n$, then $u_i(x) \geq v_i(x)$, $1 \leq i \leq n$ for all $x \geq x_0$.

18.8. Equations of the form

$$p_0 x^n y^{(n)} + p_1 x^{n-1} y^{(n-1)} + \cdots + p_n y = 0, \quad x > 0, \quad p_0 \neq 0 \quad (18.17)$$

are called *Euler's DEs*. Show that there exist $u_1(x), \ldots, u_n(x)$ such that the differential system (18.17) can be converted to $u'(x) = x^{-1}Au(x)$, where A is an $n \times n$ constant matrix.

18.9. For the given n linearly independent solutions $y_1(x), \ldots, y_n(x)$

of the homogeneous DE (17.8) in an interval J, we define

$$H(x,t) = \frac{y_1(x)\Delta_1(t) + \cdots + y_n(x)\Delta_n(t)}{W(y_1,\ldots,y_n)(t)}$$

$$= \begin{vmatrix} y_1(t) & \cdots & y_n(t) \\ y_1'(t) & \cdots & y_n'(t) \\ \cdots & & \\ y_1^{(n-2)}(t) & \cdots & y_n^{(n-2)}(t) \\ y_1(x) & \cdots & y_n(x) \end{vmatrix} \Bigg/ \begin{vmatrix} y_1(t) & \cdots & y_n(t) \\ y_1'(t) & \cdots & y_n'(t) \\ \cdots & & \\ y_1^{(n-1)}(t) & \cdots & y_n^{(n-1)}(t) \end{vmatrix} ;$$

i.e., $\Delta_i(t)$ is the cofactor of the element $y_i^{(n-1)}(t)$ in the Wronskian $W(y_1,\ldots,y_n)(t)$. Show the following:

(i) $H(x,t)$ is defined for all (x,t) in $J \times J$.

(ii) $\partial^i H(x,t)/\partial x^i$, $0 \le i \le n$, are continuous for all (x,t) in $J \times J$.

(iii) For each fixed t in J the function $z(x) = H(x,t)$ is a solution of the DE (17.8) satisfying $z^{(i)}(t) = 0$, $0 \le i \le n-2$, $z^{(n-1)}(t) = 1$.

(iv)   The function

$$y(x) = \int_{x_0}^{x} H(x,t)r(t)dt$$

is a particular solution of the nonhomogeneous DE (16.17), satisfying $y^{(i)}(x_0) = 0$, $0 \le i \le n-1$.

(v)   The general solution of (16.17) can be written as

$$y(x) = \sum_{i=1}^{n} c_i y_i(x) + \int_{x_0}^{x} H(x,t)r(t)dt,$$

where c_i, $1 \le i \le n$, are arbitrary constants.

18.10. Let $v(x)$ be the solution of the initial value problem

$$y^{(n)} + p_1 y^{(n-1)} + \cdots + p_n y = 0$$
$$y^{(i)}(0) = 0, \quad 0 \le i \le n-2, \quad y^{(n-1)}(0) = 1. \tag{18.18}$$

Show that the function

$$y(x) = \int_{x_0}^{x} v(x-t)r(t)dt$$

is the solution of the nonhomogeneous DE

$$y^{(n)} + p_1 y^{(n-1)} + \cdots + p_n y = r(x) \tag{18.19}$$

satisfying $y^{(i)}(x_0) = 0$, $0 \le i \le n-1$.

18.11. *Open-loop input–output control systems* can be written in the form
$$u' = Au + by(x), \quad z = c^T u + dy(x),$$

where the functions $y(x)$, $z(x)$ and the constant d are scalar. Here $y(x)$ is the known input and $z(x)$ is the unknown output. Show that if $u(0) = u^0$ is known, then

(i) $u(x) = e^{Ax}u^0 + \displaystyle\int_0^x e^{A(x-t)}by(t)dt$;

(ii) $z(x) = c^T e^{Ax}u^0 + dy(x) + \displaystyle\int_0^x \left(c^T e^{A(x-t)}b\right)y(t)dt.$

The function $h(t) = c^T e^{Ax}b$ is called the *impulse response function* for the given control system.

Answers or Hints

18.1. Expand $\exp\left(\int_{x_0}^x A(t)dt\right)$ and then compare $u'(x)$ and $A(x)u(x)$.

18.2. Verify that $\Psi(x) = \Phi(x, x_1)\Phi(x_1, x_0)$ is a fundamental matrix solution of (18.1). The result now follows from the uniqueness.

18.3. Use $\Phi(x, t) = \Phi(x)\Phi^{-1}(t)$.

18.4. $\left[\frac{d}{dx}\Phi(x, x_0)\right]\Phi^{-1}(x, x_0) = A(x)$ is the same as $\frac{d}{dx}\Phi(x, x_0) = A(x)$ $\times\Phi(x, x_0)$.

18.5. (i) Verify directly. (ii) From Theorem 18.2, $\Phi(x - x_0) = \Phi(x)C$.

18.6. Use Theorem 18.3 and Problems 18.2 and 18.3.

18.7. (i) If $a_{ij}(x) \geq 0$, $1 \leq i, j \leq n$, then the sequence $\{\Phi^m(x)\}$ generated by $\Phi^0(x) = I$,

$$\Phi^{m+1}(x) = I + \int_{x_0}^x A(t)\Phi^m(t)dt, \quad m = 0, 1, \ldots,$$

converges to $\Phi(x, x_0)$ and obviously each $\Phi^m(x) \geq 0$, and hence $\Phi(x, x_0) \geq 0$ for all $x \geq x_0$. Now note that (17.3) is the same as

$$u_i' - a_{ii}(x)u_i = \sum_{\substack{j=1 \\ j \neq i}}^n a_{ij}(x)u_j, \quad 1 \leq i \leq n.$$

(ii) Use the representation $u(x) = \Phi(x, x_0)u(x_0)$.

18.8. For $n = 3$ let $y = u_1$, $xu_1' = u_2$ so that $u_2' = xu_1'' + u_1'$, $u_2'' = xu_1''' + 2u_1''$. Next let $xu_2' = u_3$ so that $u_3' = xu_2'' + u_2' = x(xu_1''' + 2u_1'') + u_2'$ and hence

$$xu_3' = x^3 u_1''' + 2x^2 u_1'' + xu_2' = -\frac{p_3}{p_0} u_1 - \left(2 + \frac{p_2 - p_1}{p_0}\right) u_2 + \left(3 - \frac{p_1}{p_0}\right) u_3.$$

Now write these equations in the required system form.

18.9. Verify directly.

18.10. For each fixed t, $z(x) = v(x - t)$ is also a solution of (18.18) satisfying $z^{(i)}(t) = 0$, $0 \le i \le n - 2$, $z^{(n-1)}(t) = 1$. Now use Problem 18.9(iv).

18.11. Use (18.15).

Lecture 19
Systems with Constant Coefficients

Our discussion in Lecture 18 has restricted usage of obtaining explicit solutions of homogeneous and, in general, of nonhomogeneous differential systems. This is so because the solution (18.4) involves an infinite series with repeated integrations and (18.14) involves its inversion. In fact, even if the matrix $A(x)$ is of second order, no general method of finding the explicit form of the fundamental matrix is available. Further, if the matrix A is constant, then the computation of the elements of the fundamental matrix e^{Ax} from the series (18.4) may turn out to be difficult, if not impossible. However, in this case the notion of eigenvalues and eigenvectors of the matrix A can be used to avoid unnecessary computation. For this, the first result we prove is the following theorem.

Theorem 19.1. Let $\lambda_1, \ldots, \lambda_n$ be the distinct eigenvalues of the matrix A and $v^1, \ldots, v^n$ be the corresponding eigenvectors. Then the set

$$u^1(x) = v^1 e^{\lambda_1 x}, \quad \cdots \quad , u^n(x) = v^n e^{\lambda_n x} \tag{19.1}$$

is a fundamental set of solutions of (18.6).

Proof. Since v^i is an eigenvector of A corresponding to the eigenvalue λ_i, we find

$$(u^i(x))' = (v^i e^{\lambda_i x})' = \lambda_i v^i e^{\lambda_i x} = A v^i e^{\lambda_i x} = A u^i(x)$$

and hence $u^i(x)$ is a solution of (18.6). To show that (19.1) is a fundamental set, we note that $W(0) = \det[v^1, \ldots, v^n] \neq 0$, since $v^1, \ldots, v^n$ are linearly independent from Theorem 14.1. Thus, the result follows from Theorem 17.1. ■

Obviously, from Theorem 19.1 it follows that

$$e^{Ax} = \left[v^1 e^{\lambda_1 x}, \ldots, v^n e^{\lambda_n x} \right] \left[v^1, \ldots, v^n \right]^{-1} \tag{19.2}$$

and the general solution of (18.6) can be written as

$$u(x) = \sum_{i=1}^{n} c_i v^i e^{\lambda_i x}. \tag{19.3}$$

R.P. Agarwal and D. O'Regan, *An Introduction to Ordinary Differential Equations*,
doi: 10.1007/978-0-387-71276-5_19, © Springer Science + Business Media, LLC 2008

Example 19.1. Using the results of Example 14.1, Theorem 19.1 concludes that the set

$$u^1(x) = \begin{bmatrix} 1 \\ -1 \\ 1 \end{bmatrix} e^x, \quad u^2(x) = \begin{bmatrix} -1 \\ 0 \\ 1 \end{bmatrix} e^{2x}, \quad u^3(x) = \begin{bmatrix} 1 \\ 2 \\ 1 \end{bmatrix} e^{4x}$$

is a fundamental set of solutions of the differential system

$$u' = \begin{bmatrix} 2 & 1 & 0 \\ 1 & 3 & 1 \\ 0 & 1 & 2 \end{bmatrix} u.$$

Unfortunately, when the matrix A has only $k < n$ distinct eigenvalues, then the computation of e^{Ax} is not easy. However, among several existing methods we shall discuss only two which may be relatively easier as compared with others. The first method is given in the following result.

Theorem 19.2. Let $\lambda_1, \ldots, \lambda_k$, $k \le n$ be distinct eigenvalues of the matrix A with multiplicities $r_1, \ldots, r_k$, respectively, so that

$$p(\lambda) = (\lambda - \lambda_1)^{r_1} \cdots (\lambda - \lambda_k)^{r_k}; \tag{19.4}$$

then

$$e^{Ax} = \sum_{i=1}^{k} \left[e^{\lambda_i x} a_i(A) q_i(A) \sum_{j=0}^{r_i-1} \left\{ \frac{1}{j!} (A - \lambda_i I)^j x^i \right\} \right], \tag{19.5}$$

where

$$q_i(\lambda) = p(\lambda)(\lambda - \lambda_i)^{-r_i}, \quad 1 \le i \le k \tag{19.6}$$

and $a_i(\lambda)$, $1 \le i \le k$ are the polynomials of degree less than r_i in the expansion

$$\frac{1}{p(\lambda)} = \frac{a_1(\lambda)}{(\lambda - \lambda_1)^{r_1}} + \cdots + \frac{a_k(\lambda)}{(\lambda - \lambda_k)^{r_k}}. \tag{19.7}$$

Proof. Relations (19.6) and (19.7) imply that

$$1 = a_1(\lambda)q_1(\lambda) + \cdots + a_k(\lambda)q_k(\lambda).$$

This relation has been derived from the characteristic equation $p(\lambda) = 0$ of A, and therefore, using the Cayley–Hamilton Theorem 14.2, we have

$$I = a_1(A)q_1(A) + \cdots + a_k(A)q_k(A). \tag{19.8}$$

Since the matrices $\lambda_i I$ and $A - \lambda_i I$ commute and $e^{\lambda_i I x} = e^{\lambda_i x} I$, we have

$$e^{Ax} = e^{\lambda_i I x} e^{(A - \lambda_i I)x} = e^{\lambda_i x} \sum_{j=0}^{\infty} \left\{ \frac{1}{j!} (A - \lambda_i I)^j x^j \right\}.$$

Premultiplying both sides of this equation by $a_i(A)q_i(A)$, and observing that $q_i(A)(A - \lambda_i I)^{r_i} = p(A) = 0$, and consequently, $q_i(A)(A - \lambda_i I)^j = 0$ for all $j \geq r_i$, it follows that

$$a_i(A)q_i(A)e^{Ax} = e^{\lambda_i x}a_i(A)q_i(A) \sum_{j=0}^{r_i-1} \left\{ \frac{1}{j!}(A - \lambda_i I)^j x^j \right\}.$$

Summing this relation from $i = 1$ to k and using (19.8), we get (19.5). ∎

Corollary 19.3. If $k = n$, i.e., A has n distinct eigenvalues, then $a_i(A) = (1/q_i(\lambda_i))I$, and hence (19.5) reduces to

$$e^{Ax} = \sum_{i=1}^{n} \frac{q_i(A)}{q_i(\lambda_i)} e^{\lambda_i x}$$

$$= \sum_{i=1}^{n} \frac{(A - \lambda_1 I) \cdots (A - \lambda_{i-1} I)(A - \lambda_{i+1} I) \cdots (A - \lambda_n I)}{(\lambda_i - \lambda_1) \cdots (\lambda_i - \lambda_{i-1})(\lambda_i - \lambda_{i+1}) \cdots (\lambda_i - \lambda_n)} e^{\lambda_i x}.$$

$$(19.9)$$

Corollary 19.4. If $k = 1$, i.e., A has all the eigenvalues equal to λ_1, then $a_i(A) = q_i(A) = I$, and hence (19.5) reduces to

$$e^{Ax} = e^{\lambda_1 x} \sum_{j=0}^{n-1} \left\{ \frac{1}{j!}(A - \lambda_1 I)^j x^j \right\}. \qquad (19.10)$$

Corollary 19.5. If $k = 2$ and $r_1 = (n - 1)$, $r_2 = 1$ then we have

$$a_1(A) = \frac{1}{(\lambda_2 - \lambda_1)^{n-1}} \left[(\lambda_2 - \lambda_1)^{n-1} I - (A - \lambda_1 I)^{n-1} \right] (A - \lambda_2 I)^{-1},$$

$$q_1(A) = (A - \lambda_2 I), \quad a_2(A) = \frac{1}{(\lambda_2 - \lambda_1)^{n-1}} I, \quad q_2(A) = (A - \lambda_1 I)^{n-1},$$

and hence (19.5) reduces to

$$e^{Ax} = e^{\lambda_1 x} \left[I - \left(\frac{A - \lambda_1 I}{\lambda_2 - \lambda_1} \right)^{n-1} \right] \sum_{j=0}^{n-2} \left\{ \frac{1}{j!}(A - \lambda_1 I)^j x^j \right\}$$

$$+ e^{\lambda_2 x} \left(\frac{A - \lambda_1 I}{\lambda_2 - \lambda_1} \right)^{n-1}$$

$$= e^{\lambda_1 x} \sum_{j=0}^{n-2} \left\{ \frac{1}{j!}(A - \lambda_1 I)^j x^j \right\} - e^{\lambda_1 x} \frac{1}{(\lambda_2 - \lambda_1)^{n-1}}$$

$$\times \sum_{j=0}^{n-2} \left\{ \frac{1}{j!}(A - \lambda_1 I)^{n-1+j} x^j \right\} + e^{\lambda_2 x} \left(\frac{A - \lambda_1 I}{\lambda_2 - \lambda_1} \right)^{n-1}.$$

Now since $(A - \lambda_2 I) = (A - \lambda_1 I) - (\lambda_2 - \lambda_1)I$, we find

$$(A - \lambda_1 I)^{n-1}(A - \lambda_2 I) = (A - \lambda_1 I)^n - (\lambda_2 - \lambda_1)(A - \lambda_1 I)^{n-1}.$$

Thus, by the Cayley–Hamilton Theorem 14.2, we get $(A - \lambda_1 I)^n = (\lambda_2 - \lambda_1)(A - \lambda_1 I)^{n-1}$. Using this relation repeatedly, we obtain $(A - \lambda_1 I)^{n+j-1} = (\lambda_2 - \lambda_1)^j (A - \lambda_1 I)^{n-1}$. It, therefore, follows that

$$
\begin{aligned}
e^{Ax} &= e^{\lambda_1 x} \sum_{j=0}^{n-2} \left\{ \frac{1}{j!}(A - \lambda_1 I)^j x^j \right\} \\
&\quad + \left[e^{\lambda_2 x} - e^{\lambda_1 x} \sum_{j=0}^{n-2} \left\{ \frac{1}{j!}(\lambda_2 - \lambda_1)^j x^j \right\} \right] \left(\frac{A - \lambda_1 I}{\lambda_2 - \lambda_1} \right)^{n-1}.
\end{aligned}
$$

$$(19.11)$$

The second method is discussed in the following theorem.

Theorem 19.6 (Putzer's Algorithm). Let $\lambda_1, \ldots, \lambda_n$ be the eigenvalues of the matrix A which are arranged in some arbitrary, but specified order. Then

$$e^{Ax} = \sum_{j=0}^{n-1} r_{j+1}(x) P_j,$$

where $P_0 = I$, $P_j = \prod_{k=1}^{j}(A - \lambda_k I)$, $j = 1, \ldots, n$ and $r_1(x), \ldots, r_n(x)$ are recursively given by

$$
\begin{aligned}
r_1'(x) &= \lambda_1 r_1(x), \quad r_1(0) = 1 \\
r_j'(x) &= \lambda_j r_j(x) + r_{j-1}(x), \quad r_j(0) = 0, \quad j = 2, \ldots, n.
\end{aligned}
$$

(Note that each eigenvalue in the list is repeated according to its multiplicity. Further, since the matrices $(A - \lambda_i I)$ and $(A - \lambda_j I)$ commute, we can for convenience adopt the convention that $(A - \lambda_j I)$ follows $(A - \lambda_i I)$ if $i > j$.)

Proof. It suffices to show that $\Phi(x)$ defined by

$$\Phi(x) = \sum_{j=0}^{n-1} r_{j+1}(x) P_j$$

satisfies $\Phi' = A\Phi$, $\Phi(0) = I$. For this, we define $r_0(x) \equiv 0$. Then it follows that

$$\Phi'(x) - \lambda_n \Phi(x) = \sum_{j=0}^{n-1}(\lambda_{j+1} r_{j+1}(x) + r_j(x)) P_j - \lambda_n \sum_{j=0}^{n-1} r_{j+1}(x) P_j$$

$$= \sum_{j=0}^{n-1} (\lambda_{j+1} - \lambda_n) r_{j+1}(x) P_j + \sum_{j=0}^{n-1} r_j(x) P_j$$

$$= \sum_{j=0}^{n-2} (\lambda_{j+1} - \lambda_n) r_{j+1}(x) P_j + \sum_{j=0}^{n-2} r_{j+1}(x) P_{j+1}$$

$$= \sum_{j=0}^{n-2} \{ (\lambda_{j+1} - \lambda_n) P_j + (A - \lambda_{j+1} I) P_j \} r_{j+1}(x) \quad (19.12)$$

$$= (A - \lambda_n I) \sum_{j=0}^{n-2} P_j r_{j+1}(x)$$

$$= (A - \lambda_n I)(\Phi(x) - r_n(x) P_{n-1})$$

$$= (A - \lambda_n I)\Phi(x) - r_n(x) P_n, \quad (19.13)$$

where to obtain (19.12) and (19.13) we have used $P_{j+1} = (A - \lambda_{j+1} I) P_j$ and $P_n = (A - \lambda_n I) P_{n-1}$, respectively. Now by the Cayley–Hamilton Theorem 14.2, $P_n = p(A) = 0$, and therefore (19.13) reduces to $\Phi'(x) = A\Phi(x)$. Finally, to complete the proof we note that

$$\Phi(0) = \sum_{j=0}^{n-1} r_{j+1}(0) P_j = r_1(0) I = I. \quad \blacksquare$$

Example 19.2. Consider a 3×3 matrix A having all three eigenvalues equal to λ_1. To use Theorem 19.6, we note that $r_1(x) = e^{\lambda_1 x}$, $r_2(x) = x e^{\lambda_1 x}$, $r_3(x) = (1/2)x^2 e^{\lambda_1 x}$ is the solution set of the system

$$\begin{aligned}
r_1' &= \lambda_1 r_1, & r_1(0) &= 1 \\
r_2' &= \lambda_1 r_2 + r_1, & r_2(0) &= 0 \\
r_3' &= \lambda_1 r_3 + r_2, & r_3(0) &= 0.
\end{aligned}$$

Thus, it follows that

$$e^{Ax} = e^{\lambda_1 x} \left[I + x(A - \lambda_1 I) + \frac{1}{2} x^2 (A - \lambda_1 I)^2 \right], \quad (19.14)$$

which is exactly the same as (19.10) for $n = 3$.

In particular, the matrix

$$A = \begin{bmatrix} 2 & 1 & -1 \\ -3 & -1 & 1 \\ 9 & 3 & -4 \end{bmatrix}$$

has all its eigenvalues equal to -1, and hence from (19.14) we obtain

$$e^{Ax} = \frac{1}{2} e^{-x} \begin{bmatrix} 2 + 6x - 3x^2 & 2x & -2x + x^2 \\ -6x & 2 & 2x \\ 18x - 9x^2 & 6x & 2 - 6x + 3x^2 \end{bmatrix}.$$

Example 19.3. Consider a 3×3 matrix A with eigenvalues $\lambda_1, \lambda_1, \lambda_2$. To use Theorem 19.6, we note that $r_1(x) = e^{\lambda_1 x}$, $r_2(x) = xe^{\lambda_1 x}$,

$$r_3(x) = \frac{xe^{\lambda_1 x}}{(\lambda_1 - \lambda_2)} + \frac{e^{\lambda_2 x} - e^{\lambda_1 x}}{(\lambda_1 - \lambda_2)^2}$$

and hence

$$e^{Ax} = e^{\lambda_1 x} \left[I + x(A - \lambda_1 I) + \left\{ \frac{x}{(\lambda_1 - \lambda_2)} + \frac{e^{(\lambda_2 - \lambda_1)x} - 1}{(\lambda_1 - \lambda_2)^2} \right\} (A - \lambda_1 I)^2 \right],$$
$$(19.15)$$

which is precisely the same as (19.11) for $n = 3$.

In particular, the matrix

$$A = \begin{bmatrix} -1 & 0 & 4 \\ 0 & -1 & 2 \\ 0 & 0 & 1 \end{bmatrix}$$

has the eigenvalues $-1, -1, 1$ and hence from (19.15) we find

$$e^{Ax} = \begin{bmatrix} e^{-x} & 0 & 2(e^x - e^{-x}) \\ 0 & e^{-x} & e^x - e^{-x} \\ 0 & 0 & e^x \end{bmatrix}.$$

Problems

19.1. (i)   If $A = \begin{bmatrix} \alpha & \beta \\ -\beta & \alpha \end{bmatrix}$, show that

$$e^{Ax} = e^{\alpha x} \begin{bmatrix} \cos \beta x & \sin \beta x \\ -\sin \beta x & \cos \beta x \end{bmatrix}.$$

(ii)   If $A = \begin{bmatrix} 0 & 1 \\ -1 & -2\delta \end{bmatrix}$, show that

$$e^{Ax} = \begin{bmatrix} e^{-\delta x} \left(\cos \omega x + \dfrac{\delta}{\omega} \sin \omega x \right) & \dfrac{1}{\omega} e^{-\delta x} \sin \omega x \\ -\dfrac{1}{\omega} e^{-\delta x} \sin \omega x & e^{-\delta x} \left(\cos \omega x - \dfrac{\delta}{\omega} \sin \omega x \right) \end{bmatrix},$$

where $\omega = \sqrt{1 - \delta^2}$.

(iii) If

$$A = \begin{bmatrix} 0 & 1 & 0 & 0 \\ 3\omega^2 & 0 & 0 & 2\omega \\ 0 & 0 & 0 & 1 \\ 0 & -2\omega & 0 & 0 \end{bmatrix},$$

show that

$$e^{Ax} = \begin{bmatrix} 4 - 3\cos\omega x & \dfrac{1}{\omega}\sin\omega x & 0 & \dfrac{2}{\omega}(1 - \cos\omega x) \\ 3\omega\sin\omega x & \cos\omega x & 0 & 2\sin\omega x \\ 6(-\omega x + \sin\omega x) & -\dfrac{2}{\omega}(1 - \cos\omega x) & 1 & \dfrac{1}{\omega}(-3\omega x + 4\sin\omega x) \\ 6\omega(-1 + \cos\omega x) & -2\sin\omega x & 0 & -3 + 4\cos\omega x \end{bmatrix}.$$

(iv)   If $A^2 = \alpha A$, show that $e^{Ax} = I + [(e^{\alpha x} - 1)/\alpha]A$.

   19.2. Let A and P be $n \times n$ matrices given by

$$A = \begin{bmatrix} \lambda & 1 & 0 & \cdots & 0 \\ 0 & \lambda & 1 & \cdots & 0 \\ & \cdots & & & \\ 0 & 0 & 0 & \cdots & 1 \\ 0 & 0 & 0 & \cdots & \lambda \end{bmatrix}, \quad P = \begin{bmatrix} 0 & 1 & 0 & 0 & \cdots & 0 \\ 0 & 0 & 1 & 0 & \cdots & 0 \\ & \cdots & & & & \\ 0 & 0 & 0 & 0 & \cdots & 1 \\ 0 & 0 & 0 & 0 & \cdots & 0 \end{bmatrix}.$$

Show the following:

(i) $P^n = 0$.

(ii) $(\lambda I)P = P(\lambda I)$.

(iii) $e^{Ax} = e^{\lambda x}\left[I + xP + \dfrac{1}{2!}x^2 P^2 + \cdots + \dfrac{1}{(n-1)!}x^{n-1}P^{n-1}\right]$.

   19.3 (Kirchner's Algorithm). Let $\lambda_1, \ldots, \lambda_k$ be distinct eigenvalues of the matrix A with multiplicities $r_1, \ldots, r_k$, respectively. Define

$$p(\lambda) = \prod_{j=1}^{k}(\lambda - \lambda_j)^{r_j}, \quad q_i(\lambda) = p(\lambda)(\lambda - \lambda_i)^{-r_i}, \quad q(\lambda) = \sum_{j=1}^{k} q_j(\lambda),$$

$$f_m(x) = 1 + x + \cdots + \frac{x^{m-1}}{(m-1)!}, \quad p_i(A) = (q(A))^{-1}q_i(A).$$

Show that

$$e^{Ax} = \sum_{j=1}^{k} p_j(A)f_{r_j}((A - \lambda_j I)x)e^{\lambda_j x}.$$

Further, deduce the result (19.10) when $k = 1$.

19.4. Let A and B be two $n \times n$ matrices. We say that A and B are *similar* if and only if there exists a nonsingular matrix P such that $P^{-1}AP = B$. Show the following:

(i) $v(x)$ is a solution of the differential system $v' = Bv$ if and only if $u(x) = Pv(x)$, where $u(x)$ is a solution of the differential system (18.6).

(ii) $e^{Ax} = Pe^{Bx}P^{-1}$.

19.5. Let $u(x)$ be a solution of the differential system (18.6). Show that both the real and imaginary parts of $u(x)$ are solutions of (18.6).

19.6. Show the following:

(i)   Every solution of the differential system (18.6) tends to zero as $x \to \infty$ if and only if the real parts of the eigenvalues of the matrix A are negative.

(ii) Every solution of the differential system (18.6) is bounded in the interval $[0, \infty)$ if and only if the real parts of the multiple eigenvalues of the matrix A are negative, and the real parts of the simple eigenvalues of the matrix A are nonpositive.

19.7. Find the general solution of the differential system (18.6), where the matrix A is given by

(i) $\begin{bmatrix} 4 & -2 \\ 5 & 2 \end{bmatrix}$. (ii) $\begin{bmatrix} 7 & 6 \\ 2 & 6 \end{bmatrix}$. (iii) $\begin{bmatrix} 0 & 1 & 1 \\ 1 & 0 & 1 \\ 1 & 1 & 0 \end{bmatrix}$. (iv) $\begin{bmatrix} 1 & -1 & 4 \\ 3 & 2 & -1 \\ 2 & 1 & -1 \end{bmatrix}$.

(v) $\begin{bmatrix} -1 & 1 & 0 \\ 0 & -1 & 0 \\ 0 & 0 & 3 \end{bmatrix}$. (vi) $\begin{bmatrix} 5 & -3 & -2 \\ 8 & -5 & -4 \\ -4 & 3 & 3 \end{bmatrix}$.

19.8. Find the general solution of the nonhomogeneous differential system $u' = Au + b(x)$, where the matrix A and the vector $b(x)$ are given by

(i) $\begin{bmatrix} 0 & -1 \\ 3 & 4 \end{bmatrix}$, $\begin{bmatrix} x \\ -2 - 4x \end{bmatrix}$;

(ii) $\begin{bmatrix} -2 & -4 \\ -1 & 1 \end{bmatrix}$, $\begin{bmatrix} 1 + 4x \\ (3/2)x^2 \end{bmatrix}$;

(iii) $\begin{bmatrix} -1 & 1 & 1 \\ 1 & -1 & 1 \\ 1 & 1 & 1 \end{bmatrix}$, $\begin{bmatrix} e^x \\ e^{3x} \\ 4 \end{bmatrix}$;

(iv) $\begin{bmatrix} 2 & 1 & -2 \\ -1 & 0 & 0 \\ 1 & 1 & -1 \end{bmatrix}$, $\begin{bmatrix} 2 - x \\ 1 \\ 1 - x \end{bmatrix}$.

19.9. Find the solutions of the following initial value problems:

(i) $u' = \begin{bmatrix} 1 & 5 \\ -1 & -3 \end{bmatrix} u, \quad u_1(0) = -2, \ u_2(0) = 1.$

(ii) $u' = \begin{bmatrix} 0 & 1 \\ -1 & 0 \end{bmatrix} u, \quad u_1(\pi) = -1, \ u_2(\pi) = 0.$

(iii) $u' = \begin{bmatrix} 1 & 0 & 0 \\ 2 & 1 & -2 \\ 3 & 2 & 1 \end{bmatrix} u + \begin{bmatrix} 0 \\ 0 \\ e^x \cos 2x \end{bmatrix}, \quad u_1(0) = 0, \ u_2(0) = 1, \ u_3(0) = 1.$

(iv) $u' = \begin{bmatrix} -1 & 0 & 4 \\ 0 & -1 & 2 \\ 0 & 0 & 1 \end{bmatrix} u + \begin{bmatrix} e^x \\ e^{-x} \\ 0 \end{bmatrix}, \quad u_1(0) = 0, \ u_2(0) = 1, \ u_3(0) = 0.$

(v) $u' = \begin{bmatrix} 2 & 1 & -1 \\ -3 & -1 & 1 \\ 9 & 3 & -4 \end{bmatrix} u + \begin{bmatrix} 0 \\ x \\ 0 \end{bmatrix}, \quad u_1(0) = 0, \ u_2(0) = 3, \ u_3(0) = 0.$

(vi) $u' = \begin{bmatrix} 2 & 1 & 1 \\ 0 & 2 & 0 \\ 0 & 0 & 3 \end{bmatrix} u + \begin{bmatrix} 1 \\ 0 \\ x \end{bmatrix}, \quad u_1(0) = 1, \ u_2(0) = 1, \ u_3(0) = 1.$

19.10. Consider the DE (18.18). Show the following:

(i)   Its characteristic equation is

$$p(\lambda) = \lambda^n + p_1 \lambda^{n-1} + \cdots + p_n = 0. \tag{19.16}$$

(ii) If $\lambda_1 \neq \lambda_2 \neq \cdots \neq \lambda_n$ are the roots of (19.16), then $e^{\lambda_i x}$, $1 \leq i \leq n$ are n linearly independent solutions of the DE (18.18).

(iii) If $\lambda_1 \neq \lambda_2 \neq \cdots \neq \lambda_k$ $(k < n)$ are the roots of (19.16) with multiplicities $r_1, \ldots, r_k$, respectively, then $e^{\lambda_i x}$, $x e^{\lambda_i x}, \ldots, x^{(r_i - 1)} e^{\lambda_i x}$, $1 \leq i \leq k$ are n linearly independent solutions of the DE (18.18).

Answers or Hints

19.1. Verify directly.

19.2. (i) Observe that in each multiplication the position of 1 is shifted by one column, so in P^2 the nth and $(n-1)$th rows are 0. (ii) Obvious. (iii) Since $A = \lambda I + P$, we can use parts (i) and (ii).

19.3. Clearly, $q(\lambda)$ is a polynomial, and since $q(\lambda_i) = q_i(\lambda_i) \neq 0$, $i = 1, \ldots, k$ it follows that $p(\lambda)$ and $q(\lambda)$ have no common factor. Thus, there exist polynomials $q_*(\lambda)$ and $p_*(\lambda)$ such that $q(\lambda)q_*(\lambda) + p(\lambda)p_*(\lambda) = 1$. Hence, in view of $p(A) = 0$ we obtain $q(A)q_*(A) = I$, i.e., $q_*(A) = q(A)^{-1}$. Thus, $q(A)^{-1}$ exists and is expressible as a polynomial in A. Now

$$e^{Ax} = q(A)^{-1} q(A) e^{Ax} = q(A)^{-1} \sum_{j=1}^{k} q_j(A) e^{(A - \lambda_j I)x} \cdot e^{\lambda_j I x}.$$

Finally, use $q_j(A)(A - \lambda_j I)^i = 0$ for all $i \geq r_j$. For $k = 1$ note that $q_1(\lambda) = q(\lambda) = 1$ and hence $q(A)^{-1}q_*(A) = I$.

19.4. (i) Verify directly. (ii) Since $P^{-1}AP = B$ implies $A = PBP^{-1}$, $e^{Ax} = e^{PBP^{-1}x}$.

19.5. If $u(x) = p(x) + iq(x)$ is a solution of $u' = A(x)u$, then $p'(x) + iq'(x) = A(x)p(x) + iA(x)q(x)$.

19.6. Let $\lambda_j = \alpha_j + i\beta_j$, $1 \leq j \leq k \leq n$ be distinct eigenvalues of the matrix A with multiplicities r_j, $1 \leq j \leq k$, respectively. If $\alpha = \max_{1 \leq j \leq k} \alpha_j$ and $r = \max_{1 \leq j \leq k} r_j$, then there exists $x_1 \geq x_0 \geq 0$ sufficiently large such that for all $x \geq x_1$ the relation (19.5) gives $\|e^{Ax}\| \leq Ce^{\alpha x}x^r$.

19.7. (i) $e^{3x} \begin{bmatrix} 2c_1 \cos 3x + 2c_2 \sin 3x \\ c_1(\cos 3x + 3\sin 3x) + c_2(\sin 3x - 3\cos 3x) \end{bmatrix}$.

(ii) $c_1 e^{10x} \begin{bmatrix} 2 \\ 1 \end{bmatrix} + c_2 e^{3x} \begin{bmatrix} 3 \\ -2 \end{bmatrix}$. (iii) $\begin{bmatrix} e^{2x} & e^{-x} & 0 \\ e^{2x} & 0 & e^{-x} \\ e^{2x} & -e^{-x} & -e^{-x} \end{bmatrix} \begin{bmatrix} c_1 \\ c_2 \\ c_3 \end{bmatrix}$.

(iv) $c_1 e^x \begin{bmatrix} -1 \\ 4 \\ 1 \end{bmatrix} + c_2 e^{-2x} \begin{bmatrix} 1 \\ -1 \\ -1 \end{bmatrix} + c_3 e^{3x} \begin{bmatrix} 1 \\ 2 \\ 1 \end{bmatrix}$.

(v) $\begin{bmatrix} 0 & -e^{-x} & xe^{-x} \\ 0 & 0 & e^{-x} \\ e^{3x} & 0 & 0 \end{bmatrix} \begin{bmatrix} c_1 \\ c_2 \\ c_3 \end{bmatrix}$. (vi) $e^x \begin{bmatrix} 1 & 0 & 2x \\ 0 & 2 & 4x \\ 2 & -3 & -2x-1 \end{bmatrix} \begin{bmatrix} c_1 \\ c_2 \\ c_3 \end{bmatrix}$.

19.8. (i) $c_1 e^x \begin{bmatrix} 1 \\ -1 \end{bmatrix} + c_2 e^{3x} \begin{bmatrix} 1 \\ -3 \end{bmatrix} + \begin{bmatrix} 1 \\ x \end{bmatrix}$.

(ii) $c_1 e^{2x} \begin{bmatrix} 1 \\ -1 \end{bmatrix} + c_2 e^{-3x} \begin{bmatrix} 4 \\ 1 \end{bmatrix} + \begin{bmatrix} x + x^2 \\ -\frac{1}{2}x^2 \end{bmatrix}$.

(iii) $\frac{1}{6} \begin{bmatrix} 2 & 1 & 3 \\ 2 & 1 & -3 \\ -2 & 2 & 0 \end{bmatrix} \begin{bmatrix} c_1 e^{-x} + \frac{1}{2}e^x + \frac{1}{4}e^{3x} - 4 \\ c_2 e^{2x} - e^x + e^{3x} - 4 \\ c_3 e^{-2x} + \frac{1}{3}e^x - \frac{1}{5}e^{3x} \end{bmatrix}$.

(iv) $c_1 e^x \begin{bmatrix} 1 \\ -1 \\ 0 \end{bmatrix} + c_2 \begin{bmatrix} \sin x \\ \cos x \\ \sin x \end{bmatrix} + c_3 \begin{bmatrix} \cos x \\ -\sin x \\ \cos x \end{bmatrix} + \begin{bmatrix} 0 \\ x \\ 1 \end{bmatrix}$.

19.9. (i) $e^{-x} \begin{bmatrix} -2\cos x + \sin x \\ \cos x \end{bmatrix}$. (ii) $\begin{bmatrix} \cos x \\ -\sin x \end{bmatrix}$.

(iii) $e^x \begin{bmatrix} 0 \\ \cos 2x - (1 + \frac{1}{2}x)\sin 2x \\ (1 + \frac{1}{2}x)\cos 2x + \frac{5}{4}\sin 2x \end{bmatrix}$. (iv) $\begin{bmatrix} \frac{1}{2}(e^x - e^{-x}) \\ e^{-x}(x + 1) \\ 0 \end{bmatrix}$.

(v) $\begin{bmatrix} 2(1 + 2x)e^{-x} + x - 2 \\ 4e^{-x} + x - 1 \\ 6(1 + 2x)e^{-x} + 3(x - 2) \end{bmatrix}$.

(vi) $\frac{1}{36}$ $\begin{bmatrix} 40e^{3x} + 36xe^{2x} + 9e^{2x} + 6x - 13 \\ 36e^{2x} \\ 40e^{3x} - 12x - 4 \end{bmatrix}$.

19.10. (i) In system form the DE (18.18) is equivalent to (18.6), where

$$A = \begin{bmatrix} 0 & 1 & 0 & \cdots & 0 & 0 \\ 0 & 0 & 1 & \cdots & 0 & 0 \\ \cdots & & & & & \\ 0 & 0 & 0 & \cdots & 0 & 1 \\ -p_n & -p_{n-1} & -p_{n-2} & \cdots & -p_2 & -p_1 \end{bmatrix}.$$

Now in $\det(A - \lambda I) = 0$ perform the operation $C_1 + \lambda C_2 + \cdots + \lambda^{n-1}C_n$.
(ii) If λ_i is a simple root of $p(\lambda) = 0$, then for the above matrix A, $[1, \lambda_i, \lambda_i^2, \ldots, \lambda_i^{n-1}]^T$ is the corresponding eigenvector. In fact,

$$(A - \lambda_i I)[1, \lambda_i, \ldots, \lambda_i^{n-1}]^T = [0, \ldots, 0, -p(\lambda_i)]^T = [0, \ldots, 0]^T.$$

Thus, corresponding to λ_i the solution vector of $u' = Au$ is $u^i(x) = [e^{\lambda_i x}, \lambda_i e^{\lambda_i x}, \ldots, \lambda_i^{n-1} e^{\lambda_i x}]$.
(iii) If λ_i is a multiple root of $p(\lambda) = 0$ with multiplicity r_i, then $p^{(j)}(\lambda_i) = 0$, $0 \le j \le r_i - 1$. Let $L = \frac{d^n}{dx^n} + p_1 \frac{d^{n-1}}{dx^{n-1}} + \cdots + p_n$, so that (18.18) can be written as $L[y] = 0$. Since $L[e^{\lambda x}] = p(\lambda)e^{\lambda x}$, j times differentiation with respect to λ gives

$$\frac{\partial^j}{\partial \lambda^j} L[e^{\lambda x}] = L\left[\frac{\partial^j}{\partial \lambda^j} e^{\lambda x}\right] = L[x^j e^{\lambda x}] = \left[\sum_{i=0}^j \binom{j}{i} p^{(j)}(\lambda) x^{j-i}\right] e^{\lambda x}.$$

Now to prove linear independence suppose we have n constants c_{ij}, $1 \le i \le k$, $0 \le j \le r_i - 1$ such that

$$\sum_{i=1}^k P_i(x) e^{\lambda_i x} = 0,$$

where $P_i(x) = \sum_{j=0}^{r_i - 1} c_{ij} x^j$. If all constants c_{ij} are not zero, then there will be at least one, say, $P_k(x)$ not identically zero. The above relation can be written as

$$P_1(x) + P_2(x) e^{(\lambda_2 - \lambda_1)x} + \cdots + P_k(x) e^{(\lambda_k - \lambda_1)x} = 0.$$

Differentiating this r_1 times reduces $P_1(x)$ to 0, and we obtain

$$Q_2(x) e^{(\lambda_2 - \lambda_1)x} + \cdots + Q_k(x) e^{(\lambda_k - \lambda_1)x} = 0,$$

or

$$Q_2(x) e^{\lambda_2 x} + \cdots + Q_k(x) e^{\lambda_k x} = 0,$$

where $\deg Q_i(x) = \deg P_i(x)$, $2 \le i \le k$ and $Q_k(x) \not\equiv 0$. Continuing this process, we arrive at

$$R_k(x) e^{\lambda_k x} = 0,$$

where $\deg R_k(x) = \deg P_k(x)$ and $R_k(x) \not\equiv 0$. However, the above relation implies that $R_k(x) \equiv 0$.

Lecture 20
Periodic Linear Systems

A function $y(x)$ is called *periodic* of period $\omega > 0$ if for all x in the domain of the function

$$y(x + \omega) = y(x). \tag{20.1}$$

Geometrically, this means that the graph of $y(x)$ repeats itself in successive intervals of length ω. For example, the functions $\sin x$ and $\cos x$ are periodic of period 2π. For convenience, we shall assume that ω is the smallest positive number for which (20.1) holds. If each component $u_i(x)$, $1 \leq i \leq n$ of $u(x)$ and each element $a_{ij}(x)$, $1 \leq i, j \leq n$ of $A(x)$ are periodic of period ω, then $u(x)$ and $A(x)$ are said to be *periodic* of period ω. Periodicity of solutions of differential systems is an interesting and important aspect of qualitative study. Here we shall provide certain characterizations for the existence of such solutions of linear differential systems.

To begin with we shall provide necessary and sufficient conditions for the differential system (17.1) to have a periodic solution of period ω.

Theorem 20.1. Let the matrix $A(x)$ and the function $b(x)$ be continuous and periodic of period ω in $\mathbb{R}$. Then the differential system (17.1) has a periodic solution $u(x)$ of period ω if and only if $u(0) = u(\omega)$.

Proof. Let $u(x)$ be a periodic solution of period ω, then by definition it is necessary that $u(0) = u(\omega)$. To show sufficiency, let $u(x)$ be a solution of (17.1) satisfying $u(0) = u(\omega)$. If $v(x) = u(x + \omega)$, then it follows that $v'(x) = u'(x+\omega) = A(x+\omega)u(x+\omega)+b(x+\omega) = A(x)v(x)+b(x)$; i.e., $v(x)$ is a solution of (17.1). However, since $v(0) = u(\omega) = u(0)$, the uniqueness of the initial value problems implies that $u(x) = v(x) = u(x+\omega)$, and hence $u(x)$ is periodic of period ω. ∎

Corollary 20.2. Let the matrix $A(x)$ be continuous and periodic of period ω in $\mathbb{R}$. Further, let $\Psi(x)$ be a fundamental matrix of the differential system (17.3). Then the differential system (17.3) has a nontrivial periodic solution $u(x)$ of period ω if and only if $\det(\Psi(0) - \Psi(\omega)) = 0$.

Proof. We know that the general solution of the differential system (17.3) is $u(x) = \Psi(x)c$, where c is an arbitrary constant vector. This $u(x)$ is periodic of period ω if and only if $\Psi(0)c = \Psi(\omega)c$, i.e., the system $(\Psi(0) - \Psi(\omega))c = 0$ has a nontrivial solution vector c. But, from Theorem 13.2 this system has a nontrivial solution if and only if $\det(\Psi(0) - \Psi(\omega)) = 0$. ∎

R.P. Agarwal and D. O'Regan, *An Introduction to Ordinary Differential Equations*, doi: 10.1007/978-0-387-71276-5_20, © Springer Science + Business Media, LLC 2008

Corollary 20.3. The differential system (18.6) has a nontrivial periodic solution of period ω if and only if the matrix $(I - e^{A\omega})$ is singular.

Corollary 20.4. Let the conditions of Theorem 20.1 be satisfied. Then the differential system (17.1) has a unique periodic solution of period ω if and only if the differential system (17.3) does not have a periodic solution of period ω other than the trivial one.

Proof. Let $\Psi(x)$ be a fundamental matrix of the differential system (17.3). Then from (18.14) the general solution of (17.1) can be written as

$$u(x) = \Psi(x)c + \int_0^x \Psi(x)\Psi^{-1}(t)b(t)dt,$$

where c is an arbitrary constant. This $u(x)$ is periodic of period ω if and only if

$$\Psi(0)c = \Psi(\omega)c + \int_0^\omega \Psi(\omega)\Psi^{-1}(t)b(t)dt;$$

i.e., the system

$$(\Psi(0) - \Psi(\omega))c = \int_0^\omega \Psi(\omega)\Psi^{-1}(t)b(t)dt$$

has a unique solution vector c. But, from Theorem 13.2 this system has a unique solution if and only if $\det(\Psi(0) - \Psi(\omega)) \neq 0$. Now the conclusion follows from Corollary 20.2. ∎

In the case when the conditions of Corollary 20.2 are satisfied, the fundamental matrix $\Psi(x)$ can be represented as a product of a periodic matrix of period ω and a fundamental matrix of a differential system with constant coefficients. This basic result is known as Floquet's theorem.

Theorem 20.5 (Floquet's Theorem). Let the conditions of Corollary 20.2 be satisfied. Then the following hold:

(i)   The matrix $\chi(x) = \Psi(x + \omega)$ is also a fundamental matrix of the differential system (17.3).

(ii)   There exists a periodic nonsingular matrix $P(x)$ of period ω and a constant matrix R such that

$$\Psi(x) = P(x)e^{Rx}. \tag{20.2}$$

Proof. Since $\Psi(x)$ is a fundamental matrix of the differential system (17.3) it follows that

$$\chi'(x) = \Psi'(x + \omega) = A(x + \omega)\Psi(x + \omega) = A(x)\chi(x);$$

i.e., $\chi(x)$ is a solution matrix of the differential system (17.3). Further, since $\det \Psi(x + \omega) \neq 0$ for all x, we have $\det \chi(x) \neq 0$ for all x. Hence, we conclude that $\chi(x)$ is a fundamental matrix of the differential system (17.3). This completes the proof of part (i).

Next we shall prove part (ii), since $\Psi(x)$ and $\Psi(x+\omega)$ are both fundamental matrices of the differential system (17.3) from Theorem 18.2 there exists a nonsingular constant matrix C such that

$$\Psi(x+\omega) \;=\; \Psi(x)C. \tag{20.3}$$

Now from Theorem 14.3 there exists a constant matrix R such that $C = e^{R\omega}$. Thus, from (20.3) it follows that

$$\Psi(x+\omega) \;=\; \Psi(x)e^{R\omega}. \tag{20.4}$$

Let $P(x)$ be a matrix defined by the relation

$$P(x) \;=\; \Psi(x)e^{-Rx}. \tag{20.5}$$

Then using (20.4) we have

$$P(x+\omega) \;=\; \Psi(x+\omega)e^{-R(x+\omega)} \;=\; \Psi(x)e^{R\omega}e^{-R(x+\omega)} \;=\; \Psi(x)e^{-Rx} \;=\; P(x).$$

Hence, $P(x)$ is periodic of period ω. Further, since $\Psi(x)$ and e^{-Rx} are nonsingular $\det P(x) \neq 0$ in $\mathbb{R}$. ∎

In relation (20.3) the matrix C is in fact $\Psi^{-1}(0)\Psi(\omega)$, and hence $e^{R\omega} = \Psi^{-1}(0)\Psi(\omega)$, which gives the matrix $R = \ln(\Psi^{-1}(0)\Psi(\omega))/\omega$. Thus, in (20.5) if the matrix $\Psi(x)$ is known only in the interval $[0,\omega]$ the periodicity property of $P(x)$ can be used to determine it in the whole interval $\mathbb{R}$. Hence, from (20.2) the fundamental matrix $\Psi(x)$ can be determined in the interval $\mathbb{R}$ provided it is known only in the interval $[0,\omega]$.

Theorem 20.5 is particularly interesting because it suggests a transformation which reduces the differential system (17.3) to a differential system of the type (18.6). Precisely, we shall prove the following result.

Theorem 20.6. Let $P(x)$ and R be the matrices obtained in Theorem 20.5. Then the transformation $u(x) = P(x)v(x)$ reduces the differential system (17.3) to the system

$$v' \;=\; Rv. \tag{20.6}$$

Proof. Since $\Psi'(x) = A(x)\Psi(x)$, the relation (20.2) gives

$$\left(P(x)e^{Rx}\right)' \;=\; A(x)P(x)e^{Rx},$$

which yields

$$P'(x) + P(x)R - A(x)P(x) \;=\; 0. \tag{20.7}$$

Now using the transformation $u = P(x)v$ in the differential system (17.3), we obtain

$$P(x)v' + P'(x)v = A(x)P(x)v,$$

which is the same as

$$P(x)v' + [P'(x) - A(x)P(x)]v = 0. \tag{20.8}$$

Combining the relations (20.7) and (20.8) we get (20.6). ∎

Once again we assume that the conditions of Corollary 20.2 are satisfied, and $\Psi_1(x)$ is another fundamental matrix of the differential system (17.3). Then from Theorem 18.2 there exists a nonsingular matrix M such that $\Psi(x) = \Psi_1(x)M$ holds. Thus, from (20.4) it follows that $\Psi_1(x + \omega)M = \Psi_1(x)Me^{R\omega}$, i.e.,

$$\Psi_1(x + \omega) = \Psi_1(x)Me^{R\omega}M^{-1}. \tag{20.9}$$

Hence, we conclude that every fundamental matrix $\Psi_1(x)$ of the differential system (17.3) determines a matrix $Me^{R\omega}M^{-1}$ which is similar to $e^{R\omega}$. Conversely, if M is any constant nonsingular matrix, then there exists a fundamental matrix $\Psi_1(x)$ of the differential system (17.3) such that the relation (20.9) holds. The nonsingular matrix C associated with the fundamental matrix $\Psi(x)$ in (20.3) is called the *monodromy matrix* of the differential system (17.3). For example, monodromy matrix for $\Psi_1(x)$ is $Me^{R\omega}M^{-1}$. The eigenvalues of C are called the *multipliers* of (17.3), and the eigenvalues of R are called the *exponents* of (17.3).

Let $\sigma_1, \ldots, \sigma_n$ and $\lambda_1, \ldots, \lambda_n$, respectively, be the multipliers and exponents of (17.3), then from the relation $C = e^{R\omega}$ it follows that $\sigma_i = e^{\lambda_i \omega}$, $1 \le i \le n$. It should be noted that the exponents of (17.3) are determined only $\mod(2\pi i/\omega)$, because even though C is determined uniquely, the matrix R is not unique. Further, since the matrix C is nonsingular, none of the multipliers $\sigma_1, \ldots, \sigma_n$ of (17.3) is zero.

From the relation (20.4), we have $\Psi(\omega) = \Psi(0)e^{R\omega}$, and hence we conclude that $\sigma_1, \ldots, \sigma_n$ are the eigenvalues of $\Psi^{-1}(0)\Psi(\omega)$, or of the matrix $\Phi(\omega, 0)$ if $\Psi(x) = \Phi(x, 0)$, i.e., $\Psi(x)$ is the principal fundamental matrix of (17.3). Thus, from Theorem 17.3 and Problem 14.1 it follows that

$$\det \Phi(\omega, 0) = \prod_{i=1}^{n} \sigma_i = \det \Phi(0,0) \exp\left(\int_0^\omega \operatorname{Tr} A(t)dt\right) = \exp\left(\sum_{i=1}^{n} \lambda_i \omega\right). \quad ∎ \tag{20.10}$$

The final result is a direct consequence of Theorem 20.5.

Theorem 20.7. Let the conditions of Corollary 20.2 be satisfied. Then a complex number λ is an exponent of the differential system (17.3) if and only if there exists a nontrivial solution of (17.3) of the form $e^{\lambda x}p(x)$, where $p(x+\omega) = p(x)$. In particular, there exists a periodic solution of (17.3) of period ω (2ω) if and only if there is a multiplier 1 (-1) of (17.3).

Proof. Suppose $u(x) = e^{\lambda x}p(x)$, $p(x+\omega) = p(x)$ is a nontrivial solution of (17.3) with $u(0) = u^0$, then $u(x) = \Phi(x,0)u^0 = e^{\lambda x}p(x)$, where $\Phi(x,0)$ is the principal fundamental matrix of (17.3). From Theorem 20.5, we also have $u(x) = \Phi(x,0)u^0 = P(x)e^{Rx}u^0$, where $P(x)$ is a periodic matrix of period ω. Therefore, $e^{\lambda(x+\omega)}p(x) = P(x)e^{R(x+\omega)}u^0$, which is the same as $P(x)e^{Rx}\left(e^{\lambda\omega}I - e^{R\omega}\right)u^0 = 0$, and hence $\det\left(e^{\lambda\omega}I - e^{R\omega}\right) = 0$, i.e., λ is an exponent of the differential system (17.3). Conversely, if λ is an exponent of (17.3), then we have $e^{Rx}u^0 = e^{\lambda x}u^0$ for all x, and hence $P(x)e^{Rx}u^0 = P(x)u^0 e^{\lambda x}$. However, $u(x) = P(x)e^{Rx}u^0$ is the solution of (17.3), and hence the conclusion follows. To prove the second assertion it suffices to note that the multiplier of (17.3) is 1 (-1) provided $\lambda = 0$ ($\pi i/\omega$), and then the solution $e^{\lambda x}p(x)$ reduces to $p(x)(e^{\pi i x/\omega}p(x))$ which is periodic of period ω (2ω). ∎

Example 20.1. In system form *Hill's equation* $y'' + p(x)y = 0$, where $p(x)$ is periodic of period π, can be written as

$$u' = \begin{bmatrix} 0 & 1 \\ -p(x) & 0 \end{bmatrix} u. \tag{20.11}$$

This, as a special case, includes *Mathieu's equation* for $p(x) = \lambda + 16d\cos 2x$.

Obviously, in (20.11) the matrix $A(x)$ is periodic of period π. Thus, for the principal fundamental matrix $\Phi(x,0)$ of (20.11) the relation (20.3) gives $\Phi(x+\pi,0) = \Phi(x,0)C$, and hence $C = \Phi(\pi,0)$. Further, in the system (20.11), $\text{Tr } A(x) = 0$, and hence Theorem 17.3 gives $\det \Phi(x,0) = 1$ for all x. Thus, from Problem 14.1 the eigenvalues σ_1 and σ_2 of C are the roots of the quadratic equation $\sigma^2 - a\sigma + 1 = 0$, where $a = u_1^1(\pi) + u_2^2(\pi)$. Let λ_1 and λ_2 be the exponents of (20.11), then it follows that $\sigma_i = e^{\lambda_i \pi}$, $i = 1, 2$.

Now we shall discuss various cases.

(i)   If $a > 2$, then σ_1 and σ_2 are real, distinct, and positive, and hence the exponents are real and distinct.

(ii) If $a < -2$, then σ_1 and σ_2 are real, distinct, and negative, and hence the exponents are complex with real and imaginary parts different from zero.

(iii) If $|a| < 2$, then σ_1 and σ_2 are complex conjugates with the absolute value 1, and hence the exponents are purely imaginary.

If the roots σ_1 and σ_2 are unequal, then there exists a pair of linearly independent solutions $p_1(x)e^{\lambda_1 x}$, $p_2(x)e^{\lambda_2 x}$ where $p_1(x)$ and $p_2(x)$ are continuous periodic functions of period π.

(iv) If $|a| = 2$, then the quadratic equation has a double root. When

$a = 2$, the double root is $\sigma = 1$, and if $a = -2$, the double root is $\sigma = -1$. In this case two linearly independent solutions of (20.11) are $\pm p_1(x)$ and $\pm x p_2(x)$. One of these solutions is periodic with period π for $\sigma = 1$, period 2π for $\sigma = -1$, and the other is unbounded.

Problems

20.1. Consider the equation $y' = ay + \sin x$. Discuss the cases (i) $a = 0$, (ii) $a > 0$, and (iii) $a < 0$ separately for the existence of a unique periodic solution.

20.2. Verify that in the DE $y' = (\cos^2 x)y$ even though the function $\cos^2 x$ is periodic of period π, the solutions are not periodic.

20.3. Consider the DE $y'' + y = \cos x$.

(i)   Show that the general solution of this DE is

$$y(x) = c_1 \sin(x + c_2) + \frac{1}{2}x \sin x,$$

where c_1 and c_2 are arbitrary constants. Observe that $y(x)$ is not periodic.

(ii)   Does this example contradict Corollary 20.4?

20.4. Consider the DE $y'' + y = \sin 2x$.

(i)   Show that $y(x) = -(1/3)\sin 2x$ is a solution of this DE and it is periodic of period 2π.

(ii)   Show that the DE $y'' + y = 0$ also admits nontrivial periodic solutions of period 2π.

(iii)   Does this contradict Corollary 20.4?

20.5. The DE for the undamped mass–spring system with a given periodic external force can conveniently be written as $y'' + k_0^2 y = A \cos kx$, where k_0 is the natural frequency of the system and k the applied frequency.

If $k \neq k_0$, a particular solution of this DE is given by

$$y(x) = \frac{A}{k_0^2 - k^2} \cos kx.$$

Thus, if the applied frequency k is close to the natural frequency k_0, then this particular solution represents an oscillation with large amplitude. This phenomenon is called *resonance*. If $k = k_0$, a particular solution cannot be obtained from this solution. Show that this particular solution is given by

$$y(x) = \frac{A}{2k_0} x \sin k_0 x,$$

which is nonperiodic.

20.6. Let $y_1(x)$ and $y_2(x)$ be two solutions of the DE $y'' + p(x)y = 0$ such that $y_1(0) = 1$, $y_1'(0) = 0$; $y_2(0) = 0$, $y_2'(0) = 1$. Further, let $p(x)$ be continuous and periodic of period ω in $\mathbb{R}$. Show the following:

(i)   The Wronskian $W(y_1, y_2)(x) = 1$ for all $x \in \mathbb{R}$.

(ii)   There is at least one nontrivial periodic solution $y(x)$ of period ω if and only if $y_1(\omega) + y_2'(\omega) = 2$.

(iii)   There is at least one nontrivial *antiperiodic solution* $y(x)$, i.e., $y(x + \omega) = -y(x)$, $x \in \mathbb{R}$ if and only if $y_1(\omega) + y_2'(\omega) = -2$.

20.7. Consider the DE (6.19), where $p_1(x)$ and $p_2(x)$ are continuous and periodic of period ω in $\mathbb{R}$. Show that a nontrivial solution $y(x)$ is periodic of period ω if and only if $y(0) = y(\omega)$ and $y'(0) = y'(\omega)$. Further, if $y_1(x)$ and $y_2(x)$ are two solutions such that $y_1(0) = 1$, $y_1'(0) = 0$; $y_2(0) = 0$, $y_2'(0) = 1$, then show that there exist constants a, b, c, d such that for all x,

$$y_1(x + \omega) = ay_1(x) + by_2(x)$$
$$y_2(x + \omega) = cy_1(x) + dy_2(x).$$

20.8. Let $f(x, y)$ be a continuous function defined for all $(x, y) \in \mathbb{R}^2$. Further, let $f(x, y)$ satisfy a Lipschitz condition in y, $f(x + \omega, y) = f(x, y)$ for some $\omega > 0$, and $f(x, y_1)f(x, y_2) < 0$ for all x and some y_1, y_2. Show that the DE (1.9) has at least one periodic solution of period ω. Apply this result to the DE (5.2) where $p(x) \not\equiv 0$ and $q(x)$ are continuous periodic functions of period ω.

***20.9.** Let $p(x)$ be continuous and $p(x + \pi) = p(x) \not\equiv 0$ for all x. If

$$0 \leq \pi \int_0^\pi |p(t)| dt \leq 4,$$

then show that all solutions of the DE $y'' + p(x)y = 0$ are bounded on $\mathbb{R}$.

Answers or Hints

20.1. (i) Infinite number of periodic solutions of period 2π. (ii) and (iii) $(a \sin x - \cos x)/(1 + a^2)$ is the only periodic solution of period 2π.

20.2. The general solution is $c \exp\left(\frac{1}{2}\left(x + \frac{1}{2}\sin 2x\right)\right)$.

20.3. (i) Verify directly. (ii) The corresponding homogeneous system has nontrivial periodic solutions of period 2π.

20.4. The function $-\frac{1}{3}\sin 2x$ is periodic of period π (smallest period).

20.5. Verify directly.

20.6. (i) Use (6.3). (ii) Use Corollary 20.2. (iii) See the proof of Corollary 20.2.

20.7. Use Theorem 20.1.

20.8. A continuous function mapping a closed interval into itself has at least one fixed point.

Lecture 21
Asymptotic Behavior of Solutions of Linear Systems

In this lecture we shall begin with the study of ultimate behavior of solutions of linear differential systems. In particular, we shall provide sufficient conditions on the known quantities in a given system so that all its solutions remain bounded or tend to zero as $x \to \infty$. Thus, from the practical point of view the results we shall discuss are very important because an explicit form of the solutions is not needed.

We begin with the differential system (18.6) and note that Problem 19.6 does provide necessary and sufficient conditions for all its solutions to remain bounded or tend to zero. Further, if in Theorem 19.2 each $\lambda_j = \alpha_j + i\beta_j$ and $\alpha = \max_{1 \le j \le k} \alpha_j$, and $r = \max_{1 \le j \le k} r_j$; then there exists an $x_1 \ge x_0 \ge 0$ such that for all $x \ge x_1$ the relation (19.5) gives $\|e^{Ax}\| \le ce^{\alpha x}x^r$, where c is some suitable constant. Let $\alpha < \eta$, then there exists a $x_2 \ge x_1$ such that for all $x \ge x_2$ the inequality $e^{\alpha x}x^r \le e^{\eta x}$ holds. Thus, for $x \ge x_2$ it follows that

$$\|e^{Ax}\| \le ce^{\eta x}. \tag{21.1}$$

However, since the interval $[0, x_2]$ is finite in (21.1) we can always choose c sufficiently large so that it holds for all $x \ge 0$.

From (21.1) it is immediate that for any solution $u(x)$ of (18.6)

$$\|u(x)\| \le c_1 e^{\eta x}, \tag{21.2}$$

where c_1 is some suitable constant.

Now we shall consider the differential system

$$v' = (A + B(x))v, \tag{21.3}$$

where $B(x)$ is an $n \times n$ matrix with continuous elements $b_{ij}(x)$, $1 \le i, j \le n$ in the interval $[x_0, \infty)$. System (21.3) can be regarded as a perturbed system of (18.6). Our first result gives sufficient conditions on the matrix $B(x)$ so that all solutions of the differential system (21.3) remain bounded provided all solutions of (18.6) are bounded.

Theorem 21.1. Let all solutions of the differential system (18.6) be bounded in $[0, \infty)$. Then all solutions of the differential system (21.3) are

R.P. Agarwal and D. O'Regan, *An Introduction to Ordinary Differential Equations*, doi: 10.1007/978-0-387-71276-5_21, © Springer Science + Business Media, LLC 2008

bounded in $[x_0, \infty)$ provided

$$\int^{\infty} \|B(t)\| dt \; < \; \infty. \tag{21.4}$$

Proof. In (18.15) let the nonhomogeneous term $b(x)$ be $B(x)v$, so that each solution $v(x)$ such that $v(x_0) = v^0$ of the differential system (21.3) satisfies the integral equation

$$v(x) \; = \; e^{A(x-x_0)}v^0 + \int_{x_0}^{x} e^{A(x-t)} B(t)v(t)dt. \tag{21.5}$$

Now since all solutions of (18.6) are bounded there exists a constant c such that $\sup_{x \geq 0} \|e^{Ax}\| = c$. Hence, for all $x \geq x_0$ we have

$$\|v(x)\| \; \leq \; c_0 + c \int_{x_0}^{x} \|B(t)\|\|v(t)\| dt, \tag{21.6}$$

where $c_0 = c\|v^0\|$.

Applying Corollary 7.6 to the inequality (21.6), we obtain

$$\|v(x)\| \; \leq \; c_0 \exp \left(c \int_{x_0}^{x} \|B(t)\| dt \right)$$

for all $x \geq x_0$. The result now follows from (21.4). ∎

Our next result gives sufficient conditions on the matrix $B(x)$ so that all solutions of the differential system (21.3) tend to zero as $x \to \infty$ provided all solutions of (18.6) tend to zero as $x \to \infty$.

Theorem 21.2. Let all solutions of the differential system (18.6) tend to zero as $x \to \infty$. Then all solutions of the differential system (21.3) tend to zero as $x \to \infty$ provided

$$\|B(x)\| \; \to \; 0 \quad \text{as} \quad x \to \infty. \tag{21.7}$$

Proof. Since all solutions of (18.6) tend to zero as $x \to \infty$, Problem 19.6 ensures that all eigenvalues of A have negative real parts. Thus, there exist constants c and $\eta = -\delta$ ($\delta > 0$) such that (21.1) holds, i.e., $\|e^{Ax}\| \leq ce^{-\delta x}$ for all $x \geq 0$. Further, because of (21.7) for a given constant $c_1 > 0$ there exists a sufficiently large $x_1 \geq x_0$ such that $\|B(x)\| \leq c_1$ for all $x \geq x_1$. Hence, for all $x \geq x_1$ equation (21.5) gives

$$\|v(x)\| \; \leq \; ce^{-\delta(x-x_0)}\|v^0\| + \int_{x_0}^{x_1} ce^{-\delta(x-t)} \|B(t)\|\|v(t)\| dt$$

$$+ \int_{x_1}^{x} ce^{-\delta(x-t)} c_1 \|v(t)\| dt,$$

which is the same as

$$w(x) \leq c_0 + c_2 \int_{x_1}^{x} w(t)dt, \qquad (21.8)$$

where $w(x) = \|v(x)\|e^{\delta x}$,

$$c_0 = ce^{\delta x_0}\|v^0\| + c\int_{x_0}^{x_1} e^{\delta t}\|B(t)\|\|v(t)\|dt,$$

and $c_2 = cc_1$.

Now in view of Corollary 7.6 from inequality (21.8), we obtain

$$w(x) \leq c_0 \exp(c_2(x - x_1))$$

and hence

$$\|v(x)\| \leq c_0 \exp((c_2 - \delta)x - c_2 x_1). \qquad (21.9)$$

Finally, because of (21.7) we can always choose c_1 so small that $c_2 = cc_1 < \delta$, and then the result follows from (21.9).

Conditions (21.4) and (21.7) are restricted to a smallness property on $B(x)$ as $x \to \infty$. Obviously, condition (21.4) is stronger than (21.7) and hence in Theorem 21.2 condition (21.7) can be replaced by (21.4); however, in Theorem 21.1 condition (21.4) cannot be replaced by (21.7). For this, we have the following example.

Example 21.1. Consider the differential systems

$$\begin{bmatrix} u_1' \\ u_2' \end{bmatrix} = \begin{bmatrix} 0 & 1 \\ -1 & 0 \end{bmatrix} \begin{bmatrix} u_1 \\ u_2 \end{bmatrix} \qquad (21.10)$$

and

$$\begin{bmatrix} v_1' \\ v_2' \end{bmatrix} = \begin{bmatrix} 0 & 1 \\ -1 & 0 \end{bmatrix} \begin{bmatrix} v_1 \\ v_2 \end{bmatrix} + \begin{bmatrix} 0 & 0 \\ 0 & \dfrac{2a}{ax+b} \end{bmatrix} \begin{bmatrix} v_1 \\ v_2 \end{bmatrix}, \qquad (21.11)$$

where a and b are positive constants.

A fundamental system of solutions of (21.10) is $[\cos x, \ -\sin x]^T$, $[\sin x, \cos x]^T$ and hence all solutions of (21.10) are bounded. However, a fundamental system of solutions of (21.11) is

$$\begin{bmatrix} a\sin x - (ax+b)\cos x \\ (ax+b)\sin x \end{bmatrix}, \quad \begin{bmatrix} a\cos x + (ax+b)\sin x \\ (ax+b)\cos x \end{bmatrix}$$

and hence all nontrivial solutions of (21.11) are unbounded as $x \to \infty$. Further, we note that $\|B(x)\| \to 0$ as $x \to \infty$, while

$$\int_0^x \|B(t)\| dt = \int_0^x \frac{2a}{ax+b} dx = \ln \left(\frac{ax+b}{b}\right)^2 \to \infty$$

as $x \to \infty$.

Next we shall consider the differential system

$$v' = Av + b(x), \tag{21.12}$$

where $b(x)$ is an $n \times 1$ vector with continuous components $b_i(x)$, $1 \leq i \leq n$ in the interval $[x_0, \infty)$. Once again we shall consider (21.12) as a perturbed system of (18.6) with the perturbation term $b(x)$. From (18.15) we know that each solution $v(x)$ such that $v(x_0) = v^0$ of the differential system (21.12) satisfies the integral equation

$$v(x) = e^{A(x-x_0)} v^0 + \int_{x_0}^x e^{A(x-t)} b(t) dt.$$

Hence, for all $x \geq x_0$ inequality (21.1) gives

$$\|v(x)\| \leq c_0 e^{\eta x} + c \int_{x_0}^x e^{\eta(x-t)} \|b(t)\| dt, \tag{21.13}$$

where $c_0 = c e^{-\eta x_0} \|v^0\|$.

From (21.13) the following result is immediate.

Theorem 21.3. Suppose the function $b(x)$ is such that

$$\|b(x)\| \leq c_3 e^{\nu x} \tag{21.14}$$

for all large x, where c_3 and ν are constants with $c_3 \geq 0$. Then every solution $v(x)$ of the system (21.12) satisfies

$$\|v(x)\| \leq c_4 e^{\zeta x} \tag{21.15}$$

for all $x \geq x_0$, where c_4 and ζ are constants with $c_4 \geq 0$.

Proof. From the given hypothesis on $b(x)$ there exists an $x_1 \geq x_0$ such that (21.14) holds for all $x \geq x_1$. Hence, from (21.13) if $\nu \neq \eta$, we have

$$\|v(x)\| \leq e^{\eta x} \left[c_0 + c \int_{x_0}^{x_1} e^{-\eta t} \|b(t)\| dt + c c_3 \int_{x_1}^x e^{(\nu-\eta)t} dt \right]$$

$$= e^{\eta x} \left[c_0 + c \int_{x_0}^{x_1} e^{-\eta t} \|b(t)\| dt + \frac{c c_3}{\nu - \eta} (e^{(\nu-\eta)x} - e^{(\nu-\eta)x_1}) \right]$$

$$\leq e^{\eta x} \left[c_0 + c \int_{x_0}^{x_1} e^{-\eta t} \|b(t)\| dt + \frac{c c_3}{|\nu-\eta|} e^{(\nu-\eta)x_1} \right] + \frac{c c_3}{|\nu-\eta|} e^{\nu x}$$

$$\leq c_4 e^{\zeta x},$$

where $\zeta = \max\{\eta, \nu\}$, and

$$c_4 = \left[c_0 + c \int_{x_0}^{x_1} e^{-\eta t} \|b(t)\| dt + \frac{cc_3}{|\nu - \eta|} \left(e^{(\nu - \eta)x_1} + 1 \right) \right].$$

For the case $\nu = \eta$, the above proof requires obvious modifications. ∎

As a consequence of (21.15) we find that every solution of the system (21.12) tends to zero as $x \to \infty$ provided $\zeta < 0$.

Now we shall study the behavior of solutions of the differential system (17.3) as $x \to \infty$. We shall prove two results which involve the eigenvalues of the matrix $(A(x) + A^T(x))$, which obviously are functions of x.

Theorem 21.4. Let the matrix $A(x)$ be continuous in $[x_0, \infty)$ and $M(x)$ be the largest eigenvalue of the matrix $(A(x) + A^T(x))$. Then every solution of the differential system (17.3) tends to zero as $x \to \infty$ provided

$$\int^{\infty} M(t)dt = -\infty. \tag{21.16}$$

Proof. Let $u(x)$ be a solution of the differential system (17.3), then $|u(x)|^2 = u^T(x)u(x)$. Thus, it follows that

$$\begin{aligned}
\frac{d}{dx}|u(x)|^2 &= u^T(x)u'(x) + (u^T(x))'u(x) \\
&= u^T(x)A(x)u(x) + u^T(x)A^T(x)u(x) \\
&= u^T(x)(A(x) + A^T(x))u(x).
\end{aligned}$$

Now since the matrix $(A(x) + A^T(x))$ is symmetric and $M(x)$ is its largest eigenvalue, it is clear that

$$u^T(x)(A(x) + A^T(x))u(x) \leq M(x)|u(x)|^2.$$

Hence, for all $x \geq x_0$ it follows that

$$0 \leq |u(x)|^2 \leq |u(x_0)|^2 + \int_{x_0}^x M(t)|u(t)|^2 dt.$$

Next using Corollary 7.6, we obtain

$$|u(x)|^2 \leq |u(x_0)|^2 \exp\left(\int_{x_0}^x M(t)dt \right). \tag{21.17}$$

The result now follows from (21.16). ∎

If in Theorem 21.4 the condition (21.16) is replaced by $\int^{\infty} M(t)dt < \infty$, then (21.17) implies that the solution $u(x)$ of (17.3) remains bounded as $x \to \infty$.

Theorem 21.5. Let the matrix $A(x)$ be continuous in $[x_0, \infty)$, and $m(x)$ be the smallest eigenvalue of the matrix $(A(x) + A^T(x))$. Then every solution of the differential system (17.3) is unbounded as $x \to \infty$ provided

$$\limsup_{x \to \infty} \int^x m(t)dt = \infty. \qquad (21.18)$$

Proof. As in the proof of Theorem 21.4 for all $x \geq x_0$, it is easy to see that

$$|u(x)|^2 \geq |u(x_0)|^2 + \int_{x_0}^x m(t)|u(t)|^2 dt,$$

which implies that

$$|u(x)|^2 \geq |u(x_0)|^2 \exp\left(\int_{x_0}^x m(t)dt\right).$$

Now the conclusion follows from (21.18). ∎

Example 21.2. For the matrix

$$A(x) = \begin{bmatrix} \dfrac{1}{(1+x)^2} & x^2 \\ -x^2 & -1 \end{bmatrix},$$

we have

$$(A(x) + A^T(x)) = \begin{bmatrix} \dfrac{2}{(1+x)^2} & 0 \\ 0 & -2 \end{bmatrix},$$

and hence

$$M(x) = \frac{2}{(1+x)^2}, \quad \int_0^\infty M(t)dt = \int_0^\infty \frac{2}{(1+t)^2}dt = 2 < \infty.$$

Thus, all solutions of the differential system $u' = A(x)u$ remain bounded as $x \to \infty$.

Example 21.3. For the matrix

$$A(x) = \begin{bmatrix} -\dfrac{1}{1+x} & (1+x^2) \\ -(1+x^2) & -2 \end{bmatrix},$$

we have

$$(A(x) + A^T(x)) = \begin{bmatrix} -\dfrac{2}{1+x} & 0 \\ 0 & -4 \end{bmatrix},$$

and hence

$$M(x) = -\frac{2}{1+x}, \quad \int_0^\infty M(t)dt = \int_0^\infty -\frac{2}{1+t}dt = -\infty.$$

Thus, all solutions of the differential system $u' = A(x)u$ tend to zero as $x \to \infty$.

Lecture 22
Asymptotic Behavior of Solutions of Linear Systems (Contd.)

With respect to the differential system (17.3) we shall consider the perturbed system

$$v' = (A(x) + B(x))v, \qquad (22.1)$$

where $B(x)$ is an $n \times n$ matrix with continuous elements $b_{ij}(x)$, $1 \le i, j \le n$ in the interval $[x_0, \infty)$. We begin with an interesting example which shows that the boundedness of all solutions of (17.3), and the condition (21.4) do not imply boundedness of solutions of the differential system (22.1), i.e., when the matrix A is a function of x, then the conclusion of Theorem 21.1 need not be true.

Example 22.1. Consider the differential system

$$
\begin{aligned}
u_1' &= -au_1 \\
u_2' &= (\sin\ln x + \cos\ln x - 2a)u_2, \quad 1 < 2a < 1 + e^{-\pi}/2
\end{aligned} \qquad (22.2)
$$

whose general solution is

$$
\begin{aligned}
u_1(x) &= c_1 e^{-ax} \\
u_2(x) &= c_2 \exp((\sin\ln x - 2a)x).
\end{aligned}
$$

Since $a > 1/2$, every solution of (22.2) tends to zero as $x \to \infty$.

Now we consider the perturbed differential system

$$
\begin{aligned}
v_1' &= -av_1 \\
v_2' &= (\sin\ln x + \cos\ln x - 2a)v_2 + e^{-ax}v_1;
\end{aligned} \qquad (22.3)
$$

i.e., the perturbing matrix is

$$
B(x) = \begin{bmatrix} 0 & 0 \\ e^{-ax} & 0 \end{bmatrix}.
$$

R.P. Agarwal and D. O'Regan, *An Introduction to Ordinary Differential Equations*, doi: 10.1007/978-0-387-71276-5_22, © Springer Science + Business Media, LLC 2008

It is easily seen that $\int^\infty \|B(t)\|dt < \infty$, and the general solution of the differential system (22.3) is

$$v_1(x) = c_1 e^{-ax}$$

$$v_2(x) = \exp((\sin \ln x - 2a)x)\left(c_2 + c_1 \int_0^x \exp(-t \sin \ln t)dt\right).$$

(22.4)

Let $x = x_n = \exp((2n + 1/2)\pi)$, $n = 1, 2, \ldots$ then we have

$$\sin \ln x_n = 1$$

$$-\sin \ln t \geq 1/2 \quad \text{for all} \quad \exp((2n - 1/2)\pi) \leq t \leq \exp((2n - 1/6)\pi),$$

$$\text{i.e., for all} \quad x_n e^{-\pi} \leq t \leq x_n e^{-2\pi/3}.$$

Thus, it follows that

$$
\begin{aligned}
\int_0^{x_n} \exp(-t \sin \ln t)dt \quad &> \quad \int_{\exp((2n-1/2)\pi)}^{\exp((2n-1/6)\pi)} \exp(-t \sin \ln t)dt \\
&\geq \quad \int_{x_n e^{-\pi}}^{x_n e^{-2\pi/3}} e^{t/2}dt \\
&> \quad \exp\left(\frac{1}{2}x_n e^{-\pi}\right)(e^{-2\pi/3} - e^{-\pi})x_n.
\end{aligned}
$$

(22.5)

Therefore, if $c_1 > 0$ $(c_1 < 0)$ we have

$$
\begin{aligned}
v_2(x_n) &> (<)\ e^{(1-2a)x_n}\left(c_2 + c_1 x_n(e^{-2\pi/3} - e^{-\pi})\exp\left(\frac{1}{2}x_n e^{-\pi}\right)\right) \\
&= c_2 e^{(1-2a)x_n} + c_1 x_n(e^{-2\pi/3} - e^{-\pi})\exp\left(\left(1 - 2a + \frac{1}{2}e^{-\pi}\right)x_n\right).
\end{aligned}
$$

Since $2a < 1 + e^{-\pi}/2$, we see that $v_2(x_n) \to \infty$ $(-\infty)$ as $n \to \infty$. Thus, $v_2(x)$ remains bounded only if $c_1 = 0$.

This example also shows that for (17.3) and (22.1) Theorem 21.2 need not hold even when condition (21.7) is replaced by the stronger condition (21.4). Therefore, to prove results similar to Theorems 21.1 and 21.2, we need some additional conditions on $A(x)$. The following result is analogous to Theorem 21.1.

Theorem 22.1.   Let all solutions of the differential system (17.3) be bounded in $[x_0, \infty)$, and the condition (21.4) be satisfied. Then all solutions of the differential system (22.1) are bounded in $[x_0, \infty)$ provided

$$\liminf_{x \to \infty} \int^x \text{Tr}\, A(t)dt > -\infty, \quad \text{or} \quad \text{Tr}\, A(x) = 0. \tag{22.6}$$

Proof.   Let $\Psi(x)$ be a fundamental matrix of the differential system (17.3). Since all solutions of the differential system (17.3) are bounded, $\|\Psi(x)\|$ is bounded. Next from Theorem 17.3, we have

$$\det \Psi(x) \;=\; \det \Psi(x_0)\exp\left(\int_{x_0}^{x} \operatorname{Tr} A(t)dt\right)$$

and hence

$$\Psi^{-1}(x) \;=\; \frac{\operatorname{adj}\Psi(x)}{\det \Psi(x)} \;=\; \frac{\operatorname{adj}\Psi(x)}{\det \Psi(x_0)\exp\left(\int_{x_0}^{x} \operatorname{Tr} A(t)dt\right)}. \tag{22.7}$$

Thus, from (22.6) it follows that $\|\Psi^{-1}(x)\|$ is bounded.

Now in (18.14) let the nonhomogeneous term $b(x)$ be $B(x)v$, so that each solution $v(x)$ such that $v(x_0) = v^0$ of the differential system (22.1) satisfies the integral equation

$$v(x) \;=\; \Psi(x)\Psi^{-1}(x_0)v^0 + \int_{x_0}^{x} \Psi(x)\Psi^{-1}(t)B(t)v(t)dt. \tag{22.8}$$

Thus, if

$$c \;=\; \max\left\{\sup_{x\geq x_0}\|\Psi(x)\|, \quad \sup_{x\geq x_0}\|\Psi^{-1}(x)\|\right\} \tag{22.9}$$

it follows that

$$\|v(x)\| \;\leq\; c_0 + c^2\int_{x_0}^{x}\|B(t)\|\|v(t)\|dt,$$

where $c_0 = c\|\Psi^{-1}(x_0)v^0\|$.

This inequality immediately implies that

$$\|v(x)\| \;\leq\; c_0\exp\left(c^2\int_{x_0}^{x}\|B(t)\|dt\right).$$

The result now follows from (21.4).   ■

The next result is parallel to that of Theorem 21.2.

Theorem 22.2.   Let the fundamental matrix $\Psi(x)$ of the differential system (17.3) be such that

$$\|\Psi(x)\Psi^{-1}(t)\| \;\leq\; c, \quad x_0 \leq t \leq x < \infty \tag{22.10}$$

where c is a positive constant. Further, let condition (21.4) be satisfied. Then all solutions of the differential system (22.1) are bounded in $[x_0, \infty)$.

Moreover, if all solutions of (17.3) tend to zero as $x \to \infty$, then all solutions of the differential system (22.1) tend to zero as $x \to \infty$.

Proof. Using (22.10) in (22.8), we get

$$\|v(x)\| \leq c\|v^0\| + c \int_{x_0}^{x} \|B(t)\| \|v(t)\| dt$$

and hence

$$\|v(x)\| \leq c\|v^0\| \exp\left(c \int_{x_0}^{\infty} \|B(t)\| dt\right) =: M < \infty.$$

Thus, each solution of the differential system (22.1) is bounded in $[x_0, \infty)$.

Now since (22.8) is the same as

$$v(x) = \Psi(x)\Psi^{-1}(x_0)v^0 + \int_{x_0}^{x_1} \Psi(x)\Psi^{-1}(t)B(t)v(t)dt$$
$$+ \int_{x_1}^{x} \Psi(x)\Psi^{-1}(t)B(t)v(t)dt$$

it follows that

$$\|v(x)\| \leq \|\Psi(x)\| \|\Psi^{-1}(x_0)\| \|v^0\| + \|\Psi(x)\| \int_{x_0}^{x_1} \|\Psi^{-1}(t)\| \|B(t)\| \|v(t)\| dt$$
$$+ cM \int_{x_1}^{\infty} \|B(t)\| dt.$$

Let $\epsilon > 0$ be a given number. Then in view of (21.4), the last term in the above inequality can be made less than $\epsilon/2$ by choosing x_1 sufficiently large. Further, since all solutions of (17.3) tend to zero, it is necessary that $\|\Psi(x)\| \to 0$ as $x \to \infty$. Thus, the sum of first two terms on the right side can be made arbitrarily small by choosing x large enough, say, less than $\epsilon/2$. Hence, $\|v(x)\| < \epsilon$ for large x. But, this immediately implies that $\|v(x)\| \to 0$ as $x \to \infty$. ∎

In our next result we shall show that in Theorems 22.1 and 22.2 conditions (22.6) and (22.10) can be replaced by the periodicity of the matrix $A(x)$.

Theorem 22.3. Let the matrix $A(x)$ be periodic of period ω in $[x_0, \infty)$. Further, let the condition (21.4) be satisfied. Then the following hold:

(i) All solutions of the differential system (22.1) are bounded in $[x_0, \infty)$ provided all solutions of (17.3) are bounded in $[x_0, \infty)$.

(ii) All solutions of the differential system (22.1) tend to zero as $x \to \infty$ provided all solutions of (17.3) tend to zero as $x \to \infty$.

Proof. For a given fundamental matrix $\Psi(x)$ of (17.3), Theorem 20.5 implies that $\Psi(x) = P(x)e^{Rx}$, where $P(x)$ is a nonsingular periodic matrix of period ω, and R is a constant matrix. Using this in (22.8), we find

$$v(x) = P(x)e^{R(x-x_0)}P^{-1}(x_0)v^0 + \int_{x_0}^x P(x)e^{Rx}e^{-Rt}P^{-1}(t)B(t)v(t)dt.$$

Hence, it follows that

$$\begin{aligned}
\|v(x)\| &\leq \|P(x)\|\|e^{Rx}\|\|e^{-Rx_0}P^{-1}(x_0)v^0\| \\
&\quad + \int_{x_0}^x \|P(x)\|\|e^{R(x-t)}\|\|P^{-1}(t)\|\|B(t)\|\|v(t)\|dt.
\end{aligned} \qquad (22.11)$$

Now since $P(x)$ is nonsingular and periodic, $\det P(x)$ is periodic and does not vanish; i.e., it is bounded away from zero in $[x_0, \infty)$. Hence, $P(x)$ along with its inverse $P^{-1}(x) = [\mathrm{adj}\, P(x)/\det P(x)]$ is bounded in $[x_0, \infty)$. Thus, if

$$c_4 = \max\left\{ \sup_{x \geq x_0} \|P(x)\|, \quad \sup_{x \geq x_0} \|P^{-1}(x)\| \right\}$$

inequality (22.11) can be replaced by

$$\|v(x)\| \leq c_5\|e^{Rx}\| + c_4^2 \int_{x_0}^x \|e^{R(x-t)}\|\|B(t)\|\|v(t)\|dt, \qquad (22.12)$$

where $c_5 = c_4\|e^{-Rx_0}P^{-1}(x_0)v^0\|$.

Now if all solutions of the differential system (17.3) are bounded, then it is necessary that $\|e^{Rx}\| \leq c_6$ for all $x \geq 0$, and hence from (22.12) we have

$$\|v(x)\| \leq c_5 c_6 + c_4^2 c_6 \int_{x_0}^x \|B(t)\|\|v(t)\|dt,$$

which immediately gives that

$$\|v(x)\| \leq c_5 c_6 \exp\left(c_4^2 c_6 \int_{x_0}^x \|B(t)\|dt \right)$$

and now part (i) follows from (21.4).

On the other hand, if all solutions of (17.3) tend to zero as $x \to \infty$, then there exist positive constants c_7 and α such that $\|e^{Rx}\| \leq c_7 e^{-\alpha x}$ for all $x \geq 0$. Thus, inequality (22.12) implies that

$$\|v(x)\| \leq c_5 c_7 e^{-\alpha x} + c_4^2 c_7 \int_{x_0}^x e^{-\alpha(x-t)}\|B(t)\|\|v(t)\|dt,$$

which easily gives

$$\|v(x)\| \le c_5 c_7 \exp\left(c_4^2 c_7 \int_{x_0}^x \|B(t)\| dt - \alpha x\right).$$

Hence, in view of condition (21.4) we find that $v(x) \to 0$ as $x \to \infty$. ∎

Finally, we shall consider the differential system (17.1) as a perturbed system of (17.3) and prove the following two results.

Theorem 22.4. Suppose every solution of the differential system (17.3) is bounded in $[x_0, \infty)$. Then every solution of (17.1) is bounded provided at least one of its solutions is bounded.

Proof. Let $u^1(x)$ and $u^2(x)$ be two solutions of the differential system (17.1). Then $\phi(x) = u^1(x) - u^2(x)$ is a solution of the differential system (17.3). Hence, $u^1(x) = u^2(x) + \phi(x)$. Now since $\phi(x)$ is bounded in $[x_0, \infty)$, if $u^2(x)$ is a bounded solution of (17.1), it immediately follows that $u^1(x)$ is also a bounded solution of (17.1). ∎

From the above theorem it is also clear that if every solution of (17.3) tends to zero as $x \to \infty$, and if one solution of (17.1) tends to zero as $x \to \infty$, then every solution of (17.1) tends to zero as $x \to \infty$.

Theorem 22.5. Suppose every solution of the differential system (17.3) is bounded in $[x_0, \infty)$, and the condition (22.6) is satisfied. Then every solution of (17.1) is bounded provided

$$\int_{x_0}^{\infty} \|b(t)\| dt < \infty. \tag{22.13}$$

Proof. Let $\Psi(x)$ be a fundamental matrix of the differential system (17.3). Since each solution of (17.3) is bounded, as in Theorem 22.1 both $\|\Psi(x)\|$ and $\|\Psi^{-1}(x)\|$ are bounded in $[x_0, \infty)$. Thus, there exists a finite constant c as defined in (22.9). Hence, for any solution $u(x)$ of (17.1) such that $u(x_0) = u^0$ relation (18.14) gives

$$\|u(x)\| \le c\|\Psi^{-1}(x_0)u^0\| + c^2 \int_{x_0}^x \|b(t)\| dt.$$

The conclusion now follows from the condition (22.13). ∎

Problems

22.1. Consider the second-order DE

$$y'' + p(x)y = 0 \tag{22.14}$$

and its perturbed equation

$$z'' + (p(x) + q(x))z = 0, \qquad (22.15)$$

where $p(x)$ and $q(x)$ are continuous functions in $[x_0, \infty)$. Show that, if all solutions of (22.14) are bounded in $[x_0, \infty)$, then all solutions of (22.15) are bounded in $[x_0, \infty)$ provided $\int^{\infty} |q(t)| dt < \infty$.

22.2. Consider the second-order DE (22.14), where $p(x) \to \infty$ monotonically as $x \to \infty$. Show that all solutions of (22.14) are bounded in $[x_0, \infty)$.

22.3. Consider the second-order DE (22.14), where $\int^{\infty} t|p(t)| dt < \infty$. Show that, for any solution $y(x)$ of (22.14), $\lim_{x \to \infty} y'(x)$ exists, and every nontrivial solution is asymptotic to $d_0 x + d_1$ for some constants d_0 and d_1 not both zero.

22.4. Consider the second-order DE $y'' + (1 + p(x))y = 0$, where $p \in C^{(1)}[x_0, \infty)$, $\lim_{x \to \infty} p(x) = 0$, and $\int^{\infty} |p'(t)| dt < \infty$. Show that all solutions of this DE are bounded in $[x_0, \infty)$.

22.5. Show that all solutions of the following DEs are bounded in $[0, \infty)$:

(i) $y'' + [1 + 1/(1 + x^4)]y = 0$.

(ii) $y'' + e^x y = 0$.

(iii) $y'' + cy' + [1 + 1/(1 + x^2)]y = 0$, $c > 0$.

(iv) $y'' + cy' + [1 + 1/(1 + x^4)]y = \sin x$, $c > 0$.

22.6. Show that there are no bounded solutions of the DE

$$y'' + \left[1 + \frac{1}{1 + x^4}\right] y = \cos x, \quad x \in [0, \infty).$$

22.7. Show that all solutions of the differential system (17.3), where

(i) $A(x) = \begin{bmatrix} -x & 0 & 0 \\ 0 & -x^2 & 0 \\ 0 & 0 & -x^2 \end{bmatrix}$,   (ii) $A(x) = \begin{bmatrix} -e^x & -1 & -\cos x \\ 1 & -e^{2x} & x^2 \\ \cos x & -x^2 & -e^{3x} \end{bmatrix}$

tend to zero as $x \to \infty$.

22.8. Show that all solutions of the differential system (17.1), where

(i) $A(x) = \begin{bmatrix} -e^{-x} & 0 \\ 0 & e^{-3x} \end{bmatrix}$, $b(x) = \begin{bmatrix} \cos x \\ x \cos x^2 \end{bmatrix}$,

$$\text{(ii)} \quad A(x) = \begin{bmatrix} (1+x)^{-2} & \sin x & 0 \\ -\sin x & 0 & x \\ 0 & -x & 0 \end{bmatrix}, \quad b(x) = \begin{bmatrix} 0 \\ (1+x)^{-2} \\ (1+x)^{-4} \end{bmatrix}$$

are bounded in $[0, \infty)$.

22.9. With respect to the differential system (18.6) let the perturbed system be

$$v' = Av + g(x, v), \tag{22.16}$$

where $g \in C[[x_0, \infty) \times \mathbb{R}^n, \mathbb{R}^n]$ and $\|g(x, v)\| \leq \lambda(x)\|v\|$, where $\lambda(x)$ is a nonnegative continuous function in $[x_0, \infty)$. Show the following:

(i)   If all solutions of (18.6) are bounded, then all solutions of (22.16) are bounded provided $\int^\infty \lambda(t)dt < \infty$.

(ii) If all solutions of (18.6) tend to zero as $x \to \infty$, then all solutions of (22.16) tend to zero as $x \to \infty$ provided $\lambda(x) \to 0$ as $x \to \infty$.

22.10. With respect to the differential system (17.3) let the perturbed system be

$$v' = A(x)v + g(x, v), \tag{22.17}$$

where $g \in C[[x_0, \infty) \times \mathbb{R}^n, \mathbb{R}^n]$ and $\|g(x, v)\| \leq \lambda(x)\|v\|$, here $\lambda(x)$ is a nonnegative continuous function in $[x_0, \infty)$ such that $\int^\infty \lambda(t)dt < \infty$. Show the following:

(i)   If all solutions of (17.3) are bounded and condition (22.6) is satisfied, then all solutions of (22.17) are bounded.

(ii) If all solutions of (17.3) tend to zero as $x \to \infty$ and condition (22.10) is satisfied, then all solutions of (22.17) tend to zero as $x \to \infty$.

Answers or Hints

22.1. Write (22.14) and (22.15) in system form and then apply Theorem 22.1.

22.2. Use the fact that $p(x) \to \infty$ monotonically to get an inequality of the form

$$\frac{y^2(x)p(x)}{2} \leq |C| + \int_{x_0}^x \frac{y^2(t)p(t)}{2} \frac{dp(t)}{p(t)}.$$

Now apply Corollary 7.5.

22.3. Clearly, (22.14) is equivalent to the integral equation

$$y(x) = y(1) + (x-1)y'(1) - \int_1^x (x-t)p(t)y(t)dt.$$

Thus, for $x \geq 1$ it follows that

$$\frac{|y(x)|}{x} \le c + \int_1^x t|p(t)|\frac{|y(t)|}{t}dt,$$

where $c \ge 0$ is a constant. Now Corollary 7.5 implies $|y(x)| \le c_1 x$, where c_1 is a positive constant. We choose $x_0 \ge 1$ sufficiently large so that

$$\int_{x_0}^x |p(t)||y(t)|dt \le c_1 \int_{x_0}^x t|p(t)|dt < 1.$$

Let $y_1(x)$ be a solution of (22.14) satisfying $y_1'(x_0) = 1$, then since $y_1'(x) = 1 - \int_{x_0}^x p(t)y_1(t)dt$, the above inequality implies that $y_1'(x) \to d_1 \ne 0$ as $x \to \infty$, i.e., $y(x) \to d_1 x$ as $x \to \infty$. Finally, since $y_2(x) = y_1(x)\int_x^\infty y_1^{-2}(t)dt$ is another solution of (22.14), it follows that $y_2(x) \to d_0$ as $x \to \infty$. Hence, the general solution $y(x)$ of (22.14) is asymptotic to $d_0 + d_1 x$.

22.4. Multiply the given DE by y'. Use Corollary 7.5.

22.5. (i) Use Problem 22.1. (ii) Use Problem 22.2. (iii) Use Theorem 21.1. (iv) Write in system form. Finally, apply Corollary 7.5.

22.6. Note that for the DE $w'' + w = \cos x$ the solution $w(x) = \frac{1}{2}x\sin x$ is unbounded. Let $y(x)$ be a bounded solution of the given DE; then the function $z(x) = y(x) - w(x)$ satisfies the DE $z'' + z = -\frac{1}{1+x^4}y(x)$. Now it is easy to show that $z(x)$ is bounded, which leads to a contradiction.

22.7. Use Theorem 21.4.

22.8. (i) For the given differential system,
$u_1(x) = c_1\exp(e^{-x}) + \exp(e^{-x})\int_0^x \exp(-e^{-t})\cos tdt$
$u_2(x) = c_2\exp\left(-\frac{1}{3}e^{-3x}\right) + \exp\left(-\frac{1}{3}e^{-3x}\right)\int_0^x \exp\left(\frac{1}{3}e^{-3t}\right)t\cos t^2 dt.$
Now find upper bounds of $|u_1(x)|$ and $|u_2(x)|$. (ii) First use remark following Theorem 21.4 and then Theorem 22.5.

22.9. (i) System (22.16) satisfying $v(x_0) = v^0$ is equivalent to the integral equation

$$v(x) = e^{A(x-x_0)}v^0 + \int_{x_0}^x e^{A(x-t)}g(t,v(t))dt.$$

Now use the given conditions and Corollary 7.5. (ii) The proof is similar to that of Theorem 21.2.

22.10. System (22.17) satisfying $v(x_0) = v^0$ is equivalent to the integral equation

$$v(x) = \Phi(x)\Phi^{-1}(x_0)v^0 + \int_{x_0}^x \Phi(x)\Phi^{-1}(t)g(t,v(t))dt,$$

where $\Phi(x)$ is a fundamental matrix of (17.3). (i) Similar to Theorem 22.1. (ii) Similar to Theorem 22.2.

Lecture 23

Preliminaries to Stability of Solutions

In Lecture 16 we have provided smoothness conditions so that the solution $u(x, x_0, u^0)$ of the initial value problem (15.4) is a continuous function of x, x_0, and u^0 at the point (x, x_0, u^0), where x is in some finite interval $J = [x_0, x_0 + \alpha]$. Geometrically, this means that for all $\epsilon > 0$ there exists $\|\Delta u^0\|$ sufficiently small so that the solution $u(x, x_0, u^0 + \Delta u^0)$ remains in a strip of width 2ϵ surrounding the solution $u(x, x_0, u^0)$ for all $x \in [x_0, x_0 + \alpha]$. Thus, a small change in u^0 brings about only a small change in the solutions of (15.4) in a finite interval $[x_0, x_0 + \alpha]$. However, the situation is very much different when the finite interval $[x_0, x_0 + \alpha]$ is replaced by $[x_0, \infty)$. For example, let us consider the initial value problem $y' = ay$, $y(0) = y_0$ whose unique solution is $y(x, 0, y_0) = y_0 e^{ax}$. It follows that

$$|\Delta y| = |y(x, 0, y_0 + \Delta y_0) - y(x, 0, y_0)| = |\Delta y_0| e^{ax}$$

for all $x \geq 0$. Hence, if $a \leq 0$ then $|\Delta y| = |\Delta y_0| e^{ax} \leq \epsilon$ for all $x \geq 0$ provided $|\Delta y_0| \leq \epsilon$. But, if $a > 0$, then $|\Delta y| \leq \epsilon$ holds only if $|\Delta y_0| \leq \epsilon e^{-ax}$, which is possible only for finite values of x no matter how small $|\Delta y_0|$ is, i.e., $|\Delta y|$ becomes large for large x even for small values of $|\Delta y_0|$.

A solution $u(x, x_0, u^0)$ of the initial value problem (15.4) existing in the interval $[x_0, \infty)$ is said to be *stable* if small changes in u^0 bring only small changes in the solutions of (15.4) for all $x \geq x_0$. Otherwise, we say that the solution $u(x, x_0, u^0)$ is *unstable*. Thus, the solution $y(x) = y_0 e^{ax}$ of the problem $y' = ay$, $y(0) = y_0$ is stable only if $a \leq 0$, and unstable for $a > 0$. We shall now give a few definitions which classify various types of behavior of solutions.

Definition 23.1. A solution $u(x) = u(x, x_0, u^0)$ of the initial value problem (15.4) is said to be *stable*, if for each $\epsilon > 0$ there is a $\delta = \delta(\epsilon, x_0) > 0$ such that $\|\Delta u^0\| < \delta$ implies that $\|u(x, x_0, u^0 + \Delta u^0) - u(x, x_0, u^0)\| < \epsilon$.

Definition 23.2. A solution $u(x) = u(x, x_0, u^0)$ of the initial value problem (15.4) is said to be *unstable* if it is not stable.

Definition 23.3. A solution $u(x) = u(x, x_0, u^0)$ of the initial value problem (15.4) is said to be *asymptotically stable* if it is stable and there exists a $\delta_0 > 0$ such that $\|\Delta u^0\| < \delta_0$ implies that

$$\|u(x, x_0, u^0 + \Delta u^0) - u(x, x_0, u^0)\| \to 0 \quad \text{as} \quad x \to \infty.$$

R.P. Agarwal and D. O'Regan, *An Introduction to Ordinary Differential Equations*,
doi: 10.1007/978-0-387-71276-5_23, © Springer Science + Business Media, LLC 2008

The above definitions were introduced by A. M. Lyapunov in 1892, and hence some authors prefer to call a stable solution as *Lyapunov stable*, or stable in the sense of Lyapunov.

Example 23.1. Every solution of the DE $y' = x$ is of the form $y(x) = y(x_0) - x_0^2/2 + x^2/2$, and hence it is stable but not bounded.

Example 23.2. Every solution of the DE $y' = 0$ is of the form $y(x) = y(x_0)$, and hence stable but not asymptotically stable.

Example 23.3. Every solution of the DE $y' = p(x)y$ is of the form $y(x) = y(x_0)\exp\left(\int_{x_0}^{x} p(t)dt\right)$, and hence its trivial solution $y(x) \equiv 0$ is asymptotically stable if and only if $\int_{x_0}^{x} p(t)dt \to -\infty$ as $x \to \infty$.

From Example 23.1 it is clear that the concepts of stability and boundedness of solutions are independent. However, in the case of homogeneous linear differential system (17.3) these concepts are equivalent as seen in the following theorem.

Theorem 23.1. All solutions of the differential system (17.3) are stable if and only if they are bounded.

Proof. Let $\Psi(x)$ be a fundamental matrix of the differential system (17.3). If all solutions of (17.3) are bounded, then there exists a constant c such that $\|\Psi(x)\| \leq c$ for all $x \geq x_0$. Now given any $\epsilon > 0$, we choose $\|\Delta u^0\| < \epsilon/(c\|\Psi^{-1}(x_0)\|) = \delta(\epsilon) > 0$, so that

$$\|u(x, x_0, u^0 + \Delta u^0) - u(x, x_0, u^0)\| = \|\Psi(x)\Psi^{-1}(x_0)\Delta u^0\|$$
$$\leq c\|\Psi^{-1}(x_0)\|\|\Delta u^0\| < \epsilon,$$

i.e., all solutions of (17.3) are stable.

Conversely, if all solutions of (17.3) are stable, then, in particular, the trivial solution, i.e., $u(x, x_0, 0) \equiv 0$ is stable. Therefore, given any $\epsilon > 0$, there exists a $\delta = \delta(\epsilon) > 0$ such that $\|\Delta u^0\| < \delta$ implies that $\|u(x, x_0, \Delta u^0)\| < \epsilon$, for all $x \geq x_0$. However, since $u(x, x_0, \Delta u^0) = \Psi(x)\Psi^{-1}(x_0)\Delta u^0$, we find that $\|u(x, x_0, \Delta u^0)\| = \|\Psi(x)\Psi^{-1}(x_0)\Delta u^0\| < \epsilon$. Now let Δu^0 be a vector $(\delta/2)e^j$, then we have

$$\|\Psi(x)\Psi^{-1}(x_0)\Delta u^0\| = \|\psi^j(x)\|\frac{\delta}{2} < \epsilon,$$

where $\psi^j(x)$ is the jth column of $\Psi(x)\Psi^{-1}(x_0)$. Therefore, it follows that

$$\|\Psi(x)\Psi^{-1}(x_0)\| = \max_{1 \leq j \leq n} \|\psi^j(x)\| \leq \frac{2\epsilon}{\delta}.$$

Hence, for any solution $u(x, x_0, u^0)$ of the differential system (17.3) we have

$$\|u(x, x_0, u^0)\| \;=\; \|\Psi(x)\Psi^{-1}(x_0)u^0\| \;<\; \frac{2\epsilon}{\delta}\|u^0\|,$$

i.e., all solutions of (17.3) are bounded.   ∎

Corollary 23.2.   If the real parts of the multiple eigenvalues of the matrix A are negative, and the real parts of the simple eigenvalues of the matrix A are nonpositive, then all solutions of the differential system (18.6) are stable.

Our next result gives necessary and sufficient conditions so that all solutions of the differential system (17.3) are asymptotically stable.

Theorem 23.3.   Let $\Psi(x)$ be a fundamental matrix of the differential system (17.3). Then all solutions of the differential system (17.3) are asymptotically stable if and only if

$$\|\Psi(x)\| \;\to\; 0 \quad \text{as} \quad x \to \infty. \tag{23.1}$$

Proof.   Every solution $u(x, x_0, u^0)$ of the differential system (17.3) can be expressed as $u(x, x_0, u^0) = \Psi(x)\Psi^{-1}(x_0)u^0$. Since $\Psi(x)$ is continuous, condition (23.1) implies that there exists a constant c such that $\|\Psi(x)\| \leq c$ for all $x \geq x_0$. Thus, $\|u(x, x_0, u^0)\| \leq c\|\Psi^{-1}(x_0)\|\|u^0\|$, and hence every solution of (17.3) is bounded, and now from Theorem 23.1 it follows that every solution of (17.3) is stable. Further, since

$$\begin{aligned}
\|u(x, x_0, u^0 + \Delta u^0) - u(x, x_0, u^0)\| \;&=\; \|\Psi(x)\Psi^{-1}(x_0)\Delta u^0\| \\
&\leq\; \|\Psi(x)\|\|\Psi^{-1}(x_0)\Delta u^0\| \;\to\; 0
\end{aligned}$$

as $x \to \infty$, it follows that every solution of (17.3) is asymptotically stable.

Conversely, if all solutions of (17.3) are asymptotically stable, then, in particular, the trivial solution, i.e., $u(x, x_0, 0) \equiv 0$ is asymptotically stable. Hence, $\|u(x, x_0, \Delta u^0)\| \to 0$ as $x \to \infty$. However, this implies that $\|\Psi(x)\| \to 0$ as $x \to \infty$.   ∎

Corollary 23.4.   If the real parts of the eigenvalues of the matrix A are negative, then all solutions of the differential system (18.6) are asymptotically stable.

It is interesting to note that for the perturbed differential system (21.3) Theorems 21.1 and 23.1 can be combined, to obtain the following result.

Theorem 23.5.   Let all solutions of the differential system (18.6) be stable, and the condition (21.4) be satisfied. Then all solutions of the differential system (21.3) are stable.

Similarly, a combination of Theorems 21.2 and 23.3 gives the following result.

Theorem 23.6.   Let all solutions of the differential system (18.6) be asymptotically stable, and the condition (21.7) be satisfied. Then all solutions of the differential system (21.3) are asymptotically stable.

From Example 21.1 it is clear that in Theorem 23.5 condition (21.4) cannot be replaced by (21.7).

For our later applications we need a stronger definition of stability which is as follows.

Definition 23.4.   A solution $u(x) = u(x, x_0, u^0)$ of the initial value problem (15.4) is said to be *uniformly stable*, if for each $\epsilon > 0$ there is a $\delta = \delta(\epsilon) > 0$ such that for any solution $u^1(x) = u(x, x_0, u^1)$ of the problem $u' = g(x, u)$, $u(x_0) = u^1$ the inequalities $x_1 \geq x_0$ and $\|u^1(x_1) - u(x_1)\| < \delta$ imply that $\|u^1(x) - u(x)\| < \epsilon$ for all $x \geq x_1$.

Example 23.4.   Every solution of the DE $y' = p(x)y$ is of the form $y(x) = y(x_0) \exp\left(\int_{x_0}^{x} p(t)dt\right)$, and hence its trivial solution $y(x) \equiv 0$ is uniformly stable if and only if $\int_{x_1}^{x} p(t)dt$ is bounded above for all $x \geq x_1 \geq x_0$. In particular, if we choose $p(x) = \sin \ln x + \cos \ln x - 1.25$, then we have

$$\int_{x_0}^{x} p(t)dt = (t \sin \ln t - 1.25t)\Big|_{x_0}^{x} \to -\infty$$

as $x \to \infty$, and hence from Example 23.3 the trivial solution is asymptotically stable. But, if we choose $x = e^{(2n+1/3)\pi}$ and $x_1 = e^{(2n+1/6)\pi}$, then it can easily be seen that

$$\int_{x_1}^{x} p(t)dt = e^{2n\pi}\left[e^{\pi/3}\left(\sin\frac{\pi}{3} - 1.25\right) - e^{\pi/6}\left(\sin\frac{\pi}{6} - 1.25\right)\right]$$
$$\simeq 0.172e^{2n\pi} \to \infty \quad \text{as} \quad n \to \infty,$$

and hence the trivial solution is not uniformly stable. Thus, asymptotic stability does not imply uniform stability.

Example 23.5.   Every solution of the DE $y' = 0$ is of the form $y(x) = y(x_0)$, and hence uniformly stable but not asymptotically stable. Hence, uniform stability does not imply asymptotic stability.

Our final result provides necessary and sufficient conditions so that all solutions of the differential system (17.3) are uniformly stable.

Theorem 23.7.   Let $\Psi(x)$ be a fundamental matrix of the differential system (17.3). Then all solutions of the differential system (17.3) are uniformly stable if and only if the condition (22.10) holds.

Proof. Let $u(x) = u(x, x_0, u^0)$ be a solution of the differential system (17.3). Then for any $x_1 \geq x_0$, we have $u(x) = \Psi(x)\Psi^{-1}(x_1)u(x_1)$. If $u^1(x) = \Psi(x)\Psi^{-1}(x_1)u^1(x_1)$ is any other solution, and the condition (22.10) is satisfied, then we have

$$\|u^1(x) - u(x)\| \leq \|\Psi(x)\Psi^{-1}(x_1)\|\|u^1(x_1) - u(x_1)\| \leq c\|u^1(x_1) - u(x_1)\|$$

for all $x_0 \leq x_1 \leq x < \infty$. Thus, if $\epsilon > 0$ then $x_1 \geq x_0$ and $\|u^1(x_1) - u(x_1)\| < \epsilon/c = \delta(\epsilon) > 0$ imply that $\|u^1(x) - u(x)\| < \epsilon$, and hence the solution $u(x)$ is uniformly stable.

Conversely, if all solutions of (17.3) are uniformly stable, then, in particular, the trivial solution, i.e., $u(x, x_0, 0) \equiv 0$ is uniformly stable. Therefore, given any $\epsilon > 0$, there exists a $\delta = \delta(\epsilon) > 0$ such that $x_1 \geq x_0$ and $\|u^1(x_1)\| < \delta$ imply that $\|u^1(x)\| < \epsilon$ for all $x \geq x_1$. Thus, we have $\|\Psi(x)\Psi^{-1}(x_1)u^1(x_1)\| < \epsilon$ for all $x \geq x_1$. The rest of the proof is the same as that of Theorem 23.1. ∎

Problems

23.1. Test the stability, asymptotic stability or instability for the trivial solution of each of the following systems:

(i) $u' = \begin{bmatrix} 0 & 1 \\ -1 & 0 \end{bmatrix} u.$
    (ii) $u' = \begin{bmatrix} -1 & e^{2x} \\ 0 & -1 \end{bmatrix} u.$

(iii) $u' = \begin{bmatrix} 0 & 1 & 0 \\ 0 & 0 & 1 \\ -1 & -6 & -5 \end{bmatrix} u.$
    (iv) $u' = \begin{bmatrix} 1 & 2 & 0 \\ 0 & 1 & 1 \\ 1 & 3 & 1 \end{bmatrix} u.$

(v) $u' = \begin{bmatrix} 1 & -1 & -1 \\ 1 & 1 & -3 \\ 1 & -5 & -3 \end{bmatrix} u.$

23.2 (Hurwitz's Theorem). A necessary and sufficient condition for the negativity of the real parts of all zeros of the polynomial

$$x^n + a_1 x^{n-1} + \cdots + a_{n-1}x + a_n$$

with real coefficients is the positivity of all the leading principal minors of the *Hurwitz matrix*

$$\begin{bmatrix} a_1 & 1 & 0 & 0 & 0 & \cdots & 0 \\ a_3 & a_2 & a_1 & 1 & 0 & \cdots & 0 \\ a_5 & a_4 & a_3 & a_2 & a_1 & \cdots & 0 \\ \cdots & & & & & & \\ 0 & 0 & 0 & 0 & 0 & \cdots & a_n \end{bmatrix}.$$

Use the Hurwitz theorem to find the parameter a in the differential systems

(i) $\quad u' = \begin{bmatrix} 0 & 0 & 1 \\ -3 & 0 & 0 \\ a & 2 & -1 \end{bmatrix} u;$ (ii) $\quad u' = \begin{bmatrix} 0 & 1 & 0 \\ 0 & 0 & 1 \\ a & -3 & -2 \end{bmatrix} u$

so that the trivial solution is asymptotically stable.

23.3. In the differential system (17.1) let $A(x)$ and $b(x)$ be continuous in $[x_0, \infty)$. Show the following:

(i)　If all solutions are bounded in $[x_0, \infty)$, then they are stable.

(ii)　If all solutions are stable and one is bounded, then all solutions are bounded in $[x_0, \infty)$.

23.4. In the differential system (17.3) let $A(x)$ be continuous in $[x_0, \infty)$. System (17.3) is said to be stable if all its solutions are stable, and it is called *restrictively stable* if the system (17.3) together with its adjoint system (18.10) are stable. Show the following:

(i)　A necessary and sufficient condition for restrictive stability is that there exists a constant $c > 0$ such that $\|\Phi(x, x_0)\Phi(x_0, t)\| \leq c, \ x \geq x_0, \ t \geq x_0$ where $\Phi(x, x_0)$ is the principal fundamental matrix of (17.3).

(ii)　If the system (17.3) is stable and the condition (22.6) is satisfied, then it is restrictively stable.

(iii)　If the adjoint system (18.10) is stable and

$$\limsup_{x \to \infty} \int^x \operatorname{Tr} A(t)dt \ < \ \infty,$$

then the system (17.3) is restrictively stable.

23.5. Show that the stability of any solution of the nonhomogeneous differential system (17.1) is equivalent to the stability of the trivial solution of the homogeneous system (17.3).

23.6. If the Floquet multipliers σ_i of (17.3) satisfy $|\sigma_i| < 1, \ i = 1, \ldots, n$, then show that the trivial solution is asymptotically stable.

23.7. Show that the question of stability of the solution $u(x) = u(x, x_0, u^0)$ of (15.1) can always be reduced to the question of stability of the trivial solution of the differential system $v' = G(x, v)$, where $v = u - u(x)$ and $G(x, v) = g(x, v + u(x)) - g(x, u(x))$.

Answers or Hints

23.1. (i) Stable. (ii) Unstable. (iii) Asymptotically stable. (iv) Unstable. (v) Unstable.

23.2. (i) $a < -6$. (ii) $-6 < a < 0$.

23.3. If $u_1(x)$ and $u_2(x)$ are solutions of (17.1), then $u_1(x) - u_2(x) = \Phi(x, x_0)(u_1(x_0) - u_2(x_0))$.

23.4. (i) Condition $\|\Phi(x, x_0)\Phi(x_0, t)\| \leq c$, $x \geq x_0$, $t \geq x_0$ is equivalent to the existence of a constant c_1 such that $\|\Phi(x, x_0)\| \leq c_1$, $\|\Phi^{-1}(x, x_0)\| \leq c_1$. Now use Theorems 18.3 and 23.1. (ii) Use Theorem 17.3 to deduce that $|\det \Phi(x, x_0)| > d > 0$ for all $x \geq x_0$. Now the relation $\Phi^{-1}(x, x_0) = \operatorname{adj} \Phi(x, x_0) / \det \Phi(x, x_0)$ ensures that $\Phi^{-1}(x, x_0)$ is bounded for all $x \geq x_0$. (iii) Use part (ii) and the relation $\operatorname{Tr}(-A^T(x)) = -\operatorname{Tr} A(x)$.

23.5. See Problem 23.6.

23.6. The asymptotic stability of the trivial solution of (20.6) implies the same of the trivial solution of (17.3). Now since $e^{\omega(\text{real part of } \lambda_i)} = |\sigma_i| < 1$, the real parts of the Floquet exponents must be negative.

23.7. Let $v = u - u(x)$, then $v' = u' - u'(x) = g(x, u) - g(x, u(x)) = g(x, v + u(x)) - g(x, u(x)) = G(x, v)$. Clearly, in the new system $v' = G(x, v)$, $G(x, 0) = 0$.

Lecture 24
Stability of
Quasi-Linear Systems

In Problems 22.9 and 22.10 we have considered the differential systems (22.16) and (22.17) as the perturbed systems of (18.6) and (17.3), respectively, and provided sufficient conditions on the nonlinear perturbed function $g(x, v)$ so that the asymptotic properties of the unperturbed systems are maintained for the perturbed systems. Analogously, we expect that under certain conditions on the function $g(x, v)$ stability properties of the unperturbed systems carry through for the perturbed systems. For obvious reasons, systems (22.16) and (22.17) are called *quasi-linear differential systems*.

Let the function $g(x, v)$ satisfy the condition

$$\|g(x, v)\| = o(\|v\|) \tag{24.1}$$

uniformly in x as $\|v\|$ approaches zero. This implies that for v in a sufficiently small neighborhood of the origin, $\|g(x, v)\|/\|v\|$ can be made arbitrarily small. Condition (24.1) assures that $g(x, 0) \equiv 0$, and hence $v(x) \equiv 0$ is a solution of the perturbed differential systems.

We begin with an interesting example which shows that the asymptotic stability of the trivial solution of the unperturbed system (17.3) and the condition (24.1) do not imply the asymptotic stability of the trivial solution of the perturbed system (22.17).

Example 24.1. Consider the differential system

$$\begin{aligned}
u_1' &= -au_1 \\
u_2' &= (\sin 2x + 2x \cos 2x - 2a)u_2, \quad 1 < 2a < 3/2
\end{aligned} \tag{24.2}$$

whose general solution is

$$\begin{aligned}
u_1(x) &= c_1 e^{-ax} \\
u_2(x) &= c_2 \exp((\sin 2x - 2a)x).
\end{aligned}$$

Since $a > 1/2$, every solution of (24.2) tends to zero as $x \to \infty$, and hence the trivial solution of (24.2) is asymptotically stable.

R.P. Agarwal and D. O'Regan, *An Introduction to Ordinary Differential Equations*,
doi: 10.1007/978-0-387-71276-5_24, © Springer Science + Business Media, LLC 2008

Now we consider the perturbed differential system

$$
\begin{aligned}
v_1' &= -av_1 \\
v_2' &= (\sin 2x + 2x\cos 2x - 2a)v_2 + v_1^2,
\end{aligned}
\tag{24.3}
$$

i.e., the perturbing function $g(x, v) = [0 \ , \ v_1^2]^T$. Obviously, for this $g(x, v)$ the condition (24.1) is satisfied.

The general solution of the differential system (24.3) is

$$
\begin{aligned}
v_1(x) &= c_1 e^{-ax} \\
v_2(x) &= \left(c_2 + c_1^2 \int_0^x e^{-t\sin 2t} dt \right) \exp((\sin 2x - 2a)x).
\end{aligned}
$$

Let $x = x_n = (n + 1/4)\pi, \ n = 1, 2, \ldots$ then we have

$$
-\sin 2t \ \geq \ \frac{1}{2} \quad \text{for all} \quad x_n + \frac{\pi}{3} \leq t \leq x_n + \frac{\pi}{2}.
$$

Thus, it follows that

$$
\begin{aligned}
\int_0^{x_{n+1}} e^{-t\sin 2t} dt \ &> \ \int_{x_n + \pi/3}^{x_n + \pi/2} e^{-t\sin 2t} dt \\
&> \ \int_{x_n + \pi/3}^{x_n + \pi/2} e^{t/2} dt \ > \ 0.4 \exp\left(\frac{1}{2}x_n + \frac{\pi}{4} \right).
\end{aligned}
$$

Therefore, we have

$$
v_2(x_{n+1}) \ > \ c_2 e^{(1-2a)x_n} + 0.4 c_1^2 \exp\left(\frac{\pi}{4} + \left(\frac{3}{2} - 2a \right) x_n \right).
$$

Since $2a < 3/2$, we see that $v_2(x_{n+1}) \to \infty$ as $n \to \infty$ if $c_1 \neq 0$. Thus, the trivial solution of (24.3) is unstable.

In our first result we shall show that the asymptotic stability of the trivial solution of the differential system (18.6) and the condition (24.1) do imply the asymptotic stability of the trivial solution of (22.16).

Theorem 24.1. Suppose that the real parts of the eigenvalues of the matrix A are negative, and the function $g(x, v)$ satisfies the condition (24.1). Then the trivial solution of the differential system (22.16) is asymptotically stable.

Proof. In (18.15) let the nonhomogeneous term $b(x)$ be $g(x, v(x))$, so that each solution $v(x)$ such that $v(x_0) = v^0$ of the differential system (22.16) satisfies the integral equation

$$
v(x) \ = \ e^{A(x-x_0)} v^0 + \int_{x_0}^x e^{A(x-t)} g(t, v(t)) dt.
\tag{24.4}
$$

Now since the real parts of the eigenvalues of the matrix A are negative, there exist constants c and $\eta = -\delta$ ($\delta > 0$) such that $\|e^{Ax}\| \leq ce^{-\delta x}$ for all $x \geq 0$. Hence, from (24.4) we have

$$\|v(x)\| \leq ce^{-\delta(x-x_0)}\|v^0\| + c\int_{x_0}^{x} e^{-\delta(x-t)}\|g(t, v(t))\|dt, \quad x \geq x_0. \quad (24.5)$$

In view of the condition (24.1) for a given $m > 0$ there exists a positive number d such that

$$\|g(t, v)\| \leq m\|v\| \quad (24.6)$$

for all $x \geq x_0$, $\|v\| \leq d$.

Let us assume that $\|v^0\| < d$. Then there exists a number x_1 such that $\|v(x)\| < d$ for all $x \in [x_0, x_1)$. Using (24.6) in (24.5), we obtain

$$\|v(x)\|e^{\delta x} \leq ce^{\delta x_0}\|v^0\| + cm\int_{x_0}^{x} \|v(t)\|e^{\delta t}dt, \quad x \in [x_0, x_1). \quad (24.7)$$

Applying Corollary 7.6 to the inequality (24.7), we get

$$\|v(x)\| \leq c\|v^0\| \exp((cm - \delta)(x - x_0)), \quad x \in [x_0, x_1). \quad (24.8)$$

But since v^0 and m are at our disposal, we may choose m such that $cm < \delta$, and $v(x_0) = v^0$ so that $\|v^0\| < d/c$ implies that $\|v(x)\| < d$ for all $x \in [x_0, x_1)$.

Next since the function $g(x, v)$ is continuous in $[x_0, \infty) \times \mathbb{R}^n$, we can extend the solution $v(x)$ interval by interval by preserving the bound δ. Hence, given any solution $v(x) = v(x, x_0, v^0)$ with $\|v^0\| < d/c$, we see that v is defined in $[x_0, \infty)$ and satisfies $\|v(x)\| < d$. But d can be made as small as desired, therefore the trivial solution of the differential system (22.16) is stable. Further, $cm < \delta$ implies that it is asymptotically stable. ∎

In the above result the magnitude of $\|v^0\|$ cannot be arbitrary. For example, the solution

$$y(x) = \frac{y_0}{y_0 - (y_0 - 1)e^x}$$

of the initial value problem $y' = -y + y^2$, $y(0) = y_0 > 1$ becomes unbounded as $x \to \ln[y_0/(y_0 - 1)]$.

Example 24.2.   The motion of a *simple pendulum* with *damping* is governed by a DE of the form

$$\theta'' + \frac{k}{m}\theta' + \frac{g}{L}\sin\theta = 0,$$

which is usually approximated by a simpler DE

$$\theta'' + \frac{k}{m}\theta' + \frac{g}{L}\theta = 0.$$

In system form these equations can be written as (18.6) and (22.16), respectively, where

$$A = \begin{bmatrix} 0 & 1 \\ -\dfrac{g}{L} & -\dfrac{k}{m} \end{bmatrix} \quad \text{and} \quad g(x,v) = \begin{bmatrix} 0 \\ \dfrac{g}{L}(v_1 - \sin v_1) \end{bmatrix}.$$

Since the matrix A has eigenvalues

$$-\frac{k}{2m} \pm \left(\frac{k^2}{4m^2} - \frac{g}{L}\right)^{1/2},$$

both of which have negative real parts if k, m, g, and L are positive, and since

$$\|g(x,v)\| = \left|\frac{g}{L}(v_1 - \sin v_1)\right| = \frac{g}{L}\left|\frac{v_1^3}{3!} - \cdots\right| \leq M|v_1|^3$$

for some constant M, Theorem 24.1 is applicable. Thus, we see that when $\|v\|$ is sufficiently small the use of more refined differential system, i.e., including the nonlinear function $g(x,v)$ does not lead to a radically different behavior of the solution from that obtained from the linear differential system as $x \to \infty$.

Now we state the following result whose proof differs slightly from Theorem 24.1.

Theorem 24.2.   Suppose that the matrix A possesses at least one eigenvalue with a positive real part, and the function $g(x,v)$ satisfies the condition (24.1). Then the trivial solution of the differential system (22.16) is unstable.

Theorems 24.1 and 24.2 fail to embrace the critical case, i.e., when the real parts of all the eigenvalues of the matrix A are nonpositive, and when at least one eigenvalue is zero. In this critical case, the nonlinear function $g(x,v)$ begins to influence the stability of the trivial solution of the system (22.16), and generally it is impossible to test for stability on the basis of eigenvalues of A. For example, the trivial solution of the DE $y' = ay^3$ is asymptotically stable if $a < 0$, stable if $a = 0$, and unstable if $a > 0$.

Our final result in this lecture is for the differential system (22.17), where as in Problem 22.10 we shall assume that

$$\|g(x,v)\| \leq \lambda(x)\|v\|, \tag{24.9}$$

where $\lambda(x)$ is a nonnegative continuous function such that $\int^{\infty} \lambda(t)dt < \infty$.

Obviously, condition (24.9) implies that $v(x) \equiv 0$ is a solution of the differential system (22.17).

Theorem 24.3.   Suppose that the solutions of the differential system (17.3) are uniformly (uniformly and asymptotically) stable, and the function $g(x, v)$ satisfies condition (24.9). Then the trivial solution of the differential system (22.17) is uniformly (uniformly and asymptotically) stable.

Proof.   Since all the solutions of the differential system (17.3) are uniformly stable, by Theorem 23.7 there exists a constant c such that for any fundamental matrix $\Psi(x)$ of (17.3) we have $\|\Psi(x)\Psi^{-1}(t)\| \le c$ for all $x_0 \le t \le x < \infty$.

In (18.14) let the nonhomogeneous term $b(x)$ be $g(x, v(x))$, so that each solution $v(x)$ such that $v(x_1) = v^1$, $x_1 \ge x_0$ of the differential system (22.17) satisfies the integral equation

$$v(x) = \Psi(x)\Psi^{-1}(x_1)v^1 + \int_{x_1}^{x} \Psi(x)\Psi^{-1}(t)g(t, v(t))dt. \qquad (24.10)$$

Thus, it follows that

$$\|v(x)\| \le c\|v^1\| + c\int_{x_1}^{x} \lambda(t)\|v(t)\|dt.$$

From this, we find

$$\|v(x)\| \le c\|v^1\|\exp\left(c\int_{x_1}^{x} \lambda(t)dt\right) \le K\|v^1\|,$$

where

$$K = c\exp\left(c\int_{x_0}^{\infty} \lambda(t)dt\right).$$

Hence, for a given $\epsilon > 0$, if $\|v^1\| < K^{-1}\epsilon$ then $\|v(x)\| < \epsilon$ for all $x \ge x_1$; i.e., the trivial solution of the differential system (22.17) is uniformly stable.

Finally, if the solutions of the differential system (17.3) are, in addition, asymptotically stable, then from Theorem 23.3 it follows that $\|\Psi(x)\| \to 0$ as $x \to \infty$. Thus, given any $\epsilon > 0$ we can choose x_2 large enough so that $\|\Psi(x)\Psi^{-1}(x_0)v^0\| \le \epsilon$ for all $x \ge x_2$. For the solution $v(x) = v(x, x_0, v^0)$ we then have

$$\|v(x)\| \le \|\Psi(x)\Psi^{-1}(x_0)v^0\| + \int_{x_0}^{x} \|\Psi(x)\Psi^{-1}(t)\|\|g(t, v(t))\|dt$$

$$\le \epsilon + c\int_{x_0}^{x} \lambda(t)\|v(t)\|dt, \quad x \ge x_2.$$

From this, we get

$$\|v(x)\| \le \epsilon \exp\left(c \int_{x_0}^{x} \lambda(t)dt\right) \le L\epsilon, \quad x \ge x_2$$

where $L = \exp\left(c \int_{x_0}^{\infty} \lambda(t)dt\right)$.

Since ϵ is arbitrary and L does not depend on ϵ or x_2, we conclude that $\|v(x)\| \to 0$ as $x \to \infty$, i.e., the trivial solution of the differential system (22.17) is, in addition, asymptotically stable.

Problems

24.1. Test the stability, asymptotic stability, or unstability for the trivial solution of each of the following systems:

(i)
$$
\begin{aligned}
u_1' &= -2u_1 + u_2 + 3u_3 + 8u_1^2 + u_2^3 \\
u_2' &= -6u_2 - 5u_3 + 7u_3^4 \\
u_3' &= -u_3 + u_1^4 + u_2^2 + u_3^3.
\end{aligned}
$$

(ii)
$$
\begin{aligned}
u_1' &= 2u_1 + u_2 - u_1^2 - u_2^2 \\
u_2' &= u_1 + 3u_2 - u_1^3 \sin u_3 \\
u_3' &= u_2 + 2u_3 + u_1^2 + u_2^2.
\end{aligned}
$$

(iii)
$$
\begin{aligned}
u_1' &= \ln(1 - u_3) \\
u_2' &= \ln(1 - u_1) \\
u_3' &= \ln(1 - u_2).
\end{aligned}
$$

(iv)
$$
\begin{aligned}
u_1' &= u_1 - \cos u_2 - u_3 + 1 \\
u_2' &= u_2 - \cos u_3 - u_1 + 1 \\
u_3' &= u_3 - \cos u_1 - u_2 + 1.
\end{aligned}
$$

24.2. Test the stability, asymptotic stability or unstability for the trivial solution of each of the following equations:

(i) $y''' + 2y'' + 3y' + 9\sinh y = 0$.

(ii) $y''' + 3y'' - 4y' + 7y + y^2 = 0$.

(iii) $y'''' + y + \cosh y - 1 = 0$.

(iv) $y'''' + 2y''' + 3y'' + 11y + y \sin y = 0$.

Answers or Hints

24.1. (i) Asymptotically stable. (ii) Unstable. (iii) Unstable. (iv) Unstable.

24.2. (i) Unstable. (ii) Unstable. (iii) Unstable. (iv) Unstable.

Lecture 25
Two-Dimensional Autonomous Systems

The differential system (15.1) is said to be *autonomous* if the function $g(x, u)$ is independent of x. Thus, two-dimensional autonomous systems are of the form

$$\begin{aligned}
u_1' &= g_1(u_1, u_2) \\
u_2' &= g_2(u_1, u_2).
\end{aligned} \tag{25.1}$$

Throughout, we shall assume that the functions g_1 and g_2 together with their first partial derivatives are continuous in some domain D of the u_1u_2-plane. Thus, for all $(u_1^0, u_2^0) \in D$ the differential system (25.1) together with $u_1(x_0) = u_1^0$, $u_2(x_0) = u_2^0$ has a unique solution in some interval J containing x_0. The main interest in studying (25.1) is twofold:

1. A large number of dynamic processes in applied sciences are governed by such systems.

2. The qualitative behavior of its solutions can be illustrated through the geometry in the u_1u_2-plane.

For the autonomous differential system (25.1) the following result is fundamental.

Theorem 25.1. If $u(x) = (u_1(x), u_2(x))$ is a solution of the differential system (25.1) in the interval (α, β), then for any constant c the function $v(x) = (u_1(x + c), u_2(x + c))$ is also a solution of (25.1) in the interval $(\alpha - c, \beta - c)$.

Proof. Since $v'(x) = u'(x + c)$ and $u'(x) = g(u(x))$ it follows that $v'(x) = u'(x + c) = g(u(x + c)) = g(v(x))$, i.e., $v(x)$ is also a solution of (25.1). ∎

Obviously, the above property does not usually hold for nonautonomous differential systems, e.g., a solution of $u_1' = u_1$, $u_2' = xu_1$ is $u_1(x) = e^x$, $u_2(x) = xe^x - e^x$, and $u_2'(x + c) = (x + c)e^{x+c} \neq xu_1(x + c)$ unless $c = 0$.

In the domain D of the u_1u_2-plane any solution of the differential system (25.1) may be regarded as a parametric curve given by $(u_1(x), u_2(x))$ with x as the parameter. This curve $(u_1(x), u_2(x))$ is called a *trajectory* or an *orbit* or a *path* of (25.1), and the u_1u_2-plane is called the *phase plane*. Thus,

R.P. Agarwal and D. O'Regan, *An Introduction to Ordinary Differential Equations*, doi: 10.1007/978-0-387-71276-5_25, © Springer Science + Business Media, LLC 2008

from Theorem 25.1 for any constant c both $(u_1(x), u_2(x))$, $x \in (\alpha, \beta)$ and $(u_1(x+c), u_2(x+c))$, $x \in (\alpha - c, \beta - c)$ represent the same trajectory. For the trajectories of the differential system (25.1) the following property is very important.

Theorem 25.2. Through each point $(u_1^0, u_2^0) \in D$ there passes one and only one trajectory of the differential system (25.1).

Proof.   Suppose, on the contrary, there are two different trajectories $(u_1(x), u_2(x))$ and $(v_1(x), v_2(x))$ passing through (u_1^0, u_2^0), i.e., $u_1(x_0) = u_1^0 = v_1(x_1)$ and $u_2(x_0) = u_2^0 = v_2(x_1)$, where $x_0 \neq x_1$ follows by the uniqueness of solutions of the initial value problems. By Theorem 25.1, $u_1^1(x) = u_1(x - x_1 + x_0)$, $u_2^1(x) = u_2(x - x_1 + x_0)$ is also a solution of (25.1). Note $u_1^1(x_1) = u_1(x_0) = u_1^0 = v_1(x_1)$, and $u_1^1(x_1) = u_2(x_0) = u_2^0 = v_2(x_1)$. Hence, from the uniqueness of solutions of the initial value problems we find that $u_1^1(x) \equiv v_1(x)$ and $u_2^1(x) \equiv v_2(x)$. Thus, $(u_1(x), u_2(x))$ and $(v_1(x), v_2(x))$ are the same trajectories with different parameterizations.

Example 25.1.   For the differential system $u_1' = u_2$, $u_2' = -u_1$ there are an infinite number of solutions $u_1(x) = \sin(x+c)$, $u_2(x) = \cos(x+c)$, $0 \leq c < 2\pi$, $-\infty < x < \infty$. However, they represent the same trajectory, i.e., the circle $u_1^2 + u_2^2 = 1$.

Thus, it is important to note that a trajectory is a curve in D that is represented parametrically by more than one solution. Hence, in conclusion, $u(x) = (u_1(x), u_2(x))$ and $v(x) = (u_1(x+c), u_2(x+c))$, $c \neq 0$ are distinct solutions of (25.1), but they represent the same curve parametrically.

Definition 25.1.   Any point $(u_1^0, u_2^0) \in D$ at which both g_1 and g_2 vanish simultaneously is called a *critical point* of (25.1). A critical point is also referred to as a *point of equilibrium* or *stationary point* or *rest point* or *singular point*.

If (u_1^0, u_2^0) is a critical point of (25.1), then obviously $u_1(x) = u_1^0$, $u_2(x) = u_2^0$ is a solution of (25.1), and from Theorem 25.2 no trajectory can pass through the point (u_1^0, u_2^0).

A critical point (u_1^0, u_2^0) is said to be *isolated* if there exists no other critical point in some neighborhood of (u_1^0, u_2^0). By a critical point, we shall hereafter mean an isolated critical point.

Example 25.2.   For the differential system

$$\begin{aligned} u_1' &= a_{11}u_1 + a_{12}u_2 \\ u_2' &= a_{21}u_1 + a_{22}u_2, \quad a_{11}a_{22} - a_{21}a_{12} \neq 0 \end{aligned} \quad (25.2)$$

there is only one critical point, namely, $(0,0)$ in $D = \mathbb{R}^2$.

Example 25.3. For the simple undamped pendulum system $u_1' = u_2$, $u_2' = -(g/L)\sin u_1$ there are an infinite number of critical points $(n\pi, 0)$, $n = 0, \pm 1, \pm 2, \ldots$ in $D = \mathbb{R}^2$.

If (u_1^0, u_2^0) is a critical point of (25.1), then the substitution $v_1 = u_1 - u_1^0$, $v_2 = u_2 - u_2^0$ transforms (25.1) into an equivalent system with $(0,0)$ as a critical point, thus without loss of generality we can assume that $(0,0)$ is a critical point of the system (25.1). An effective technique in studying the differential system (25.1) near the critical point $(0,0)$ is to approximate it by a linear system of the form (25.2), and expect that a "good" approximation (25.2) will provide solutions which themselves are "good" approximations to the solutions of the system (25.1). For example, if the system (25.1) can be written as

$$
\begin{aligned}
u_1' &= a_{11}u_1 + a_{12}u_2 + h_1(u_1, u_2) \\
u_2' &= a_{21}u_1 + a_{22}u_2 + h_2(u_1, u_2),
\end{aligned}
\tag{25.3}
$$

where $h_1(0,0) = h_2(0,0) = 0$ and

$$
\lim_{u_1, u_2 \to 0} \frac{h_1(u_1, u_2)}{(u_1^2 + u_2^2)^{1/2}} = \lim_{u_1, u_2 \to 0} \frac{h_2(u_1, u_2)}{(u_1^2 + u_2^2)^{1/2}} = 0,
$$

then the results of the previous lecture immediately give the following theorem.

Theorem 25.3. (i) If the zero solution of the system (25.2) is asymptotically stable, then the zero solution of the system (25.3) is asymptotically stable.

(ii) If the zero solution of the system (25.2) is unstable, then the zero solution of the system (25.3) is unstable.

(iii) If the zero solution of the system (25.2) is stable, then the zero solution of the system (25.3) may be asymptotically stable, stable, or unstable.

Of course, if the functions $g_1(u_1, u_2)$ and $g_2(u_1, u_2)$ possess continuous second-order partial derivatives in the neighborhood of the critical point $(0,0)$, then by Taylor's formula, differential system (25.1) can always be written in the form (25.3) with

$$
a_{11} = \frac{\partial g_1}{\partial u_1}(0,0), \quad a_{12} = \frac{\partial g_1}{\partial u_2}(0,0), \quad a_{21} = \frac{\partial g_2}{\partial u_1}(0,0), \quad a_{22} = \frac{\partial g_2}{\partial u_2}(0,0).
$$

The picture of all trajectories of a system is called the *phase portrait* of the system. Since the solutions of (25.2) can be determined explicitly, a complete description of its phase portrait can be given. However, as we have seen earlier the nature of the solutions of (25.2) depends on the eigenvalues of the matrix

$$A = \begin{bmatrix} a_{11} & a_{12} \\ a_{21} & a_{22} \end{bmatrix};$$

i.e., the roots of the equation

$$\lambda^2 - (a_{11} + a_{22})\lambda + a_{11}a_{22} - a_{21}a_{12} = 0. \qquad (25.4)$$

The phase portrait of (25.2) depends almost entirely on the roots λ_1 and λ_2 of (25.4). For this, there are several different cases which must be studied separately.

Case 1. λ_1 and λ_2 are real, distinct, and of the same sign.
If v^1, v^2 are the corresponding eigenvectors of A, then from Theorem 19.1 the general solution of (25.2) can be written as

$$\begin{bmatrix} u_1(x) \\ u_2(x) \end{bmatrix} = c_1 \begin{bmatrix} v_1^1 \\ v_2^1 \end{bmatrix} e^{\lambda_1 x} + c_2 \begin{bmatrix} v_1^2 \\ v_2^2 \end{bmatrix} e^{\lambda_2 x}, \qquad (25.5)$$

where c_1 and c_2 are arbitrary constants.

For simplicity, we can always assume that $\lambda_1 > \lambda_2$. Thus, if $\lambda_2 < \lambda_1 < 0$ then all solutions of (25.2) tend to $(0,0)$ as $x \to \infty$. Therefore, the critical point $(0,0)$ of (25.2) is asymptotically stable. In case $c_1 = 0$ and $c_2 \neq 0$, we have $u_2 = (v_2^2/v_1^2)u_1$, i.e., the trajectory is a straight line with slope v_2^2/v_1^2. Similarly, if $c_1 \neq 0$ and $c_2 = 0$ we obtain the straight line $u_2 = (v_2^1/v_1^1)u_1$. To obtain other trajectories, we assume that c_1 and c_2 both are different from zero. Then since

$$\frac{u_2(x)}{u_1(x)} = \frac{c_1 v_2^1 e^{\lambda_1 x} + c_2 v_2^2 e^{\lambda_2 x}}{c_1 v_1^1 e^{\lambda_1 x} + c_2 v_1^2 e^{\lambda_2 x}} = \frac{c_1 v_2^1 + c_2 v_2^2 e^{(\lambda_2 - \lambda_1)x}}{c_1 v_1^1 + c_2 v_1^2 e^{(\lambda_2 - \lambda_1)x}}, \qquad (25.6)$$

which tends to v_2^1/v_1^1 as $x \to \infty$, all trajectories tend to $(0,0)$ with the slope v_2^1/v_1^1. Similarly, as $x \to -\infty$ all trajectories become asymptotic to the line with the slope v_2^2/v_1^2. This situation is illustrated in Figure 25.1 for two different values of the slope v_2^1/v_1^1 and v_2^2/v_1^2. Here the critical point $(0,0)$ is called a *stable node*.

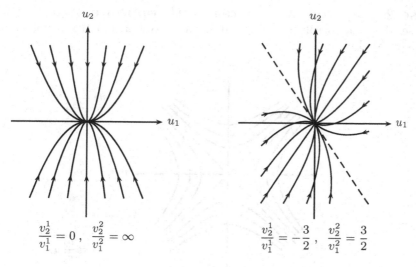

$$\frac{v_2^1}{v_1^1} = 0 \,, \quad \frac{v_2^2}{v_1^2} = \infty \qquad\qquad \frac{v_2^1}{v_1^1} = -\frac{3}{2} \,, \quad \frac{v_2^2}{v_1^2} = \frac{3}{2}$$

Figure 25.1

If $\lambda_1 > \lambda_2 > 0$, then all nontrivial solutions tend to infinity as x tends to ∞. Therefore, the critical point $(0,0)$ is unstable. The trajectories are the same as for $\lambda_2 < \lambda_1 < 0$ except that the direction of the motion is reversed as depicted in Figure 25.2. As $x \to -\infty$, the trajectories tend to $(0,0)$ with the slope v_2^2/v_1^2, and as $x \to \infty$ trajectories become asymptotic to the line with the slope v_2^1/v_1^1. Here the critical point $(0,0)$ is called an *unstable node*.

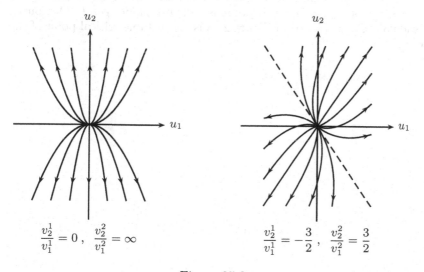

$$\frac{v_2^1}{v_1^1} = 0 \,, \quad \frac{v_2^2}{v_1^2} = \infty \qquad\qquad \frac{v_2^1}{v_1^1} = -\frac{3}{2} \,, \quad \frac{v_2^2}{v_1^2} = \frac{3}{2}$$

Figure 25.2

Case 2. λ_1 **and** λ_2 **are real with opposite signs.** Of course, the general solution of the differential system (25.2) remains the same (25.5). Let

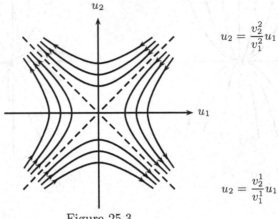

$$u_2 = \frac{v_2^2}{v_1^2} u_1$$

$$u_2 = \frac{v_2^1}{v_1^1} u_1$$

Figure 25.3

$\lambda_1 > 0 > \lambda_2$. If $c_1 = 0$ and $c_2 \neq 0$, then as in Case 1 we have $u_2 = (v_2^2/v_1^2)u_1$, and as $x \to \infty$ both $u_1(x)$ and $u_2(x)$ tend to zero. If $c_1 \neq 0$ and $c_2 = 0$, then $u_2 = (v_2^1/v_1^1)u_1$ and both $u_1(x)$ and $u_2(x)$ tend to infinity as $x \to \infty$, and approach zero as $x \to -\infty$. If c_1 and c_2 both are different from zero, then from (25.6) it follows that u_2/u_1 tends to v_2^1/v_1^1 as $x \to \infty$. Hence, all trajectories are asymptotic to the line with slope v_2^1/v_1^1 as $x \to \infty$. Similarly, as $x \to -\infty$ all trajectories are asymptotic to the line with slope v_2^2/v_1^2. Also, it is obvious that both $u_1(x)$ and $u_2(x)$ tend to infinity as $x \to \pm\infty$. This type of critical point is called a *saddle point*. Obviously, the saddle point displayed in Figure 25.3 is an unstable critical point of the system.

Lecture 26

Two-Dimensional Autonomous Systems (Contd.)

We shall continue our study of the phase portrait of the differential system (25.2).

Case 3. $\lambda_1 = \lambda_2 = \lambda.$ In this case from Corollary 19.4 the general solution of the differential system (25.2) can be written as

$$
\begin{bmatrix} u_1(x) \\ u_2(x) \end{bmatrix} = c_1 \begin{bmatrix} 1 + (a_{11} - \lambda)x \\ a_{21}x \end{bmatrix} e^{\lambda x} + c_2 \begin{bmatrix} a_{12}x \\ 1 + (a_{22} - \lambda)x \end{bmatrix} e^{\lambda x},
$$

$$(26.1)$$

where c_1 and c_2 are arbitrary constants.

If $\lambda < 0$, both $u_1(x)$ and $u_2(x)$ tend to 0 as $x \to \infty$ and hence the critical point $(0,0)$ of (25.2) is asymptotically stable. Further, from (26.1) it follows that

$$
\frac{u_2}{u_1} = \frac{c_2 + [a_{21}c_1 + (a_{22} - \lambda)c_2]x}{c_1 + [a_{12}c_2 + (a_{11} - \lambda)c_1]x}. \tag{26.2}
$$

Thus, in particular, if $a_{12} = a_{21} = 0$, $a_{11} = a_{22} \neq 0$, then equation (25.4) gives $\lambda = a_{11} = a_{22}$, and (26.2) reduces to $u_2/u_1 = c_2/c_1$. Therefore, all trajectories are straight lines with slope c_2/c_1. The phase portrait in this case is illustrated in Figure 26.1(a). Here, the origin is called *stable proper (star-shaped) node.* In the general case as $x \to \pm\infty$, (26.2) tends to

$$
\frac{a_{21}c_1 + (a_{22} - \lambda)c_2}{a_{12}c_2 + (a_{11} - \lambda)c_1}.
$$

But, since $(a_{11} - \lambda)(a_{22} - \lambda) = a_{12}a_{21}$ this ratio is the same as $a_{21}/(a_{11} - \lambda)$. Thus, as $x \to \pm\infty$, all trajectories are asymptotic to the line $u_2 = (a_{21}/(a_{11} - \lambda))u_1$. The origin $(0,0)$ here is called *stable improper node*; see Figure 26.1(b).

If $\lambda > 0$, all solutions tend to ∞ as $x \to \infty$ and hence the critical point $(0,0)$ of (25.2) is unstable. The trajectories are the same as for $\lambda < 0$ except that the direction of the motion is reversed (see Figure 26.2(a)–(b)).

R.P. Agarwal and D. O'Regan, *An Introduction to Ordinary Differential Equations,*
doi: 10.1007/978-0-387-71276-5_26, © Springer Science + Business Media, LLC 2008

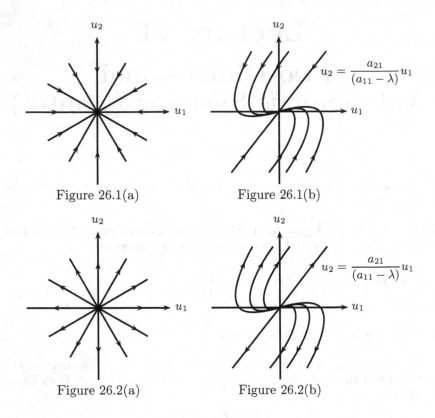

Figure 26.1(a)   Figure 26.1(b)

Figure 26.2(a)   Figure 26.2(b)

Case 4. λ_1 and λ_2 are complex conjugates.

Let $\lambda_1 = \mu + i\nu$ and $\lambda_2 = \mu - i\nu$, where we can assume that $\nu > 0$. If $v = v^1 + iv^2$ is the eigenvector of A corresponding to the eigenvalue $\lambda_1 = \mu + i\nu$, then a solution of (25.2) can be written as

$$
\begin{aligned}
u(x) &= e^{(\mu+i\nu)x}(v^1 + iv^2) = e^{\mu x}(\cos\nu x + i\sin\nu x)(v^1 + iv^2) \\
&= e^{\mu x}[v^1 \cos\nu x - v^2 \sin\nu x] + ie^{\mu x}[v^1 \sin\nu x + v^2 \cos\nu x].
\end{aligned}
$$

Therefore, from Problem 19.5,

$$
u^1(x) = e^{\mu x}[v^1 \cos\nu x - v^2 \sin\nu x] \quad \text{and} \quad u^2(x) = e^{\mu x}[v^1 \sin\nu x + v^2 \cos\nu x]
$$

are two real-valued linearly independent solutions of (25.2), and every solution $u(x)$ of (25.2) is of the form $u(x) = c_1 u^1(x) + c_2 u^2(x)$. This expression can easily be rewritten as

$$
\begin{aligned}
u_1(x) &= r_1 e^{\mu x} \cos(\nu x - \delta_1) \\
u_2(x) &= r_2 e^{\mu x} \cos(\nu x - \delta_2),
\end{aligned}
\tag{26.3}
$$

where $r_1 \geq 0$, $r_2 \geq 0$, δ_1 and δ_2 are some constants.

If $\mu = 0$, then both $u_1(x) = r_1 \cos(\nu x - \delta_1)$ and $u_2(x) = r_2 \cos(\nu x - \delta_2)$ are periodic of period $2\pi/\nu$. The function $u_1(x)$ varies between $-r_1$ and r_1, while $u_2(x)$ varies between $-r_2$ and r_2. Thus, each trajectory beginning at the point (u_1^*, u_2^*) when $x = x^*$ will return to the same point when $x = x^* + 2\pi/\nu$. Thus, the trajectories are closed curves, and the phase portrait of (25.2) has the form described in Figure 26.3(a). In this case the critical point $(0, 0)$ is stable but not asymptotically stable, and is called a *center*.

If $\mu < 0$, then the effect of the factor $e^{\mu x}$ in (26.3) is to change the simple closed curves of Figure 26.3(a) into the spirals of Figure 26.3(b). This is because the point

$$\left(u_1\left(\frac{2\pi}{\nu}\right), u_2\left(\frac{2\pi}{\nu}\right) \right) = \exp\left(\frac{2\pi\mu}{\nu}\right) (u_1(0), u_2(0))$$

is closer to the origin $(0, 0)$ than $(u_1(0), u_2(0))$. In this case the critical point $(0, 0)$ is asymptotically stable, and is called a *stable focus*.

If $\mu > 0$, then all trajectories of (25.2) spiral away from the origin $(0, 0)$ as $x \to \infty$ and are illustrated in Figure 26.3(c). In this case the critical point $(0, 0)$ is unstable, and is named an *unstable focus*.

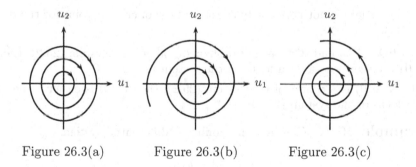

Figure 26.3(a)   Figure 26.3(b)   Figure 26.3(c)

We summarize the above analysis in the following theorem.

Theorem 26.1. For the differential system (25.2), let λ_1 and λ_2 be the eigenvalues of the matrix A. Then the behavior of its trajectories near the critical point $(0, 0)$ is as follows:

(i)   stable node, if λ_1 and λ_2 are real, distinct, and negative;

(ii)   unstable node, if λ_1 and λ_2 are real, distinct, and positive;

(iii) saddle point (unstable), if λ_1 and λ_2 are real, distinct, and of opposite sign;

(iv)   stable node, if λ_1 and λ_2 are real, equal, and negative;

(v)   unstable node, if λ_1 and λ_2 are real, equal, and positive;

(vi)   stable center, if λ_1 and λ_2 are pure imaginary;

(vii)   stable focus, if λ_1 and λ_2 are complex conjugates, with negative real part;

(viii) unstable focus, if λ_1 and λ_2 are complex conjugates with positive real part.

The behavior of the linear system (25.2) near the origin also determines the nature of the trajectories of the nonlinear system (25.3) near the critical point $(0,0)$. We state the following result whose proof can be found in advanced books.

Theorem 26.2.   For the differential system (25.2), let λ_1 and λ_2 be the eigenvalues of the matrix A. Then we have the following

(a) The nonlinear system (25.3) has the same type of critical point at the origin as the linear system (25.2) whenever

(i) $\lambda_1 \neq \lambda_2$ and $(0,0)$ is a node of the system (25.2);

(ii) $(0,0)$ is a saddle point of the system (25.2);

(iii) $\lambda_1 = \lambda_2$ and $(0,0)$ is not a star-shaped node of the system (25.2);

(iv) $(0,0)$ is a focus of the system (25.2).

(b) The origin is not necessarily the same type of critical point for the two systems:

(i)   If $\lambda_1 = \lambda_2$ and $(0,0)$ is a star-shaped node of the system (25.2), then $(0,0)$ is either a node or a focus of the system (25.3).

(ii)   If $(0,0)$ is a center of the system (25.2), then $(0,0)$ is either a center or a focus of the system (25.3).

Example 26.1.   Consider the nonlinear differential system

$$\begin{aligned}
u_1' &= 1 - u_1 u_2 \\
u_2' &= u_1 - u_2^3.
\end{aligned}$$
(26.4)

The equations $1 - u_1 u_2 = 0$ and $u_1 - u_2^3 = 0$ imply that $(1,1)$ and $(-1,-1)$ are the only critical points of the system (26.4).

For the point $(1,1)$, in (26.4) we use the substitution $v_1 = u_1 - 1$, $v_2 = u_2 - 1$ to obtain the new system

$$\begin{aligned}
v_1' &= 1 - (v_1 + 1)(v_2 + 1) = -v_1 - v_2 - v_1 v_2 \\
v_2' &= (v_1 + 1) - (v_2 + 1)^3 = v_1 - 3v_2 - 3v_2^2 - v_2^3,
\end{aligned}$$
(26.5)

which plays the role of the differential system (25.3). Obviously, in (26.5) the functions $h_1(v_1, v_2) = -v_1 v_2$ and $h_2(v_1, v_2) = -3v_2^2 - v_2^3$ are such that

$$\lim_{v_1, v_2 \to 0} \frac{h_1(v_1, v_2)}{(v_1^2 + v_2^2)^{1/2}} = 0 = \lim_{v_1, v_2 \to 0} \frac{h_2(v_1, v_2)}{(v_1^2 + v_2^2)^{1/2}}.$$

Corresponding to (26.5) the associated linear system is

$$\begin{aligned} v_1' &= -v_1 - v_2 \\ v_2' &= v_1 - 3v_2. \end{aligned} \tag{26.6}$$

Since for the matrix

$$\begin{bmatrix} -1 & -1 \\ 1 & -3 \end{bmatrix}$$

the eigenvalues $\lambda_1 = \lambda_2 = -2$, the zero solution of the system (26.6) is asymptotically stable. Thus, from Theorem 25.3 the zero solution of (26.5) is asymptotically stable. Hence, the critical point $(1, 1)$ of the differential system (26.4) is asymptotically stable. Further, from Theorem 26.1 the zero solution of the differential system (26.6) is a stable node. Thus, from Theorem 26.2 the zero solution of (26.5) is a stable node. Hence the critical point $(1, 1)$ of (26.4) is a stable node.

Similarly, for the point $(-1, -1)$, we use the substitution $v_1 = u_1 + 1$, $v_2 = u_2 + 1$ to obtain the new system

$$\begin{aligned} v_1' &= 1 - (v_1 - 1)(v_2 - 1) = v_1 + v_2 - v_1 v_2 \\ v_2' &= (v_1 - 1) - (v_2 - 1)^3 = v_1 - 3v_2 + 3v_2^2 - v_2^3. \end{aligned} \tag{26.7}$$

Corresponding to (26.7) the associated linear system is

$$\begin{aligned} v_1' &= v_1 + v_2 \\ v_2' &= v_1 - 3v_2. \end{aligned} \tag{26.8}$$

Since for the matrix

$$\begin{bmatrix} 1 & 1 \\ 1 & -3 \end{bmatrix}$$

the eigenvalues $\lambda_1 = -1 + \sqrt{5} > 0$ and $\lambda_2 = -1 - \sqrt{5} < 0$, the zero solution of the system (26.8) is an unstable saddle point. Therefore, for the nonlinear system (26.7) the zero solution is an unstable saddle point. Hence, the critical point $(-1, -1)$ of the differential system (26.4) is an unstable saddle point.

Remark 26.1. For the nonlinear systems

$$\begin{aligned} u_1' &= -u_2 - u_1^2 \\ u_2' &= u_1 \end{aligned} \tag{26.9}$$

and

$$u_1' = -u_2 - u_1^3$$
$$u_2' = u_1$$

(26.10)

the corresponding homogeneous system is

$$u_1' = -u_2$$
$$u_2' = u_1.$$

(26.11)

Clearly, for (26.11) the critical point $(0,0)$ is a center. It is known that the critical point $(0,0)$ of (26.9) is a center, while the same point for (26.10) is a focus.

Remark 26.2. If the general nonlinear system (25.1) does not contain linear terms, then an infinite number of critical points are possible. Further, the nature of these points depend on the nonlinearity in (25.1), and hence it is rather impossible to classify these critical points.

Problems

26.1. Show that any solution $u(x) = (u_1(x), u_2(x))$ of the differential system

$$u_1' = u_2 (e^{u_1} - 1)$$
$$u_2' = u_1 + e^{u_2}$$

which starts in the right half-plane $(u_1 > 0)$ must remain there for all x.

26.2. Let $u(x) = (u_1(x), u_2(x))$ be a solution of the differential system (25.1). Show that if $u(x_0+\omega) = u(x_0)$ for some $\omega > 0$, then $u(x+\omega) = u(x)$ for all $x \geq x_0$, i.e., $u(x)$ is periodic of period ω.

26.3. Let $u(x) = (u_1(x), u_2(x))$ be a solution of the differential system (25.1). Show the following:

(i)   If $u(x)$ is periodic of period ω, then the trajectory of this solution is a closed curve in the $u_1 u_2$-plane.

(ii)   If the trajectory of $u(x)$ is a closed curve containing no critical points of (25.1), then this solution is periodic.

26.4. Prove that all solutions of the following DEs are periodic:

(i) $y'' + a^2 y + b y^3 = 0$, $\quad b > 0$ $\quad$ (*Duffing's equation*).
(ii) $y'' + y^3/(1 + y^4) = 0$.
(iii) $y'' + e^{y^2} - 1 = 0$.
(iv) $y'' + y + y^7 = 0$.

26.5. Show that all solutions of the following differential systems are periodic:

(i) $\begin{aligned} u_1' &= u_2(1 + u_1^2 + u_2^2) \\ u_2' &= -2u_1(1 + u_1^2 + u_2^2). \end{aligned}$

(ii) $\begin{aligned} u_1' &= u_2 \exp(1 + u_1^2) \\ u_2' &= -u_1 \exp(1 + u_1^2). \end{aligned}$

26.6. Find all critical points of each of the following differential systems and determine whether they are stable or unstable:

(i) $\begin{aligned} u_1' &= u_1 + 4u_2 \\ u_2' &= u_1 + u_2 - u_1^2. \end{aligned}$

(ii) $\begin{aligned} u_1' &= u_1 - u_2 + u_1^2 \\ u_2' &= 12u_1 - 6u_2 + u_1u_2. \end{aligned}$

(iii) $\begin{aligned} u_1' &= 8u_1 - u_2^2 \\ u_2' &= u_2 - u_1^2. \end{aligned}$

(iv) $\begin{aligned} u_1' &= u_1 - u_1^3 - u_1u_2^2 \\ u_2' &= 2u_2 - u_2^5 - u_1^4u_2. \end{aligned}$

26.7. Determine the type of stability of the critical point $(0,0)$ of each of the following linear systems and sketch the phase portraits:

(i) $\begin{aligned} u_1' &= -2u_1 + u_2 \\ u_2' &= -5u_1 - 6u_2. \end{aligned}$

(ii) $\begin{aligned} u_1' &= 4u_1 + u_2 \\ u_2' &= 3u_1 + 6u_2. \end{aligned}$

(iii) $\begin{aligned} u_1' &= u_2 \\ u_2' &= 2u_1 - u_2. \end{aligned}$

(iv) $\begin{aligned} u_1' &= u_1 + u_2 \\ u_2' &= 3u_1 - u_2. \end{aligned}$

(v) $\begin{aligned} u_1' &= 3u_1 + u_2 \\ u_2' &= u_2 - u_1. \end{aligned}$

(vi) $\begin{aligned} u_1' &= -2u_1 - 5u_2 \\ u_2' &= 2u_1 + 2u_2. \end{aligned}$

(vii) $\begin{aligned} u_1' &= -u_1 - u_2 \\ u_2' &= u_1 - u_2. \end{aligned}$

(viii) $\begin{aligned} u_1' &= 7u_1 + u_2 \\ u_2' &= -3u_1 + 4u_2. \end{aligned}$

26.8. Find all critical points of each of the following differential systems and determine their nature:

(i) $\begin{aligned} u_1' &= 4u_2^2 - u_1^2 \\ u_2' &= 2u_1u_2 - 4u_2 - 8. \end{aligned}$

(ii) $\begin{aligned} u_1' &= u_1u_2 \\ u_2' &= 4 - 4u_1 - 2u_2. \end{aligned}$

(iii) $\begin{aligned} u_1' &= 2u_1(u_1 - u_2) \\ u_2' &= 2 + u_2 - u_1^2. \end{aligned}$

(iv) $\begin{aligned} u_1' &= u_1(2u_2 - u_1 + 5) \\ u_2' &= u_1^2 + u_2^2 - 6u_1 - 8u_2. \end{aligned}$

26.9. Consider the *Van der Pol equation*

$$y'' + \mu(y^2 - 1)y' + y = 0, \qquad (26.12)$$

which is equivalent to the system $u_1' = u_2$, $u_2' = \mu(1 - u_1^2)u_2 - u_1$. Show that $(0,0)$ is the only critical point of the system. Determine the nature of the critical point when $\mu < 2$, $\mu = 2$, and $\mu > 2$.

26.10. *Rayleigh's equation* is

$$y'' + \mu\left(\frac{1}{3}(y')^2 - 1\right)y' + y = 0,$$

where μ is a constant. Show that differentiation of this equation and setting $y' = z$ reduces Rayleigh's equation to the Van der Pol equation.

26.11. Consider the DE $y'' + py' + qy = 0$, $q \neq 0$ which is equivalent to the system $u_1' = u_2$, $u_2' = -pu_2 - qu_1$. Determine the nature of the critical point $(0, 0)$ in each of the following cases:

(i) $p^2 > 4q$, $q > 0$, and $p < 0$.

(ii) $p^2 > 4q$, $q > 0$, and $p > 0$.

(iii) $p^2 > 4q$ and $q < 0$.

(iv) $p^2 = 4q$ and $p > 0$.

(v) $p^2 = 4q$ and $p < 0$.

(vi) $p^2 < 4q$ and $p > 0$.

(vii) $p^2 < 4q$ and $p < 0$.

(viii) $p = 0$ and $q > 0$.

Answers or Hints

26.1. Use uniqueness property.

26.2. Use uniqueness property.

26.3. (i) In every time interval $x_0 \leq x \leq x_0 + \omega$ (x_0 fixed) the trajectory C of this solution moves once around C. (ii) Such a solution moves along its trajectory C with velocity $[g_1^2(u_1, u_2) + g_2^2(u_1, u_2)]^{1/2}$, which has a positive minimum for (u_1, u_2) on C. Hence, the trajectory must return to its starting point $u_1^0 = u_1(x_0)$, $u_2^0 = u_2(x_0)$ in some finite time ω. But this implies that $u(x + \omega) = u(x)$ (cf. Problem 26.2).

26.4. (i) In system form the given DE is the same as $u_1' = u_2$, $u_2' = -a^2 u_1 - bu_1^3$. The trajectories of this system are the solution curves $\frac{1}{2}u_2^2 + \frac{a^2}{2}u_1^2 + \frac{b}{4}u_1^4 = c^2$ of the scalar equation $\frac{du_2}{du_1} = -\frac{a^2 u_1 + bu_1^3}{u_2}$. Obviously these curves are closed in $u_1 u_2$-plane. (ii) $\frac{1}{2}u_2^2 + \frac{1}{4}\ln(1 + u_1^4) = c^2$. (iii) $\frac{1}{2}u_2^2 + u_1 - \int_0^{u_1} e^{t^2}\, dt = c^2$. (iv) $\frac{1}{2}u_2^2 + \frac{1}{2}u_1^2 + \frac{1}{8}u_1^8 = c^2$.

26.5. (i) $\frac{1}{2}u_2^2 + u_1^2 = c^2$. (ii) $u_2^2 + u_1^2 = c^2$.

26.6. (i) $(0, 0)$ unstable, $(3/4, -3/16)$ unstable. (ii) $(0, 0)$ stable, $(2, 6)$ unstable, $(3, 12)$ unstable. (iii) $(0, 0)$ unstable, $(2, 4)$ unstable. (iv) $(0, 0)$ unstable, $(1, 0)$ unstable, $(-1, 0)$ unstable, $(0, 2^{1/4})$ stable, $(0, -2^{1/4})$ stable.

26.7. (i) Stable focus. (ii) Unstable node. (iii) Saddle point. (iv) Saddle point. (v) Unstable node. (vi) Center. (vii) Stable focus. (viii) Unstable focus.

26.8. (i) $(4, 2)$ saddle point, $(-2, -1)$ saddle point. (ii) $(0, 2)$ saddle point, $(1, 0)$ stable focus. (iii) $(0, -2)$ unstable node, $(2, 2)$ saddle point, $(-1, -1)$ saddle point. (iv) $(0, 0)$ saddle point, $(0, 8)$ unstable node, $(7, 1)$ saddle point, $(3, -1)$ stable node.

26.9. $\mu > 0 :$ $\mu < 2$ unstable focus, $\mu = 2$ unstable node, $\mu > 2$ unstable node. $\mu < 0 : -2 < \mu$ stable focus, $\mu = -2$ stable node, $\mu < -2$ stable node.

26.10. Verify directly.

26.11. (i) Unstable node. (ii) Stable node. (iii) Saddle point. (iv) Stable node. (v) Unstable node. (vi) Stable focus. (vii) Unstable focus. (viii) Stable center.

Lecture 27
Limit Cycles and
Periodic Solutions

We begin this lecture with the following well-known example.

Example 27.1. Consider the nonlinear differential system

$$
\begin{aligned}
u_1' &= -u_2 + u_1(1 - u_1^2 - u_2^2) \\
u_2' &= u_1 + u_2(1 - u_1^2 - u_2^2).
\end{aligned}
\tag{27.1}
$$

Since the term $u_1^2 + u_2^2$ appears in both the equations, we introduce polar coordinates (r, θ), where $u_1 = r\cos\theta$, $u_2 = r\sin\theta$ to obtain

$$
\begin{aligned}
\frac{d}{dx}r^2 = 2r\frac{dr}{dx} &= 2u_1\frac{du_1}{dx} + 2u_2\frac{du_2}{dx} \\
&= 2(u_1^2 + u_2^2) - 2(u_1^2 + u_2^2)^2 \\
&= 2r^2(1 - r^2)
\end{aligned}
$$

and hence

$$
\frac{dr}{dx} = r(1 - r^2).
$$

Similarly, we find

$$
\frac{d\theta}{dx} = \frac{d}{dx}\tan^{-1}\frac{u_2}{u_1} = \frac{1}{u_1^2}\frac{u_1\dfrac{du_2}{dx} - u_2\dfrac{du_1}{dx}}{1 + (u_2/u_1)^2} = \frac{u_1^2 + u_2^2}{u_1^2 + u_2^2} = 1.
$$

Thus, the system (27.1) is equivalent to the differential system

$$
\begin{aligned}
\frac{dr}{dx} &= r(1 - r^2) \\
\frac{d\theta}{dx} &= 1,
\end{aligned}
\tag{27.2}
$$

which can be solved easily to obtain the general solution

$$
\begin{aligned}
r(x) &= \frac{r_0}{[r_0^2 + (1 - r_0^2)e^{-2x}]^{1/2}} \\
\theta(x) &= x + \theta_0,
\end{aligned}
$$

R.P. Agarwal and D. O'Regan, *An Introduction to Ordinary Differential Equations*,
doi: 10.1007/978-0-387-71276-5_27, © Springer Science + Business Media, LLC 2008

where $r_0 = r(0)$ and $\theta_0 = \theta(0)$. Hence, the general solution of (27.1) appears as

$$
\begin{aligned}
u_1(x) &= \frac{r_0}{[r_0^2 + (1 - r_0^2)e^{-2x}]^{1/2}} \cos(x + \theta_0) \\
u_2(x) &= \frac{r_0}{[r_0^2 + (1 - r_0^2)e^{-2x}]^{1/2}} \sin(x + \theta_0).
\end{aligned}
\tag{27.3}
$$

Obviously, (27.3) defines the trajectories of (27.1) in the $u_1 u_2$-plane. Examining these trajectories, we note the following:

(i) If $r_0 = 1$, the trajectory defined by (27.3) is the unit circle

$$
\begin{aligned}
u_1(x) &= \cos(x + \theta_0) \\
u_2(x) &= \sin(x + \theta_0)
\end{aligned}
\tag{27.4}
$$

described in the anticlockwise direction. This solution is periodic of period 2π.

(ii) If $r_0 \neq 1$, the trajectories defined by (27.3) are not closed (and hence from Problem 26.3 are not the periodic solutions) but rather have a spiral behavior. If $r_0 < 1$, the trajectories are spirals lying inside the circle (27.4). As $x \to +\infty$, they approach this circle, while as $x \to -\infty$, they approach the only critical point $(0,0)$ of (27.1). If $r_0 > 1$, the trajectories are spirals lying outside the circle (27.4). These outer trajectories also approach this circle as $x \to +\infty$; while as

$$
x \to \frac{1}{2}\ln\left(1 - \frac{1}{r_0^2}\right),
$$

both u_1 and u_2 become infinite. This situation is depicted in Figure 27.1.

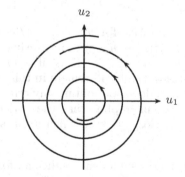

Figure 27.1

The differential system (27.1) shows that the trajectories of a nonlinear system of DEs may spiral into a simple closed curve. This, of course, is not possible for linear systems. This leads to the following important definition, which is due to Poincaré.

Definition 27.1. A closed trajectory of the differential system (25.1) which is approached spirally from either the inside or the outside by a nonclosed trajectory of (25.1) either as $x \to +\infty$ or as $x \to -\infty$ is called a *limit cycle* of (25.1).

The following result provides sufficient conditions for the existence of limit cycles of the differential system (25.1).

Theorem 27.1 (Poincaré–Bendixson Theorem). Suppose that a solution $u(x) = (u_1(x), u_2(x))$ of the differential system (25.1) remains in a bounded region of the u_1u_2-plane which contains no critical points of (25.1). Then its trajectory must spiral into a simple closed curve, which itself is the trajectory of a periodic solution of (25.1).

While the proof of this celebrated theorem is not given here, we give an example illustrating the importance of this result.

Example 27.2. Consider the DE

$$y'' + (2y^2 + 3y'^2 - 1)y' + y = 0, \qquad (27.5)$$

which is equivalent to the system

$$\begin{aligned} u_1' &= u_2 \\ u_2' &= -u_1 + (1 - 2u_1^2 - 3u_2^2)u_2. \end{aligned} \qquad (27.6)$$

For any given solution $u(x) = (u_1(x), u_2(x))$ of (27.6) we note that

$$\begin{aligned} \frac{d}{dx}(u_1^2(x) + u_2^2(x)) &= 2u_1(x)u_1'(x) + 2u_2(x)u_2'(x) \\ &= 2(1 - 2u_1^2(x) - 3u_2^2(x))u_2^2(x). \end{aligned}$$

Since $(1 - 2u_1^2 - 3u_2^2)$ is positive for $u_1^2 + u_2^2 < 1/3$, and negative for $u_1^2 + u_2^2 > 1/2$, the function $u_1^2(x) + u_2^2(x)$ is increasing when $u_1^2 + u_2^2 < 1/3$ and decreasing when $u_1^2 + u_2^2 > 1/2$. Thus, if $u(x)$ starts in the annulus $1/3 < u_1^2 + u_2^2 < 1/2$ at $x = x_0$, it will remain in this annulus for all $x \geq x_0$. Further, since this annulus does not contain any critical point of (27.6), the Poincaré–Bendixson theorem implies that the trajectory of this solution must spiral into a simple closed curve, which itself is a nontrivial periodic solution of (27.6).

The next result provides sufficient conditions for the nonexistence of closed trajectories, and hence, in particular, limit cycles of the differential system (25.1).

Theorem 27.2 (Bendixson's Theorem). If

$$\frac{\partial g_1(u_1, u_2)}{\partial u_1} + \frac{\partial g_2(u_1, u_2)}{\partial u_2}$$

has the same sign throughout the domain D, then the differential system (25.1) has no closed trajectory in D.

Proof.   Let S be a region in D which is bounded by a closed curve C. The Green's theorem states that

$$\int_C [g_1(u_1, u_2)du_2 - g_2(u_1, u_2)du_1] = \iint_S \left[\frac{\partial g_1(u_1, u_2)}{\partial u_1} + \frac{\partial g_2(u_1, u_2)}{\partial u_2} \right] du_1 du_2.$$

$$(27.7)$$

Let $u(x) = (u_1(x), u_2(x))$ be a solution of (25.1) whose trajectory is a closed curve C in D, and let ω denote the period of this solution. Then it follows that

$$\int_C [g_1(u_1, u_2)du_2 - g_2(u_1, u_2)du_1]$$

$$= \int_0^\omega \left[g_1(u_1(x), u_2(x)) \frac{du_2(x)}{dx} - g_2(u_1(x), u_2(x)) \frac{du_1(x)}{dx} \right] dx = 0.$$

Thus, from (27.7) we have

$$\iint_S \left[\frac{\partial g_1(u_1, u_2)}{\partial u_1} + \frac{\partial g_2(u_1, u_2)}{\partial u_2} \right] du_1 du_2 = 0.$$

But this double integral can be zero only if its integrand changes sign. This contradiction completes the proof.   ∎

Example 27.3.   Consider the nonlinear differential system

$$\begin{aligned} u_1' &= u_1(u_1^2 + u_2^2 - 2u_1 - 3) - u_2 \\ u_2' &= u_2(u_1^2 + u_2^2 - 2u_1 - 3) + u_1. \end{aligned}$$

$$(27.8)$$

Since for this system

$$\frac{\partial g_1}{\partial u_1} + \frac{\partial g_2}{\partial u_2} = 4u_1^2 + 4u_2^2 - 6u_1 - 6 = 4\left[\left(u_1 - \frac{3}{4} \right)^2 + u_2^2 - \frac{33}{16} \right]$$

we find that

$$\frac{\partial g_1}{\partial u_1} + \frac{\partial g_2}{\partial u_2} < 0$$

in the disc D of radius 1.436 centered at $(3/4, 0)$. Thus, the Bendixson's theorem implies that the system (27.8) has no closed trajectory in this disc D.

Finally, in this lecture we shall state the following theorem.

Theorem 27.3 (Liénard–Levinson–Smith Theorem).
Consider the DE

$$y'' + f(y)y' + g(y) = 0,$$

$$(27.9)$$

where we assume the following:

(i) f is even and continuous for all y.

(ii)   There exists a number $y_0 > 0$ such that

$$F(y) = \int_0^y f(t)dt < 0$$

for $0 < y < y_0$, and $F(y) > 0$ and monotonically increasing for $y > y_0$, also $F(y) \to \infty$ as $y \to \infty$.

(iii) g is odd, has a continuous derivative for all y, and is such that $g(y) > 0$ for $y > 0$,

(iv) $G(y) = \int_0^y g(t)dt \to \infty$ as $y \to \infty$.

Then the DE (27.9) possesses an essentially unique nontrivial periodic solution.

By "essentially unique" in the above result we mean that if $y = y(x)$ is a nontrivial periodic solution of (27.9), then all other nontrivial periodic solutions of (27.9) are of the form $y = y(x - x_1)$ where x_1 is a real number. This, of course, implies that the equivalent system $u_1' = u_2$, $u_2' = -f(u_1)u_2 - g(u_1)$ has a unique closed trajectory in the u_1u_2-plane.

Example 27.4.   In Van der Pol's equation (26.12), $f(y) = \mu(y^2 - 1)$ and $g(y) = y$. Thus, it is easy to check the following:

(i) $f(-y) = \mu(y^2 - 1) = f(y)$, the function f is even and continuous for all y.

(ii) $F(y) = \mu(y^3/3 - y) < 0$ for $0 < y < \sqrt{3}$, $F(y) > 0$ and monotonically increasing for $y > \sqrt{3}$, also $F(y) \to \infty$ as $y \to \infty$.

(iii) $g(-y) = -y = -g(y)$, the function g is odd; $dg/dy = 1$, and the derivative of g is continuous for all y; $g(y) > 0$ for $y > 0$.

(iv) $G(y) = y^2/2 \to \infty$ as $y \to \infty$.

Hence, all the conditions of Theorem 27.3 for the DE (26.12) are satisfied. In conclusion we find that the DE (26.12) possesses an essentially unique nontrivial periodic solution. In other words, the equivalent system $u_1' = u_2$, $u_2' = \mu(1 - u_1^2)u_2 - u_1$ has a unique closed trajectory in the u_1u_2-plane. For $\mu = 0.1$, 1, and 10, these trajectories are illustrated in Figure 27.2.

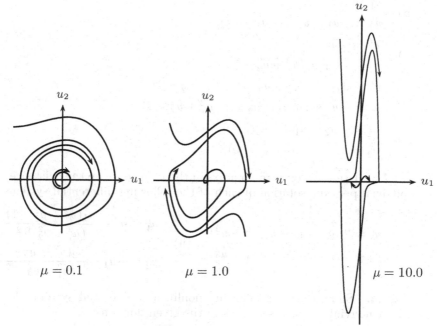

Figure 27.2

Problems

27.1. Show that the differential system

$$u_1' = u_2 + u_1\frac{f(r)}{r}$$

$$u_2' = -u_1 + u_2\frac{f(r)}{r},$$

where $r^2 = u_1^2 + u_2^2$, has limit cycles corresponding to the zeros of $f(r)$.

27.2. Find all limit cycles of each of the following differential systems:

(i)
$$u_1' = u_2 + u_1(u_1^2 + u_2^2)^{1/2}(u_1^2 + u_2^2 - 3)$$
$$u_2' = -u_1 + u_2(u_1^2 + u_2^2)^{1/2}(u_1^2 + u_2^2 - 3).$$

(ii)
$$u_1' = -u_2 + \frac{u_1(u_1^2 + u_2^2 - 2)}{(u_1^2 + u_2^2)^{1/2}}$$
$$u_2' = u_1 + \frac{u_2(u_1^2 + u_2^2 - 2)}{(u_1^2 + u_2^2)^{1/2}}.$$

(iii)
$$u_1' = u_2$$
$$u_2' = -u_1 + u_2(1 - u_1^2 - u_2^2).$$

(iv)
$$u_1' = -u_2$$
$$u_2' = u_1 + u_2(u_1^2 + u_2^2 - 1).$$

(v)
$$u_1' = u_2 + u_1(u_1^2 + u_2^2 - 1)(u_1^2 + u_2^2 - 2)$$
$$u_2' = -u_1 + u_2(u_1^2 + u_2^2 - 1)(u_1^2 + u_2^2 - 2).$$

(vi)
$$u_1' = -u_2 - u_1(u_1^2 + u_2^2 - 9)^2$$
$$u_2' = u_1 - u_2(u_1^2 + u_2^2 - 9)^2.$$

27.3. Use the Poincaré–Bendixson theorem to prove the existence of a nontrivial periodic solution of each of the following differential systems:

(i)
$$u_1' = 2u_1 - 2u_2 - u_1(u_1^2 + u_2^2)$$
$$u_2' = 2u_1 + 2u_2 - u_2(u_1^2 + u_2^2).$$

(ii)
$$u_1' = u_2 - \frac{u_1(u_1^2 + u_2^2 - 1)}{(u_1^2 + u_2^2)^{1/2}}$$
$$u_2' = -u_1 - \frac{u_2(u_1^2 + u_2^2 - 1)}{(u_1^2 + u_2^2)^{1/2}}.$$

27.4. Show that the following nonlinear differential systems do not have nontrivial periodic solutions in the given domains:

(i)
$$u_1' = u_1 + 7u_2^2 + 2u_1^3$$
$$u_2' = -u_1 + 3u_2 + u_2u_1^2, \qquad D = \mathbb{R}^2.$$

(ii)
$$u_1' = u_1 - u_1u_2^2 + u_2^3$$
$$u_2' = 3u_2 - u_2u_1^2 + u_1^3, \qquad D = \{(u_1, u_2) : u_1^2 + u_2^2 < 4\}.$$

27.5. Show that the following DEs do not have nontrivial periodic solutions:

(i) $y'' + y' + 2y'^3 + y = 0.$
(ii) $y'' + (2y^2 + 3y^4)y' - (y + 5y^3) = 0.$

27.6. Use the Liénard–Levinson–Smith theorem to prove the existence of an essentially nontrivial periodic solution of each of the following DEs:

(i) $y'' + (y^4 - 1)y' + y = 0.$
(ii) $y'' + (5y^4 - 6y^2)y' + y^3 = 0.$

Answers or Hints

27.1. Let $r_0 < r_1 < \cdots$ be the zeros of $f(r)$. Transforming the given differential system into polar coordinates (r, θ), we obtain $\frac{dr}{dx} = f(r)$, $\frac{d\theta}{dx} = -1$. Thus, if $f(r) > 0$ for $0 < r < r_0$, then $\frac{dr}{dx} > 0$ if $0 < r < r_0$, i.e., as x increases r increases, and the path curves spiral outward as long as they lie within the circle $r = r_0$. However, if $f(r) < 0$ for $0 < r < r_0$, then $\frac{dr}{dx} < 0$ if $0 < r < r_0$, i.e., as x increases r decreases, and the path curves tend to the critical point $(0,0)$. Now if $f(r) < 0$ for $r_0 < r < r_1$, then $\frac{dr}{dx} < 0$ if $r_0 < r < r_1$, i.e., as x increases r decreases, and the path curves spiral inward as long as they lie within the annulus $r_0 < r < r_1$. However, if $f(r) > 0$ for $r_0 < r < r_1$ then $\frac{dr}{dx} > 0$ if $r_0 < r < r_1$, i.e., as x increases r increases, and the path curves spiral outward as long as they lie within the annulus. The points $r_2, r_3, \ldots$ can be treated similarly.

27.2. (i) $r = \sqrt{3}$. (ii) $r = \sqrt{2}$. (iii) $r = 1$. (iv) $r = 1$. (v) $r = 1,\ \sqrt{2}$. (vi) $r = 3$.

27.3. Use Theorem 27.1.

27.4. Use Theorem 27.2.

27.5. Use Theorem 27.2.

27.6. Use Theorem 27.3.

Lecture 28
Lyapunov's Direct Method for Autonomous Systems

It is well known that a mechanical system is stable if its total energy, which is the sum of potential energy and the kinetic energy, continuously decreases. These two energies are always positive quantities and are zero when the system is completely at rest. Lyapunov's direct method uses a generalized energy function to study the stability of the solutions of the differential systems. This function is called *Lyapunov function* and is, generally, denoted by V. The main advantage of this approach is that the stability can be discussed without any prior knowledge of the solutions. In this lecture we shall study this fruitful technique for the autonomous differential system

$$u' = g(u), \tag{28.1}$$

where the function $g = (g_1, \ldots, g_n)$ and its partial derivatives $\partial g_i / \partial u_j$, $1 \leq i, j \leq n$, are assumed to be continuous in an open set $\Omega \subseteq \mathbb{R}^n$ containing the origin. Thus, for all $u^0 \in \Omega$ the initial value problem (28.1), (15.3) has a unique solution in some interval containing x_0. Further, we shall assume that $g(0) = 0$ and $g(u) \neq 0$ for $u \neq 0$ in some neighborhood of the origin so that (28.1) admits the trivial solution, and the origin is an isolated critical point of the differential system (28.1).

Let $V(u)$ be a scalar continuous function defined in Ω, i.e., $V \in C[\Omega, \mathbb{R}]$ and $V(0) = 0$. For this function we need the following.

Definition 28.1. $V(u)$ is said to be *positive definite* in Ω if and only if $V(u) > 0$ for $u \neq 0$, $u \in \Omega$.

Definition 28.2. $V(u)$ is said to be *positive semidefinite* in Ω if $V(u) \geq 0$ (with equality only at certain points) for all $u \in \Omega$.

Definition 28.3. $V(u)$ is said to be *negative definite* (*negative semidefinite*) in Ω if and only if $-V(u)$ is positive definite (positive semidefinite) in Ω.

In this and the next lecture, for the sake of easy geometric interpretation, instead of the absolute norm we shall use the Euclidean norm $\| \cdot \|_2$, and the subscript 2 will be dropped.

Definition 28.4. A function $\phi(r)$ is said to belong to the class $\mathcal{K}$ if

R.P. Agarwal and D. O'Regan, *An Introduction to Ordinary Differential Equations*,   204
doi: 10.1007/978-0-387-71276-5_28, © Springer Science + Business Media, LLC 2008

and only if $\phi \in C[[0, \rho), \mathbb{R}^+]$, $\phi(0) = 0$, and $\phi(r)$ is strictly monotonically increasing in r.

Since $V(u)$ is continuous, for sufficiently small r, $0 < c \leq r \leq d$ we have

$$V(u) \leq \max_{\|v\| \leq r} V(v), \quad V(u) \geq \min_{r \leq \|v\| \leq d} V(v) \qquad (28.2)$$

on the hypersphere $\|u\| = r$. In (28.2) the right sides are monotonic functions of r and can be estimated in terms of functions belonging to the class $\mathcal{K}$. Thus, there exist two functions $a, b \in \mathcal{K}$ such that

$$a(\|u\|) \leq V(u) \leq b(\|u\|). \qquad (28.3)$$

The left side of (28.3) provides an alternative definition for the positive definiteness of $V(u)$ as follows.

Definition 28.5. The function $V(u)$ is said to be positive definite in Ω if and only if $V(0) = 0$ and there exists a function $a(r) \in \mathcal{K}$ such that $a(r) \leq V(u)$, $\|u\| = r$, $u \in \Omega$.

Example 28.1. The function $V(u_1, u_2) = c_1 u_1^2 + c_2 u_2^2$, where $c_1 > 0$, $c_2 > 0$ is positive definite in $\Omega = \mathbb{R}^2$.

Example 28.2. The function $V(u_1, u_2, u_3) = c_1 u_1^2 + c_2 u_2^2$, where $c_1 > 0$, $c_2 > 0$ is positive semidefinite in $\Omega = \mathbb{R}^3$ since it vanishes on the u_3-axis.

Example 28.3. The function $V(u_1, u_2, u_3) = c_1 u_1^2 + (c_2 u_2 + c_3 u_3)^2$, where $c_1 > 0$ is positive semidefinite in $\Omega = \mathbb{R}^3$ because it vanishes not only at the origin but also on the line $c_2 u_2 + c_3 u_3 = 0$, $u_1 = 0$.

Example 28.4. The function $V(u_1, u_2) = u_1^2 + u_2^2 - (u_1^4 + u_2^4)$ is positive definite in the interior of the unit circle because $V(u_1, u_2) \geq \|u\|^2 - \|u\|^4$, $\|u\| < 1$.

Let S_ρ be the set $\{u \in \mathbb{R}^n : \|u\| < \rho\}$, and $u(x) = u(x, x_0, u^0)$ be any solution of (28.1), (15.3) such that $\|u(x)\| < \rho$ for all $x \geq x_0$. Since (28.1) is autonomous we can always assume that $x_0 = 0$. If the function $V \in C^{(1)}[S_\rho, \mathbb{R}]$, then the chain rule can be used to obtain

$$\frac{dV(u)}{dx} = V^*(u) = \frac{\partial V(u)}{\partial u_1} \frac{du_1}{dx} + \cdots + \frac{\partial V(u)}{\partial u_n} \frac{du_n}{dx}$$

$$= \sum_{i=1}^{n} \frac{\partial V(u)}{\partial u_i} g_i(u) = \text{grad } V(u) \cdot g(u). \qquad (28.4)$$

Thus, the derivative of $V(u)$ with respect to x along the solution $u(x)$ of (28.1) is known, although we do not have the explicit solution.

Now we shall provide three theorems regarding the stability, asymptotic stability, and unstability of the trivial solution of the differential system (28.1).

Theorem 28.1. If there exists a positive definite scalar function $V(u) \in C^{(1)}[S_\rho, \mathbb{R}^+]$ (called a *Lyapunov function*) such that $V^*(u) \leq 0$ in S_ρ, then the trivial solution of the differential system (28.1) is stable.

Proof. Since $V(u)$ is positive definite, there exists a function $a \in \mathcal{K}$ such that $a(\|u\|) \leq V(u)$ for all $u \in S_\rho$. Let $0 < \epsilon < \rho$ be given. Since $V(u)$ is continuous and $V(0) = 0$, we can find a $\delta = \delta(\epsilon) > 0$ such that $\|u^0\| < \delta$ implies that $V(u^0) < a(\epsilon)$. If the trivial solution of (28.1) is unstable, then there exists a solution $u(x) = u(x, 0, u^0)$ of (28.1) such that $\|u^0\| < \delta$ satisfying $\|u(x_1)\| = \epsilon$ for some $x_1 > 0$. However, since $V^*(u) \leq 0$ in S_ρ, we have $V(u(x_1)) \leq V(u^0)$, and hence

$$a(\epsilon) = a(\|u(x_1)\|) \leq V(u(x_1)) \leq V(u^0) < a(\epsilon),$$

which is not true. Thus, if $\|u^0\| < \delta$ then $\|u(x)\| < \epsilon$ for all $x \geq 0$. This implies that the trivial solution of (28.1) is stable. ∎

Theorem 28.2. If there exists a positive definite scalar function $V(u) \in C^{(1)}[S_\rho, \mathbb{R}^+]$ such that $V^*(u)$ is negative definite in S_ρ, then the trivial solution of the differential system (28.1) is asymptotically stable.

Proof. Since all the conditions of Theorem 28.1 are satisfied, the trivial solution of (28.1) is stable. Therefore, given $0 < \epsilon < \rho$, suppose that there exist $\delta > 0$, $\lambda > 0$ and a solution $u(x) = u(x, 0, u^0)$ of (28.1) such that

$$\lambda \leq \|u(x)\| < \epsilon, \quad x \geq 0, \quad \|u^0\| < \delta. \tag{28.5}$$

Since $V^*(u)$ is negative definite, there exists a function $a \in \mathcal{K}$ such that

$$V^*(u(x)) \leq -a(\|u(x)\|).$$

Furthermore, since $\|u(x)\| \geq \lambda > 0$ for $x \geq 0$, there exists a constant $d > 0$ such that $a(\|u(x)\|) \geq d$ for $x \geq 0$. Hence, we have

$$V^*(u(x)) \leq -d < 0, \quad x \geq 0.$$

This implies that

$$V(u(x)) = V(u^0) + \int_0^x V^*(u(t))dt \leq V(u^0) - x\,d$$

and for sufficiently large x the right side will become negative, which contradicts $V(u)$ being positive definite. Hence, no such λ exists for which (28.5) holds. Further, since $V(u(x))$ is positive and a decreasing function of x, it follows that $\lim_{x \to \infty} V(u(x)) = 0$. Therefore, $\lim_{x \to \infty} \|u(x)\| = 0$, and this implies that the trivial solution of (28.1) is asymptotically stable.   ∎

Theorem 28.3. If there exists a scalar function $V(u) \in C^{(1)}[S_\rho, \mathbb{R}]$, $V(0) = 0$ such that $V^*(u)$ is positive definite in S_ρ, and if in every neighborhood N of the origin, $N \subset S_\rho$, there is a point u^0 where $V(u^0) > 0$, then the trivial solution of the differential system (28.1) is unstable.

Proof. Let $r > 0$ be sufficiently small so that the hypersphere

$$S_r = \{u \in \mathbb{R}^n : \|u\| \leq r\} \subset S_\rho.$$

Let $M = \max_{\|u\| \leq r} V(u)$, where M is finite since V is continuous. Let r_1 be such that $0 < r_1 < r$; then by the hypotheses there exists a point $u^0 \in \mathbb{R}^n$ such that $0 < \|u^0\| < r_1$ and $V(u^0) > 0$. Along the solution $u(x) = u(x, 0, u^0)$, $x \geq 0$, $V^*(u)$ is positive, and therefore $V(u(x))$, $x \geq 0$ is an increasing function and $V(u(0)) = V(u^0) > 0$. This implies that this solution $u(x)$ cannot approach the origin. Thus, it follows that

$$\inf_{x \geq 0} V^*(u(x)) = m > 0,$$

and therefore, $V(u(x)) \geq V(u^0) + mx$ for $x \geq 0$. But the right side of this inequality can be made greater than M for x sufficiently large, which implies that $u(x)$ must leave the hypersphere S_r. Thus, the trivial solution of (28.1) is unstable.   ∎

Example 28.5. For the differential system

$$\begin{aligned} u_1' &= u_2 + u_1(r^2 - u_1^2 - u_2^2) \\ u_2' &= -u_1 + u_2(r^2 - u_1^2 - u_2^2) \end{aligned} \tag{28.6}$$

we consider the positive definite function $V(u_1, u_2) = u_1^2 + u_2^2$ in $\Omega = \mathbb{R}^2$. A simple computation gives

$$V^*(u_1, u_2) = -2(u_1^2 + u_2^2)(u_1^2 + u_2^2 - r^2).$$

Obviously, $V^*(u)$ is negative definite when $r = 0$, and hence the trivial solution of (28.6) is asymptotically stable. On the other hand, when $r \neq 0$, $V^*(u)$ is positive definite in the region $u_1^2 + u_2^2 < r^2$. Therefore, the trivial solution of (28.6) is unstable.

Example 28.6. For the differential system

$$\begin{aligned} u_1' &= -u_1 u_2^4 \\ u_2' &= u_2 u_1^4 \end{aligned} \tag{28.7}$$

we choose the positive definite function $V(u_1, u_2) = u_1^4 + u_2^4$ in $\Omega = \mathbb{R}^2$. For this function we find $V^*(u_1, u_2) \equiv 0$ and hence the trivial solution of (28.7) is stable.

Example 28.7. Consider the differential system (17.3) with $a_{ij}(x) = -a_{ji}(x)$, $i \neq j$ and $a_{ii}(x) \leq 0$ for all $x \geq 0$, and $i, j = 1, 2, \ldots, n$. Let $V(u) = u_1^2 + \cdots + u_n^2$, which is obviously positive definite in $\Omega = \mathbb{R}^n$, and

$$
\begin{aligned}
V^*(u) &= 2\sum_{i=1}^{n} u_i(x) u_i'(x) = 2\sum_{i=1}^{n} u_i(x) \left[\sum_{j=1}^{n} a_{ij}(x) u_j(x) \right] \\
&= 2\sum_{i=1}^{n}\sum_{j=1}^{n} a_{ij}(x) u_i(x) u_j(x) = 2\sum_{i=1}^{n} a_{ii}(x) u_i^2(x),
\end{aligned}
$$

i.e., $V^*(u)$ is negative semidefinite. Therefore, the trivial solution of (17.3) is stable. If $a_{ii}(x) < 0$ for all $x \geq 0$, then $V^*(u)$ is negative definite and the trivial solution of (17.3) is asymptotically stable.

Example 28.8. Consider the Liénard equation (27.9), where the functions $f(y)$ and $g(y)$ are continuously differentiable for all $y \in \mathbb{R}$, $g(0) = 0$ and $yg(y) > 0$ for all $y \neq 0$, and for some positive constant k, $yF(y) > 0$ for $y \neq 0$, $-k < y < k$, where

$$
F(y) = \int_0^y f(t)dt.
$$

Equation (27.9) is equivalent to the system

$$
\begin{aligned}
u_1' &= u_2 - F(u_1) \\
u_2' &= -g(u_1).
\end{aligned}
\tag{28.8}
$$

A suitable Lyapunov function for the differential system (28.8) is the total energy function $V(u_1, u_2) = (1/2)u_2^2 + G(u_1)$, where

$$
G(u_1) = \int_0^{u_1} g(t)dt.
$$

Obviously, $V(u_1, u_2)$ is positive definite in $\Omega = \mathbb{R}^2$. Further, it is easy to find $V^*(u) = -g(u_1)F(u_1)$. Thus, from our hypotheses it follows that $V^*(u) \leq 0$ in the strip $\{(u_1, u_2) \in \mathbb{R}^2 : -k < u_1 < k, \ -\infty < u_2 < \infty\}$. Therefore, the trivial solution of (28.8) is stable.

Problems

28.1. Show that the function $V(u_1, u_2) = c_1 u_1^2 + c_2 u_1 u_2 + c_3 u_2^2$ in $\Omega = \mathbb{R}^2$ is positive definite if and only if $c_1 > 0$ and $c_2^2 - 4c_1 c_3 < 0$, and is negative definite if and only if $c_1 < 0$ and $c_2^2 - 4c_1 c_3 < 0$.

28.2. For each of the following systems construct a Lyapunov function of the form $c_1 u_1^2 + c_2 u_2^2$ to determine whether the trivial solution is stable, asymptotically stable, or unstable:

(i)
$$u_1' = -u_1 + e^{u_1} u_2$$
$$u_2' = -e^{u_1} u_1 - u_2.$$

(ii)
$$u_1' = -u_1^3 + u_1^2 u_2$$
$$u_2' = -u_2^3 - u_1^2 u_2.$$

(iii)
$$u_1' = 2u_2 - u_1(1 + 4u_2^2)$$
$$u_2' = (1/2)u_1 - u_2 + u_1^2 u_2.$$

(iv)
$$u_1' = 2u_2 - 2u_2 \sin u_1$$
$$u_2' = -u_1 - 4u_2 + u_1 \sin u_1.$$

(v)
$$u_1' = -u_1 + 2u_1^2 + u_2^2$$
$$u_2' = -u_2 + u_1 u_2.$$

(vi)
$$u_1' = -u_1 - u_2 - u_1^3$$
$$u_2' = u_1 - u_2 - u_2^3.$$

(vii)
$$u_1' = 2u_1 u_2 + u_1^3$$
$$u_2' = -u_1^2 + u_2^5.$$

(viii)
$$u_1' = u_1^3 - u_2$$
$$u_2' = u_1 + u_2^3.$$

28.3. Consider the system

$$u_1' = u_2 - u_1 f(u_1, u_2)$$
$$u_2' = -u_1 - u_2 f(u_1, u_2),$$

where $f(u_1, u_2)$ has a convergent power series expansion in $\Omega \subseteq \mathbb{R}^2$ containing the origin, and $f(0,0) = 0$. Show that the trivial solution of the above system is

(i)   stable if $f(u_1, u_2) \geq 0$ in some region around the origin;

(ii)   asymptotically stable if $f(u_1, u_2)$ is positive definite in some region around the origin;

(iii)   unstable if in every region around the origin there are points (u_1, u_2) such that $f(u_1, u_2) < 0$.

28.4. Use Problem 28.3 to determine whether the trivial solution is stable, asymptotically stable or unstable for each of the following systems:

(i)
$$u_1' = u_2 - u_1(e^{u_1} \sin^2 u_2)$$
$$u_2' = -u_1 - u_2(e^{u_1} \sin^2 u_2).$$

(ii)
$$u_1' = u_2 - u_1(u_1^4 + u_2^6 + 2u_1^2 u_2^2 \sin^2 u_1)$$
$$u_2' = -u_1 - u_2(u_1^4 + u_2^6 + 2u_1^2 u_2^2 \sin^2 u_1).$$

(iii)
$$u_1' = u_2 - u_1(u_2^3 \sin^2 u_1)$$
$$u_2' = -u_1 - u_2(u_2^3 \sin^2 u_1).$$

28.5. Consider the equation of *undamped simple pendulum* $y'' + \omega^2 \sin y = 0$, $-\pi/2 \leq y \leq \pi/2$. Prove by means of an appropriate Lyapunov function that its trivial solution is stable.

28.6. Prove by means of an appropriate Lyapunov function that the trivial solution of the Van der Pol equation (26.12) is asymptotically stable when $\mu < 0$.

Answers or Hints

28.1. Completing the squares, we have

$$V(u_1, u_2) = c_1 \left[\left(u_1 + \frac{c_2}{2c_1} u_2 \right)^2 + \frac{1}{4c_1^2} \left(4c_1 c_3 - c_2^2 \right) u_2^2 \right].$$

28.2. (i) Asymptotically stable. (ii) Stable. (iii) Stable. (iv) Stable. (v) Asymptotically stable. (vi) Asymptotically stable. (vii) Unstable. (viii) Unstable.

28.3. Let $V(u_1, u_2) = u_1^2 + u_2^2$, then $V^*(u_1, u_2) = -2f(u_1, u_2)(u_1^2 + u_2^2)$.

28.4. (i) Stable. (ii) Asymptotically stable. (iii) Unstable.

28.5. Write the equation in system form and then use Lyapunov's function $V(u_1, u_2) = g(u_1) + \frac{1}{2} u_2^2$, where $g(u_1) = \int_0^{u_1} \omega^2 \sin t \, dt$.

28.6. Use $V(u_1, u_2) = u_1^2 + u_2^2$.

Lecture 29
Lyapunov's Direct Method for Nonautonomous Systems

In this lecture we shall extend the method of Lyapunov functions to study the stability properties of the solutions of the differential system (15.1). For this, we shall assume that the function $g(x, u)$ is continuous for all $(x, u) \in [x_0, \infty) \times S_\rho$, $x_0 \geq 0$ and smooth enough so that the initial value problem (15.4) has a unique solution in $[x_0, \infty)$ for all $u^0 \in S_\rho$. Further, we shall assume that $g(x, 0) \equiv 0$ so that the differential system (15.1) admits the trivial solution. It is clear that a Lyapunov function for the system (15.1) must depend on x and u both, i.e., $V = V(x, u)$.

Definition 29.1. A real-valued function $V(x, u)$ defined in $[x_0, \infty) \times S_\rho$ is said to be *positive definite* if and only if $V(x, 0) \equiv 0$, $x \geq x_0$, and there exists a function $a(r) \in \mathcal{K}$ such that $a(r) \leq V(x, u)$, $\|u\| = r$, $(x, u) \in [x_0, \infty) \times S_\rho$. It is *negative definite* if $V(x, u) \leq -a(r)$.

Definition 29.2. A real-valued function $V(x, u)$ defined in $[x_0, \infty) \times S_\rho$ is said to be *decrescent* if and only if $V(x, 0) \equiv 0$, $x \geq x_0$ and there exists an h, $0 < h \leq \rho$ and a function $b(r) \in \mathcal{K}$ such that $V(x, u) \leq b(\|u\|)$ for $\|u\| < h$ and $x \geq x_0$.

Example 29.1. The function

$$V(x, u_1, u_2) = (1 + \sin^2 x)u_1^2 + (1 + \cos^2 x)u_2^2$$

is positive definite in $[0, \infty) \times \mathbb{R}^2$ since $V(x, 0, 0) \equiv 0$ and $a(r) = r^2 \in \mathcal{K}$ satisfies the inequality $a(r) \leq V(x, u_1, u_2)$. This function is also decrescent since $b(r) = 2r^2 \in \mathcal{K}$ satisfies $V(x, u_1, u_2) \leq b(r)$.

Example 29.2. The function

$$V(x, u_1, u_2) = u_1^2 + (1 + x^2)u_2^2$$

is positive definite in $[0, \infty) \times \mathbb{R}^2$ but not decrescent since it can be arbitrarily large for sufficiently small $\|u\|$.

Example 29.3. The function

$$V(x, u_1, u_2) = u_1^2 + \frac{1}{(1 + x^2)} u_2^2$$

R.P. Agarwal and D. O'Regan, *An Introduction to Ordinary Differential Equations*,
doi: 10.1007/978-0-387-71276-5_29, © Springer Science + Business Media, LLC 2008

is decrescent. However, since

$$V(x, 0, u_2) = \frac{1}{(1+x^2)} u_2^2 \to 0 \quad \text{as} \quad x \to \infty,$$

we cannot find a suitable function $a(r) \in \mathcal{K}$ such that $V(x, u_1, u_2) \geq a(\|u\|)$, i.e., it is not positive definite.

Now we shall assume that $V(x, u) \in C^{(1)}[[x_0, \infty) \times S_\rho, \mathbb{R}]$ so that the chain rule can be used, to obtain

$$\frac{dV(x, u)}{dx} = V^*(x, u) = \frac{\partial V}{\partial x} + \sum_{i=1}^{n} \frac{\partial V}{\partial u_i} \frac{du_i}{dx}.$$

Our interest is in the derivative of $V(x, u)$ along a solution $u(x) = u(x, x_0, u^0)$ of the differential system (15.1). Indeed, we have

$$V^*(x, u) = \frac{\partial V}{\partial x} + \sum_{i=1}^{n} \frac{\partial V}{\partial u_i} g_i(x, u) = \frac{\partial V}{\partial x} + \text{grad}\, V(x, u) \cdot g(x, u).$$

The following two theorems regarding the stability and asymptotic stability of the trivial solution of the differential system (15.1) are parallel to the results in the autonomous case.

Theorem 29.1.   If there exists a positive definite scalar function $V(x, u) \in C^{(1)}[[x_0, \infty) \times S_\rho, \mathbb{R}^+]$ (called a *Lyapunov function*) such that $V^*(x, u) \leq 0$ in $[x_0, \infty) \times S_\rho$, then the trivial solution of the differential system (15.1) is stable.

Theorem 29.2. If there exists a positive definite and decrescent scalar function $V(x, u) \in C^{(1)}[[x_0, \infty) \times S_\rho, \mathbb{R}^+]$ such that $V^*(x, u)$ is negative definite in $[x_0, \infty) \times S_\rho$, then the trivial solution of the differential system (15.1) is asymptotically stable.

Example 29.4.   Consider the DE $y'' + p(x)y = 0$, where $p(x) \geq \delta > 0$ and $p'(x) \leq 0$ for all $x \in [0, \infty)$. This is equivalent to the system

$$
\begin{aligned}
u_1' &= u_2 \\
u_2' &= -p(x)u_1.
\end{aligned}
\tag{29.1}
$$

For the differential system (29.1) we consider the scalar function $V(x, u_1, u_2) = p(x)u_1^2 + u_2^2$, which is positive definite in $[0, \infty) \times \mathbb{R}^2$. Further, since

$$V^*(x, u) = p'(x)u_1^2 + 2p(x)u_1 u_2 + 2u_2(-p(x)u_1) = p'(x)u_1^2 \leq 0$$

in $[0, \infty) \times \mathbb{R}^2$, the trivial solution of (29.1) is stable.

Example 29.5. Consider the system

$$u_1' = -a_{11}(x)u_1 - a_{12}(x)u_2$$
$$u_2' = a_{21}(x)u_1 - a_{22}(x)u_2, \tag{29.2}$$

where $a_{21}(x) = a_{12}(x)$, and $a_{11}(x) \geq \delta > 0$, $a_{22}(x) \geq \delta > 0$ for all $x \in [0, \infty)$. For the differential system (29.2) we consider the scalar function $V(x, u_1, u_2) = u_1^2 + u_2^2$ which is obviously positive definite and decrescent in $[0, \infty) \times \mathbb{R}^2$. Further, since

$$
\begin{aligned}
V^*(x, u) &= 2u_1(-a_{11}(x)u_1 - a_{12}(x)u_2) + 2u_2(a_{21}(x)u_1 - a_{22}(x)u_2) \\
&= -2a_{11}(x)u_1^2 - 2a_{22}(x)u_2^2 \leq -2\delta(u_1^2 + u_2^2)
\end{aligned}
$$

in $[0, \infty) \times \mathbb{R}^2$, the trivial solution of (29.2) is asymptotically stable.

We shall now formulate a result which provides sufficient conditions for the trivial solution of the differential system (15.1) to be unstable.

Theorem 29.3. If there exists a scalar function $V(x, u) \in C^{(1)}[[x_0, \infty) \times S_\rho, \mathbb{R}]$ such that

(i) $|V(x, u)| \leq b(\|u\|)$ for all $(x, u) \in [x_0, \infty) \times S_\rho$, where $b \in \mathcal{K}$;

(ii) for every $\delta > 0$ there exists an u^0 with $\|u^0\| < \delta$ such that $V(x_0, u^0) < 0$;

(iii) $V^*(x, u) \leq -a(\|u\|)$ for $(x, u) \in [x_0, \infty) \times S_\rho$, where $a \in \mathcal{K}$,

then the trivial solution of the differential system (15.1) is unstable.

Proof. Let the trivial solution of (15.1) be stable. Then for every $\epsilon > 0$ such that $\epsilon < \rho$, there exists a $\delta = \delta(\epsilon) > 0$ such that $\|u^0\| < \delta$ implies that $\|u(x)\| = \|u(x, x_0, u^0)\| < \epsilon$ for all $x \geq x_0$. Let u^0 be such that $\|u^0\| < \delta$ and $V(x_0, u^0) < 0$. Since $\|u^0\| < \delta$, we have $\|u(x)\| < \epsilon$. Hence, condition (i) gives

$$|V(x, u(x))| \leq b(\|u(x)\|) < b(\epsilon) \quad \text{for all} \quad x \geq x_0. \tag{29.3}$$

Now from condition (iii), it follows that $V(x, u(x))$ is a decreasing function, and therefore for every $x \geq x_0$, we have $V(x, u(x)) \leq V(x_0, u^0) < 0$. This implies that $|V(x, u(x))| \geq |V(x_0, u^0)|$. Hence, from condition (i) we get $\|u(x)\| \geq b^{-1}(|V(x_0, u^0)|)$.

From condition (iii) again, we have $V^*(x, u(x)) \leq -a(\|u(x)\|)$, and hence on integrating this inequality between x_0 and x, we obtain

$$V(x, u(x)) \leq V(x_0, u^0) - \int_{x_0}^x a(\|u(t)\|)dt.$$

However, since $\|u(x)\| \geq b^{-1}(|V(x_0, u^0)|)$, it is clear that $a(\|u(x)\|) \geq a(b^{-1}(|V(x_0, u^0)|))$. Thus, we have

$$V(x, u(x)) \leq V(x_0, u^0) - (x - x_0)a(b^{-1}(|V(x_0, u^0)|)).$$

But this shows that $\lim_{x\to\infty} V(x, u(x)) = -\infty$, which contradicts (29.3). Hence, the trivial solution of (15.1) is unstable. ∎

Example 29.6. Consider the system

$$\begin{aligned} u_1' &= a_{11}(x)u_1 - a_{12}(x)u_2 \\ u_2' &= a_{21}(x)u_1 + a_{22}(x)u_2, \end{aligned} \qquad (29.4)$$

where $a_{21}(x) = a_{12}(x)$, and $a_{11}(x) \geq \delta > 0$, $a_{22}(x) \geq \delta > 0$ for all $x \in [0, \infty)$. For the differential system (29.4) we consider the scalar function $V(x, u_1, u_2) = -(u_1^2 + u_2^2)$, and note that for all $(x, u) \in [0, \infty) \times \mathbb{R}^2$, $|V(x, u_1, u_2)| \leq (u_1^2 + u_2^2) = r^2 = b(r)$ for $(u_1, u_2) \neq (0, 0)$, $V(x, u_1, u_2) < 0$, and $V^*(x, u_1, u_2) = -2(a_{11}(x)u_1^2 + a_{22}(x)u_2^2) \leq -2\delta(u_1^2 + u_2^2) = -2\delta r^2 = -a(r)$. Thus, the conditions of Theorem 29.3 are satisfied and the trivial solution of (29.4) is unstable.

Our final result in this lecture gives sufficient conditions for the trivial solution of the differential system (15.1) to be uniformly stable.

Theorem 29.4.   If there exists a positive definite and decrescent scalar function $V(x, u) \in C^{(1)}[[x_0, \infty) \times S_\rho, \mathbb{R}^+]$ such that $V^*(x, u) \leq 0$ in $[x_0, \infty) \times S_\rho$, then the trivial solution of the differential system (15.1) is uniformly stable.

Proof.   Since $V(x, u)$ is positive definite and decrescent, there exist functions $a, b \in \mathcal{K}$ such that

$$a(\|u\|) \leq V(x, u) \leq b(\|u\|) \qquad (29.5)$$

for all $(x, u) \in [x_0, \infty) \times S_\rho$. For each ϵ, $0 < \epsilon < \rho$, we choose a $\delta = \delta(\epsilon) > 0$ such that $b(\delta) < a(\epsilon)$. We now claim that the trivial solution of (15.1) is uniformly stable, i.e., if $x_1 \geq x_0$ and $\|u(x_1)\| < \delta$, then $\|u(x)\| < \epsilon$ for all $x \geq x_1$. Suppose this is not true. Then there exists some $x_2 > x_1$ such that $x_1 \geq x_0$ and $\|u(x_1)\| < \delta$ imply that

$$\|u(x_2)\| = \epsilon. \qquad (29.6)$$

Integrating $V^*(x, u(x)) \leq 0$ from x_1 to x, we get $V(x, u(x)) \leq V(x_1, u(x_1))$, and hence for $x = x_2$, we have

$$\begin{aligned} a(\epsilon) = a(\|u(x_2)\|) &\leq V(x_2, u(x_2)) \leq V(x_1, u(x_1)) \\ &\leq b(\|u(x_1)\|) \leq b(\delta) < a(\epsilon). \end{aligned}$$

This contradicts relation (29.6), and therefore no such x_2 exists. Hence, the trivial solution of (15.1) is uniformly stable. $\qquad$ ■

Corollary 29.5. If there exists a scalar function $V(x, u) \in C[[x_0, \infty) \times S_\rho, \mathbb{R}^+]$ such that the inequality (29.5) holds, and $V(x, u(x))$ is nonincreasing in x for every solution $u(x)$ of (15.1) with $\|u(x)\| < \rho$, then the trivial solution of (15.1) is uniformly stable.

Example 29.7. For the differential system (29.2) once again we consider the scalar function $V(x, u) = u_1^2 + u_2^2$ and note that the inequality (29.5) with $a(r) = b(r) = r^2$ is satisfied. Further, since $V^*(x, u(x)) \leq 0$ for all solutions of (29.2), the trivial solution of (29.2) is uniformly stable.

To conclude, we remark that the main drawback of Lyapunov's direct method is that there is no sufficiently general constructive method for finding the function $V(x, u)$. Nevertheless, for a series of important classes of differential systems such a construction is possible.

Problems

29.1. Show the following:

(i) $(u_1^2 + u_2^2) \cos^2 x$ is decrescent.

(ii) $u_1^2 + e^x u_2^2$ is positive definite but not decrescent.

(iii) $u_1^2 + e^{-x} u_2^2$ is decrescent.

(iv) $(1 + \cos^2 x + e^{-2x})(u_1^4 + u_2^4)$ is positive definite and decrescent.

29.2. For the DE

$$y' = (\sin \ln x + \cos \ln x - 1.25)y, \qquad (29.7)$$

consider the function

$$V(x, y) = y^2 \exp(2(1.25 - \sin \ln x)x).$$

Show the following:

(i) $V(x, y)$ is positive definite but not decrescent.

(ii) The trivial solution of (29.7) is stable.

29.3. Show that the trivial solution of the differential system

$$u_1' = p(x)u_2 + q(x)u_1(u_1^2 + u_2^2)$$
$$u_2' = -p(x)u_1 + q(x)u_2(u_1^2 + u_2^2),$$

where p, $q \in C[0, \infty)$ is stable if $q(x) \leq 0$, asymptotically stable if $q(x) \leq \delta < 0$, and unstable if $q(x) \geq \delta > 0$.

29.4. If there exists a positive definite scalar function $V(x, u) \in C^{(1)}[[x_0, \infty) \times S_\rho, \mathbb{R}^+]$ such that $V^*(x, u) \le -a(V(x, u))$, where $a \in \mathcal{K}$ in $[x_0, \infty) \times S_\rho$, then show that the trivial solution of the differential system (15.1) is asymptotically stable.

29.5. For the DE

$$y' = (x \sin x - 2x)y, \tag{29.8}$$

consider the function

$$V(x, y) = y^2 \exp\left(\int_0^x (2t - t \sin t)dt\right).$$

Show the following:

(i) $V(x, y)$ is positive definite but not decrescent.

(ii) $V^*(x, y) \le -\lambda V(x, y)$ for all $x \ge \lambda > 0$.

(iii) The trivial solution of (29.8) is asymptotically stable.

Answers or Hints

29.1. Verify directly.

29.2. Verify directly.

29.3. Use functions $V(x, u_1, u_2) = u_1^2 + u_2^2$ and $V(x, u_1, u_2) = -(u_1^2 + u_2^2)$.

29.4. The stability of the trivial solution of (15.1) follows from Theorem 29.1. Hence, given $\epsilon > 0$ $(0 < \epsilon < \rho)$ there exists a $\delta = \delta(\epsilon) > 0$ such that $\|u^0\| < \delta$ implies that $\|u(x, x_0, u^0)\| < \epsilon$ for all $x \ge x_0$. Since $V^*(x, u) \le -a(V(x, u))$ along the trajectory through (x_0, u^0), we have for all $x \ge x_0$

$$\int_{V(x_0,u^0)}^{V(x,u)} \frac{dV}{a(V)} \le -(x - x_0).$$

Thus, as $x \to \infty$, the integral tends to $-\infty$. But this is possible only if $V(x, u) \to 0$ as $x \to \infty$. Now use the fact that $V(x, u)$ is positive definite.

29.5. Verify directly.

Lecture 30
Higher-Order Exact
and Adjoint Equations

The concept of exactness which was discussed for the first-order DEs in Lecture 3 can be extended to higher-order DEs. The nth-order DE (1.5) is called *exact* if the function $F(x, y, y', \ldots, y^{(n)})$ is a derivative of some differential expression of $(n-1)$th order, say, $\phi(x, y, y', \ldots, y^{(n-1)})$. Thus, in particular, the second-order DE (6.1) is exact if

$$p_0(x)y'' + p_1(x)y' + p_2(x)y = (p(x)y' + q(x)y)', \qquad (30.1)$$

where the functions $p(x)$ and $q(x)$ are differentiable in J.

Expanding (30.1), we obtain

$$p_0(x)y'' + p_1(x)y' + p_2(x)y = p(x)y'' + (p'(x) + q(x))y' + q'(x)y,$$

and hence it is necessary that $p_0(x) = p(x)$, $p_1(x) = p'(x) + q(x)$, and $p_2(x) = q'(x)$ for all $x \in J$. These equations in turn imply that

$$p_0''(x) - p_1'(x) + p_2(x) = 0. \qquad (30.2)$$

Thus, the DE (6.1) is exact if and only if condition (30.2) is satisfied.

Similarly, the second-order nonhomogeneous DE (6.6) is exact if the expression $p_0(x)y'' + p_1(x)y' + p_2(x)y$ is exact, and in such a case (6.6) is

$$[p_0(x)y' + (p_1(x) - p_0'(x))y]' = r(x). \qquad (30.3)$$

On integrating (30.3), we find

$$p_0(x)y' + (p_1(x) - p_0'(x))y = \int^x r(t)dt + c, \qquad (30.4)$$

which is a first-order linear DE and can be integrated to find the general solution of (6.6).

Example 30.1. For the DE

$$x^2 y'' + xy' - y = x^4, \quad x > 0$$

we have $p_0''(x) - p_1'(x) + p_2(x) = 2 - 1 - 1 = 0$, and hence it is an exact DE. Using (30.4), we get

$$x^2 y' - xy = \frac{1}{5}x^5 + c,$$

R.P. Agarwal and D. O'Regan, *An Introduction to Ordinary Differential Equations*, doi: 10.1007/978-0-387-71276-5_30, © Springer Science + Business Media, LLC 2008

which is a first-order linear DE. The general solution of this DE is

$$y(x) = \frac{1}{15}x^4 + \frac{c_1}{x} + c_2 x.$$

If the DE (6.1) is not exact, then we may seek an integrating factor $z(x)$ that will make it exact. Such an integrating factor must satisfy the condition

$$(p_0(x)z(x))'' - (p_1(x)z(x))' + p_2(x)z(x) = 0. \tag{30.5}$$

Equation (30.5) is a second-order DE in $z(x)$ and it can be written as

$$q_0(x)z'' + q_1(x)z' + q_2(x)z = 0, \tag{30.6}$$

where

$$q_0(x) = p_0(x), \quad q_1(x) = 2p_0'(x) - p_1(x), \quad q_2(x) = p_0''(x) - p_1'(x) + p_2(x). \tag{30.7}$$

Equation (30.5), or equivalently (30.6), is in fact the *adjoint equation* of (6.1). To show this, we note that the DE (6.1) is equivalent to the system

$$\begin{bmatrix} u_1' \\ u_2' \end{bmatrix} = \begin{bmatrix} 0 & 1/p_0(x) \\ -p_2(x) & (p_0'(x) - p_1(x))/p_0(x) \end{bmatrix} \begin{bmatrix} u_1 \\ u_2 \end{bmatrix},$$

and its adjoint system is

$$\begin{bmatrix} v_1' \\ v_2' \end{bmatrix} = \begin{bmatrix} 0 & p_2(x) \\ -1/p_0(x) & -(p_0'(x) - p_1(x))/p_0(x) \end{bmatrix} \begin{bmatrix} v_1 \\ v_2 \end{bmatrix},$$

which is the same as

$$v_1' = p_2(x)v_2 \tag{30.8}$$

$$p_0(x)v_2' = -v_1 - (p_0'(x) - p_1(x))v_2. \tag{30.9}$$

Now using (30.8) in (30.9), we obtain

$$p_0(x)v_2'' + p_0'(x)v_2' = -p_2(x)v_2 - (p_0''(x) - p_1'(x))v_2 - (p_0'(x) - p_1(x))v_2',$$

or

$$p_0(x)v_2'' + (2p_0'(x) - p_1(x))v_2' + (p_0''(x) - p_1'(x) + p_2(x))v_2 = 0,$$

which is exactly the same as (30.6).

Obviously, relations (30.7) can be rewritten as

$$p_0(x) = q_0(x), \quad p_0'(x) - p_1(x) = q_1(x) - q_0'(x)$$
$$2p_2(x) - p_1'(x) = 2q_2(x) - q_1'(x) \tag{30.10}$$

and hence when p_i's and q_i's are interchanged these equations are unaltered; thus the relation between equations (6.1) and (30.6) is of a reciprocal nature and each equation is the adjoint of the other.

When an equation and its adjoint are the same, it is said to be *self-adjoint*. Thus, the DE (6.1) is self-adjoint if $p_0(x) = q_0(x)$, $p_1(x) = q_1(x)$, $p_2(x) = q_2(x)$. In such a situation, relations (30.7) give

$$p_1(x) = p_0'(x). \tag{30.11}$$

Thus, the self-adjoint equation takes the form

$$p_0(x)y'' + p_0'(x)y' + p_2(x)y = 0,$$

which is the same as

$$(p_0(x)y')' + p_2(x)y = 0. \tag{30.12}$$

Further, any self-adjoint equation can be written in the form (30.12).

The condition (30.11) shows that the DE (6.1) will become self-adjoint after multiplication by a function $\sigma(x)$ that satisfies the relation

$$\sigma(x)p_1(x) = (\sigma(x)p_0(x))',$$

which is the same as

$$\frac{(\sigma(x)p_0(x))'}{\sigma(x)p_0(x)} = \frac{p_1(x)}{p_0(x)}. \tag{30.13}$$

On integrating (30.13), we easily find the function $\sigma(x)$ which appears as

$$\sigma(x) = \frac{1}{p_0(x)} \exp\left(\int^x \frac{p_1(t)}{p_0(t)} dt\right). \tag{30.14}$$

Thus, with this choice of $\sigma(x)$ the DE

$$\sigma(x)p_0(x)y'' + \sigma(x)p_1(x)y' + \sigma(x)p_2(x)y = 0 \tag{30.15}$$

is a self-adjoint equation.

Since (30.15) is self-adjoint its solutions are also its integrating factors. But since (30.15) is the same as (6.1) multiplied by $\sigma(x)$, the solutions of (6.1) are the solutions of (30.15). Hence the statement "the solutions of (30.15) are the integrating factors of (30.15)" is equivalent to saying that "$\sigma(x)$ times the solutions of (6.1) are the integrating factors of (6.1)." But, since the integrating factors of (6.1) are the solutions of its adjoint equation (30.6), i.e., $z(x) = \sigma(x)y(x)$ it follows that

$$\frac{z(x)}{y(x)} = \sigma(x) = \frac{1}{p_0(x)} \exp\left(\int^x \frac{p_1(t)}{p_0(t)} dt\right), \tag{30.16}$$

where $y(x)$ and $z(x)$ are solutions of (6.1) and (30.6), respectively.

Since (6.1) is adjoint of (30.6) it is also clear from (30.16) that

$$\frac{y(x)}{z(x)} = \frac{1}{q_0(x)} \exp\left(\int^x \frac{q_1(t)}{q_0(t)} dt\right). \tag{30.17}$$

Example 30.2. The DE

$$xy'' + 2y' + a^2 xy = 0, \quad x > 0$$

is not exact since $p_0'' - p_1' + p_2 = a^2 x \neq 0$. Its adjoint equation is

$$(xz)'' - 2z' + a^2 xz = x(z'' + a^2 z) = 0,$$

whose general solution can be written as $z(x) = c_1 \cos ax + c_2 \sin ax$. The function

$$\sigma(x) = \frac{1}{x} \exp\left(\int^x \frac{2}{t} dt\right) = x.$$

Therefore, the general solution of the given DE is

$$y(x) = \frac{z(x)}{x} = \frac{c_1 \cos ax}{x} + \frac{c_2 \sin ax}{x}.$$

Now let the linear operator

$$\mathcal{P}_2 = p_0(x) \frac{d^2}{dx^2} + p_1(x) \frac{d}{dx} + p_2(x)$$

and its adjoint

$$\mathcal{Q}_2 = q_0(x) \frac{d^2}{dx^2} + q_1(x) \frac{d}{dx} + q_2(x),$$

so that the DEs (6.1) and (30.6), respectively, can be written as

$$\mathcal{P}_2[y] = p_0(x) \frac{d^2 y}{dx^2} + p_1(x) \frac{dy}{dx} + p_2(x)y = 0 \tag{30.18}$$

and

$$\mathcal{Q}_2[z] = q_0(x) \frac{d^2 z}{dx^2} + q_1(x) \frac{dz}{dx} + q_2(x)z = 0. \tag{30.19}$$

Multiplying (30.18) by z and (30.19) by y and subtracting, we get

$$zP_2[y] - yQ_2[z]$$
$$= z(p_0(x)y'' + p_1(x)y' + p_2(x)y) - y(q_0(x)z'' + q_1(x)z' + q_2(x)z)$$
$$= z(p_0(x)y'' + p_1(x)y' + p_2(x)y) - y((p_0(x)z)'' - (p_1(x)z)' + p_2(x)z)$$
$$= (p_0(x)z)y'' - (p_0(x)z)''y + (p_1(x)z)y' + (p_1(x)z)'y$$
$$= \frac{d}{dx}[(p_0(x)z)y' - (p_0(x)z)'y] + \frac{d}{dx}(p_1(x)yz)$$
$$= \frac{d}{dx}[p_0(x)(zy' - z'y) + (p_1(x) - p_0'(x))yz].$$
$$(30.20)$$

Relation (30.20) is known as *Lagrange's identity.* Further, the expression in square brackets on the right side is called the *bilinear concomitant* of the functions y and z.

Integrating (30.20), we find

$$\int_\alpha^\beta (zP_2[y] - yQ_2[z])dx = \left[p_0(x)(zy' - z'y) + (p_1(x) - p_0'(x))yz\right]\Big|_\alpha^\beta .$$
$$(30.21)$$

The relation (30.21) is called *Green's identity.*

In the special case when the operator P_2 is self-adjoint, i.e., $p_0'(x) = p_1(x)$, the Lagrange identity (30.20) reduces to

$$zP_2[y] - yP_2[z] = \frac{d}{dx}[p_0(x)(zy' - z'y)]. \qquad (30.22)$$

Thus, if y and z both are solutions of the DE (30.12), then (30.22) gives

$$p_0(x)W(y, z)(x) = \text{constant}. \qquad (30.23)$$

Further, in this case Green's identity (30.21) becomes

$$\int_\alpha^\beta (zP_2[y] - yP_2[z])dx = \left[p_0(x)(zy' - z'y)\right]\Big|_\alpha^\beta . \qquad (30.24)$$

Problems

30.1. Verify that the following DEs are exact and find their general solutions:

(i) $y'' + xy' + y = 0.$

(ii) $yy'' + y'^2 = 0$.

(iii) $x^2 y'' + (1 + 4x)y' + 2y = 0$, $x > 0$.

(iv) $(1 - x)y'' + xy' + y = 0$, $x \neq 1$.

(v) $(x + 2xy)y'' + 2xy'^2 + 2y' + 4yy' = 0$, $x > 0$.

(vi) $\sin x \, y''' + \cos x \, y'' + \sin x \, y' + \cos x \, y = 0$, $(0 < x < \pi)$.

30.2. If $p_1(x) = 2p_0'(x)$, $p_2(x) = p_0''(x)$, show that

$$p_0(x)y'' + p_1(x)y' + p_2(x)y = (p_0(x)y)''$$

and hence find the general solution of the DE

$$(x^2 + 3)y'' + 4xy' + 2y = e^{-x}.$$

30.3. Show that the DE

$$p_0(x)y^{(n)} + p_1(x)y^{(n-1)} + \cdots + p_n(x)y = 0 \qquad (30.25)$$

is exact if and only if

$$p_0^{(n)}(x) - p_1^{(n-1)}(x) + \cdots + (-1)^n p_n(x) = 0.$$

30.4. Show that the Euler equation (18.17) is exact if and only if

$$p_0 - \frac{p_1}{n} + \frac{p_2}{n(n-1)} - \cdots + (-1)^n \frac{p_n}{n!} = 0.$$

30.5. Transform the following equations to their self-adjoint forms:

(i) $xy'' - y' + x^3 y = 0$.

(ii) $x^2 y'' + xy' + (x^2 - 1)y = 0$.

(iii) $xy'' + (1 - x)y' + y = 0$.

30.6. Solve the following equations with the help of their respective adjoint equations:

(i) $xy'' + (2x - 1)y' + (x - 1)y = x^2 + 2x - 2$.

(ii) $x^2 y'' + (3x^2 + 4x)y' + (2x^2 + 6x + 2)y = 0$.

30.7. If $z_1(x)$ and $z_2(x)$ are linearly independent solutions of (30.6), show that $y_1(x) = z_1(x)/\sigma(x)$ and $y_2(x) = z_2(x)/\sigma(x)$ are linearly independent solutions of (6.1).

30.8. Show that the DE (6.19) has two solutions whose product is a constant provided $2p_1(x)p_2(x) + p_2'(x) = 0$. Hence, solve the DE

$$(x + 1)x^2 y'' + xy' - (x + 1)^3 y = 0.$$

30.9. Show that a necessary and sufficient condition for transforming (6.19) into an equation with constant coefficients by a change of the independent variable

$$\phi = \phi(x) = \int^x [p_2(t)]^{1/2} dt$$

is that the function $(p_2'(x) + 2p_1(x)p_2(x))/p_2^{3/2}(x)$ is a constant.

30.10. Show that if $z(x)$ is an integrating factor of the DE (30.25); then $z(x)$ satisfies the adjoint DE

$$q_0(x)z^{(n)} - q_1(x)z^{(n-1)} + \cdots + (-1)^n q_n(x)z = 0,$$

where

$$q_k(x) = \sum_{i=0}^{k} (-1)^i \binom{n-k+i}{i} p_{k-i}^{(i)}(x)$$

$$= p_k(x) - \binom{n-k+1}{1} p_{k-1}'(x) + \binom{n-k+2}{2} p_{k-2}''(x) - \cdots,$$

$$k = 0, 1, \ldots, n.$$

Answers or Hints

30.1. (i) $y'' + xy' + y = (y' + xy)'$, $y = c_1 \exp(-x^2/2) + c_2 \exp(-x^2/2) \times \int^x \exp(t^2/2) dt$. (ii) $yy'' + (y')^2 = (yy')'$, $y^2 = c_1 x + c_2$. (iii) $x^2 y'' + (1 + 4x)y' + 2y = (x^2 y' + (1 + 2x)y)'$, $y = (c_1/x^2) \exp(1/x) + (c_2/x^2) \exp(1/x) \times \int^x \exp(-1/t) dt$. (iv) $(1-x)y'' + xy' + y = ((1-x)y' + (1+x)y)'$, $y = c_1 e^x (1 - x)^2 + c_2 e^x (1-x)^2 \int^x [e^{-t}/(1-t)^3] dt$. (v) $(x + 2xy)y'' + 2x(y')^2 + 2y' + 4yy' = (xy(y+1))''$, $xy(y+1) = c_1 x + c_2$. (vi) $\sin xy''' + \cos xy'' + \sin xy' + \cos xy = (\sin xy'' + \sin xy)'$, $y = c_1 \cos x + c_2 \sin x + c_3(-x \cos x + \sin x \ln \sin x)$.

30.2. $y = (c_1 + c_2 x + e^{-x})/(x^2 + 3)$.

30.3. $\sum_{k=0}^{n} p_k(x)y^{(n-k)} = (\sum_{k=0}^{n-1} q_k(x)y^{(n-1-k)})'$ gives the relations $p_0 = q_0$, $p_k = q_k + q_{k-1}'$, $k = 1, 2, \ldots, n-1$, $p_n = q_{n-1}'$, which are equivalent to $\sum_{k=0}^{n} (-1)^k p_k^{(n-k)}(x) = 0$.

30.4. Use Problem 30.3.

30.5. (i) $\sigma(x) = 1/x^2$, $(1/x)y'' - (1/x^2)y' + xy = 0$. (ii) $\sigma(x) = 1/x$, $xy'' + y' + [x - (1/x)]y = 0$. (iii) $\sigma(x) = e^{-x}$, $xe^{-x}y'' + (1 - x)e^{-x}y' + e^{-x}y = 0$.

30.6. (i) $e^{-x}(c_1 + c_2 x^2) + 1 + x$. (ii) $e^{-x}(c_1 + c_2 e^{-x})/x^2$.

30.7. If $z_1(x)$, $z_2(x)$ are linearly independent solutions of (30.6), then $y_1(x) = z_1(x)/\sigma(x)$, $y_2(x) = z_2(x)/\sigma(x)$ are solutions of (6.1). Drive $W(y_1, y_2)(x) = W(z_1, z_2)(x)/\sigma^2(x)$.

30.8. Use $y(x) = c/y_1(x)$ in (6.19) to get the relation $p_2(x)y_1^2 + y'^2 = 0$. Differentiation of this relation and the fact that $y_1(x)$ is a solution gives $p_2'(x)y_1^2 - 2p_1(x)y_1'^2 = 0$, $c_1 x e^x + c_2 e^{-x}/x$.

30.9. Verify directly.

30.10. If $z(x)$ is an integrating factor of (30.25), then it is necessary that

$$zp_0 y^{(n)} + zp_1 y^{(n-1)} + \cdots + zp_n y = \left(q_0 y^{(n-1)} + q_1 y^{(n-2)} + \cdots + q_{n-1} y\right)'$$
$$= q_0 y^{(n)} + (q_1 + q_0')y^{(n-1)} + (q_2 + q_1')y^{(n-2)} + \cdots + (q_{n-1} + q_{n-2}')y' + q_{n-1}'y,$$

which implies that $zp_0 = q_0$, $zp_k = q_k + q_{k-1}'$, $1 \le k \le n-1$, $zp_n = q_{n-1}$ and hence

$$(zp_0)^{(n)} - (zp_1)^{(n-1)} + (zp_2)^{(n-2)} - \cdots + (-1)^n (zp_n) = 0.$$

Lecture 31
Oscillatory Equations

In this lecture we shall consider the following second-order linear DE

$$(p(x)y')' + q(x)y = 0 \tag{31.1}$$

and its special case

$$y'' + q(x)y = 0, \tag{31.2}$$

where the functions $p, \ q \in C(J)$, and $p(x) > 0$ for all $x \in J$. By a solution of (31.1) we mean a nontrivial function $y \in C^{(1)}(J)$ and $py' \in C^{(1)}(J)$. A solution $y(x)$ of (31.1) is said to be *oscillatory* if it has no last zero, i.e., if $y(x_1) = 0$, then there exists an $x_2 > x_1$ such that $y(x_2) = 0$. Equation (31.1) itself is said to be oscillatory if every solution of (31.1) is oscillatory. A solution $y(x)$ which is not oscillatory is called *nonoscillatory*. For example, the DE $y'' + y = 0$ is oscillatory, whereas $y'' - y = 0$ is nonoscillatory in $J = [0, \infty)$.

From the practical point of view the following result is fundamental.

Theorem 31.1 (Sturm's Comparison Theorem). If $\alpha, \ \beta \in J$ are the consecutive zeros of a nontrivial solution $y(x)$ of (31.2), and if $q_1(x)$ is continuous and $q_1(x) \geq q(x)$, $q_1(x) \not\equiv q(x)$ in $[\alpha, \beta]$, then every nontrivial solution $z(x)$ of the DE

$$z'' + q_1(x)z = 0 \tag{31.3}$$

has a zero in (α, β).

Proof. Multiplying (31.2) by $z(x)$ and (31.3) by $y(x)$ and subtracting, we obtain

$$z(x)y''(x) - y(x)z''(x) + (q(x) - q_1(x))y(x)z(x) = 0,$$

which is the same as

$$(z(x)y'(x) - y(x)z'(x))' + (q(x) - q_1(x))y(x)z(x) = 0.$$

Since $y(\alpha) = y(\beta) = 0$, an integration yields

$$z(\beta)y'(\beta) - z(\alpha)y'(\alpha) + \int_{\alpha}^{\beta}(q(x) - q_1(x))y(x)z(x)dx = 0. \tag{31.4}$$

R.P. Agarwal and D. O'Regan, *An Introduction to Ordinary Differential Equations*, doi: 10.1007/978-0-387-71276-5_31, © Springer Science + Business Media, LLC 2008

From the linearity of (31.2) we can assume that $y(x) > 0$ in (α, β), then $y'(\alpha) > 0$ and $y'(\beta) < 0$. Thus, from (31.4) it follows that $z(x)$ cannot be of fixed sign in (α, β), i.e., it has a zero in (α, β). ∎

Corollary 31.2. If $q(x) \geq (1 + \epsilon)/4x^2$, $\epsilon > 0$ for all $x > 0$, then the DE (31.2) is oscillatory in $J = (0, \infty)$.

Proof. For $\epsilon > 0$, all nontrivial solutions of the DE

$$y'' + \frac{\epsilon}{4}y = 0 \tag{31.5}$$

are oscillatory. Let $t = e^x$ in (31.5), to obtain

$$t^2 \frac{d^2y}{dt^2} + t\frac{dy}{dt} + \frac{\epsilon}{4}y = 0. \tag{31.6}$$

Now using the substitution $y = z/\sqrt{t}$ in (31.6), we find

$$\frac{d^2z}{dt^2} + \frac{1 + \epsilon}{4t^2}z = 0. \tag{31.7}$$

Since $z(t) = e^{x/2}y(e^x)$ the equation (31.7) is also oscillatory. Therefore, from Theorem 31.1 between any two zeros of a solution of (31.7) there is a zero of every solution of (31.2); i.e., DE (31.2) is oscillatory in $J = (0, \infty)$. ∎

Example 31.1. Obviously the DE $y'' = 0$ is nonoscillatory. Thus, if the function $q(x) \leq 0 \ (\not\equiv 0)$ in J, Theorem 31.1 immediately implies that each solution of the DE (31.2) cannot have more than one zero in J. Thus, in particular, the DE $y'' - x^2y = 0$ is nonoscillatory in $J = \mathbb{R}$.

Example 31.2. Obviously the DE $y'' + y = 0$ is oscillatory. Thus, by Theorem 31.1 it follows that the DE $y'' + (1 + x)y = 0$ is also oscillatory in $J = [0, \infty)$.

Example 31.3. From Corollary 31.2 it is obvious that each solution of the DE $y'' + (c/x^2)y = 0$ has an infinite number of positive zeros if $c > 1/4$, and only a finite number of zeros if $c < 1/4$.

Example 31.4. Using the substitution $y = u/\sqrt{x}$ it is easy to see that the Bessel DE

$$x^2y'' + xy' + (x^2 - a^2)y = 0 \tag{31.8}$$

can be transformed into a simple DE

$$u'' + \left(1 + \frac{1 - 4a^2}{4x^2}\right)u = 0. \tag{31.9}$$

We will use Theorem 31.1 to find the behavior of zeros of the solutions $u(x)$ of the DE (31.9) in the interval $J = (0, \infty)$. For this, three cases are to be discussed:

Case 1. If $0 \le a < 1/2$, then $1 + [(1 - 4a^2)/4x^2] > 1$ and hence from Theorem 31.1 it follows that every solution $u(x)$ of (31.9) must vanish at least once between any two zeros of a nontrivial solution of $z'' + z = 0$. Since for a given p, $\sin(x - p)$ is a solution of $z'' + z = 0$, we find that the zeros of this solution are $p,\ p \pm \pi,\ p \pm 2\pi, \ldots$. Thus, every solution $u(x)$ of (31.9) has at least one zero in every subinterval of the positive x-axis of length π, i.e., the distance between successive zeros of $u(x)$ does not exceed π.

Case 2. If $a = 1/2$, then the DE (31.9) reduces to $u'' + u = 0$ and hence the zeros of every solution $u(x)$ of (31.9) are equally spaced by the distance π on the positive x-axis.

Case 3. If $a > 1/2$, then $1 + [(1 - 4a^2)/4x^2] < 1$ and comparison with $z'' + z = 0$ implies that every solution $u(x)$ of (31.9) has at most one zero in any subinterval of the positive x-axis of length π. To prove the existence of zeros we reason as follows: For any fixed a, $(1 - 4a^2)/4x^2 \to 0$ as $x \to \infty$. Hence, there exists a $x_0 > 0$ such that $1 + [(1 - 4a^2)/4x^2] > 1/2$ whenever $x > x_0$, and now comparison with $z'' + (1/2)z = 0$ implies that every solution $u(x)$ of (31.9) has infinitely many zeros and that the distance between successive zeros eventually becomes less than $\sqrt{2}\pi$.

We leave the proof of the fact that the distance between the successive zeros of every solution of (31.9) for every a tends to π (Problem 31.4).

Next we shall extend Sturm's comparison theorem for the DE (31.1). For this, we need the following lemma.

Lemma 31.3 (Picone's Identity). Let the functions y, z, py', $p_1 z'$ be differentiable and $z(x) \ne 0$ in J. Then the following identity holds:

$$\left[\frac{y}{z}(zpy' - yp_1 z') \right]' = y(py')' - \frac{y^2}{z}(p_1 z')' + (p - p_1)y'^2 + p_1 \left(y' - \frac{y}{z}z' \right)^2.$$
$$(31.10)$$

Proof. Expanding the left side, we have

$$\frac{y}{z}(z(py')' + z'py' - y(p_1 z')' - y'p_1 z') + \left(\frac{y'}{z} - \frac{y}{z^2}z' \right)(zpy' - yp_1 z')$$

$$= y(py')' - \frac{y^2}{z}(p_1 z')' + py'^2 - \frac{2yy'p_1 z'}{z} + \frac{y^2 p_1 z'^2}{z^2}$$

$$= y(py')' - \frac{y^2}{z}(p_1 z')' + (p - p_1)y'^2 + p_1 \left(y' - \frac{y}{z}z' \right)^2. \quad \blacksquare$$

Theorem 31.4 (Sturm–Picone's Theorem). If α, $\beta \in J$ are the consecutive zeros of a nontrivial solution $y(x)$ of (31.1), and if

$p_1(x)$, $q_1(x)$ are continuous and $0 < p_1(x) \le p(x)$, $q_1(x) \ge q(x)$ in $[\alpha, \beta]$, then every nontrivial solution $z(x)$ of the DE

$$(p_1(x)z')' + q_1(x)z = 0 \qquad (31.11)$$

has a zero in $[\alpha, \beta]$.

Proof. Let $z(x) \neq 0$ in $[\alpha, \beta]$, then Lemma 31.3 is applicable and from (31.10) and the DEs (31.1) and (31.11), we find

$$\left[\frac{y}{z}(zpy' - yp_1z') \right]' = (q_1 - q)y^2 + (p - p_1)y'^2 + p_1 \left(y' - \frac{y}{z}z' \right)^2.$$

Integrating the above identity and using $y(\alpha) = y(\beta) = 0$, we obtain

$$\int_\alpha^\beta \left[(q_1 - q)y^2 + (p - p_1)y'^2 + p_1 \left(y' - \frac{y}{z}z' \right)^2 \right] dx = 0,$$

which is a contradiction unless $q_1(x) \equiv q(x)$, $p_1(x) \equiv p(x)$ and $y' - (y/z)z' \equiv 0$. The last identity is the same as $d(y/z)/dx \equiv 0$, and hence $y(x)/z(x) \equiv$ constant. However, since $y(\alpha) = 0$ this constant must be zero, and so $y(x)/z(x) \equiv 0$, or $y(x) \equiv 0$. This contradiction implies that z must have a zero in $[\alpha, \beta]$. ∎

Corollary 31.5 (Sturm's Separation Theorem). If $y_1(x)$ and $y_2(x)$ are two linearly independent solutions of the DE (31.1) in J, then their zeros are interlaced, i.e., between two consecutive zeros of one there is exactly one zero of the other.

Proof. Since $y_1(x)$ and $y_2(x)$ cannot have common zeros, Theorem 31.4 ($p_1(x) \equiv p(x)$, $q_1(x) \equiv q(x)$) implies that the solution $y_2(x)$ has at least one zero between two consecutive zeros of $y_1(x)$. Interchanging $y_1(x)$ and $y_2(x)$ we see that $y_2(x)$ has at most one zero between two consecutive zeros of $y_1(x)$. ∎

Example 31.5. It is easy to see that the functions $y_1(x) = c_1 \cos x + c_2 \sin x$ and $y_2(x) = c_3 \cos x + c_4 \sin x$ are linearly independent solutions of the DE $y'' + y = 0$ if and only if $c_1c_4 - c_2c_3 \neq 0$. Thus, from Corollary 31.5 it follows that these functions $y_1(x)$ and $y_2(x)$ have alternating zeros.

In a finite interval $J = [\alpha, \beta]$ the DE (31.1) can have at most finite number of zeros, and this we shall prove in the following result.

Theorem 31.6. The only solution of the DE (31.1) which vanishes infinitely often in $J = [\alpha, \beta]$ is the trivial solution.

Proof. We assume that the solution $y(x)$ of the DE (31.1) has an infinite number of zeros in J. The set of zeros will then have a limit point $x^* \in J$, and there will exist a sequence $\{x_m\}$ of zeros converging to x^* with

$x_m \neq x^*$, $m = 0, 1, \ldots$. We shall show that $y(x^*) = y'(x^*) = 0$, then from the uniqueness of solutions it will follow that $y(x) \equiv 0$ in J. For this, the continuity of the solution $y(x)$ implies that $y(x^*) = \lim_{m \to \infty} y(x_m) = 0$. Next from the differentiability of the solution $y(x)$, we have

$$y'(x^*) = \lim_{m \to \infty} \frac{y(x_m) - y(x^*)}{x_m - x^*} = 0. \quad \blacksquare$$

The final result in this lecture gives an easier test for the DE (31.1) to be oscillatory in $J = (0, \infty)$.

Theorem 31.7 (Leighton's Oscillation Theorem). If $\int^\infty (1/p(x))dx = \infty$ and $\int^\infty q(x)dx = \infty$, then the DE (31.1) is oscillatory in $J = (0, \infty)$.

Proof. Let $y(x)$ be a nonoscillatory solution of the DE (31.1) which we assume to be positive in $[x_0, \infty)$, where $x_0 > 0$. Then from Problem 31.7 the Riccati equation

$$z' + q(x) + \frac{z^2}{p(x)} = 0 \qquad (31.12)$$

has a solution $z(x)$ in $[x_0, \infty)$. This solution obviously satisfies the equation

$$z(x) = z(x_0) - \int_{x_0}^x q(t)dt - \int_{x_0}^x \frac{z^2(t)}{p(t)}dt. \qquad (31.13)$$

Since $\int^\infty q(t)dt = \infty$, we can always find an $x_1 > x_0$ such that

$$z(x_0) - \int_{x_0}^x q(t)dt < 0$$

for all x in $[x_1, \infty)$. Thus, from (31.13) it follows that

$$z(x) < - \int_{x_0}^x \frac{z^2(t)}{p(t)}dt$$

for all $x \in [x_1, \infty)$. Let

$$r(x) = \int_{x_0}^x \frac{z^2(t)}{p(t)}dt, \quad x \in [x_1, \infty)$$

then $z(x) < -r(x)$ and

$$r'(x) = \frac{z^2(x)}{p(x)} > \frac{r^2(x)}{p(x)} \qquad (31.14)$$

for all x in $[x_1, \infty)$. Integrating (31.14) from $x_1 > x_0$ to ∞, we obtain

$$-\frac{1}{r(\infty)} + \frac{1}{r(x_1)} > \int_{x_1}^\infty \frac{1}{p(t)}dt$$

and hence

$$\int_{x_1}^{\infty} \frac{1}{p(t)} dt \ < \ \frac{1}{r(x_1)} \ < \ \infty,$$

which is a contradiction. Thus, the solution $y(x)$ is oscillatory. ∎

Example 31.6.   Once again we consider the DE (31.9). For all a, there exists a sufficiently large x_0 such that $1 + [(1 - 4a^2)/4x^2] > 1/2$ for all $x \geq x_0$, and hence

$$\int^{\infty} \left(1 + \frac{1 - 4a^2}{4x^2} \right) dx \ = \ \infty.$$

Thus, Theorem 31.7 implies that (31.9) is oscillatory for all a.

Problems

31.1. Let the function $q_1(x)$ be continuous and $q_1(x) \geq q(x)$ in J. Show the following:

(i)   If DE (31.2) is oscillatory, then the DE (31.3) is oscillatory.

(ii)   If DE (31.3) is nonoscillatory, then the DE (31.2) is nonoscillatory.

31.2. Show that the DE (31.2) is oscillatory if any one of the following conditions is satisfied:

(i) $q(x) \geq m^2 > 0$ eventually.

(ii) $q(x) = 1 + \phi(x)$, where $\phi(x) \to 0$ as $x \to \infty$.

(iii) $q(x) \to \infty$ as $x \to \infty$.

31.3. Let the function $q(x)$ be such that $0 < m \leq q(x) \leq M$ in $[\alpha, \beta]$. Further, let $\alpha \leq x_1 < x_2 < \cdots < x_n \leq \beta$ be the zeros of a solution $y(x)$ of the DE (31.2). Then show the following:

(i) $\pi/\sqrt{m} \geq x_{i+1} - x_i \geq \pi/\sqrt{M}$, $i = 1, 2, \ldots, n - 1$.

(ii) $n > (\beta - \alpha)(\sqrt{m}/\pi) - 1$.

31.4. Use Problem 31.3 to show that the distance between the successive zeros of every solution of (31.9) for each a tends to π.

31.5. Let the function $q_1(x)$ be continuous and $q_1(x) \geq q(x)$, $q_1(x) \neq q(x)$ in $[\alpha, \beta]$. Further, let $y(x)$ and $z(x)$ be respective solutions of (31.2) and (31.3) such that $y'(\alpha)/y(\alpha) \geq z'(\alpha)/z(\alpha)$, $y(\alpha) \neq 0$, $z(\alpha) \neq 0$, or $y(\alpha) = z(\alpha) = 0$.

(i)   Use Sturm's comparison theorem to show that $z(x)$ has at least as many zeros in $[\alpha, \beta]$ as $y(x)$.

(ii) Further, if $y(x)$ and $z(x)$ have the same number of zeros, then show that $y'(\beta)/y(\beta) > z'(\beta)/z(\beta)$ provided $y(\beta) \neq 0$.

31.6. Give an example to show that Sturm's separation theorem need not be true for DEs of order higher than two.

31.7. Show that the DE (31.1) has a solution without zeros in an interval J if and only if the Riccati equation (31.12) has a solution defined throughout J.

31.8. Show that every solution of the Hermite DE $y'' - 2xy' + 2ay = 0$ $(a \geq 0)$ has at most finitely many zeros in the interval $J = \mathbb{R}$.

31.9. Let $p, q \in C^{(1)}(J)$, $q(x) \neq 0$ in J and $p(x)q(x)$ be nonincreasing (nondecreasing) in J. Then show that the absolute values of the relative maxima and minima of every solution $y(x)$ of the DE (31.1) are nondecreasing (nonincreasing) as x increases.

31.10. Use Problem 31.9 to show that the magnitude of the oscillations of each solution of the Bessel equation (31.8) is nonincreasing in the interval (a, ∞).

Answers or Hints

31.1. (i) Use Theorem 31.1. (ii) Use part (i).

31.2. Use Problem 31.1.

31.3. (i) Compare (31.2) with $y'' + my = 0$ and $y'' + My = 0$ and apply Theorem 31.1. (ii) Note that each of the intervals $[\alpha, x_1]$ and $[x_n, \beta]$ is of length less than $\pi/\sqrt{m}$.

31.4. Note that $1 - \epsilon \leq q(x) = 1 + \frac{1-4a^2}{4x^2} \leq 1 + \epsilon$ for all sufficiently large x, say, $x \geq x^*$. Thus, in view of Problem 31.3(i), if $x_i, x_{i+1}(\geq x^*)$ are two consecutive zeros of the solution $u(x)$ of (31.9), then

$$\frac{\pi}{\sqrt{(1+\epsilon)}} \leq x_{i+1} - x_i \leq \frac{\pi}{\sqrt{(1-\epsilon)}}.$$

31.5. (i) If the zeros of $y(x)$ are $(\alpha \leq)x_1 < x_2 < \cdots < x_n(\leq \beta)$, then by Theorem 31.1, $z(x)$ has at least $(n-1)$ zeros in (x_1, x_n). Thus, it suffices to show that $z(x)$ has a zero in $[\alpha, x_1]$. If $y(\alpha) = z(\alpha) = 0$, then the proof is complete, otherwise, as in Theorem 31.1 we have

$$(z(x)y'(x) - y(x)z'(x))|_\alpha^{x_1} = \int_\alpha^{x_1} (q_1(x) - q(x))y(x)z(x)dx.$$

Thus, if $y(x) > 0$, $x \in [\alpha, x_1)$ and $z(x) > 0$, $x \in [\alpha, x_1]$, then $z(x_1)y'(x_1) - (z(\alpha)y'(\alpha) - y(\alpha)z'(\alpha)) > 0$. But, now $y'(x_1) < 0$ and $\frac{y'(\alpha)}{y(\alpha)} \geq \frac{z'(\alpha)}{z(\alpha)}$ leads to a contradiction. (ii) Use part (i).

31.6. Consider $y''' + y = 0$.

31.7. Use $py' = zy$.

31.8. Use $z(x) = e^{-x^2/2}y(x)$ to obtain $z'' + (1 + 2a - x^2)z = 0$.

31.9. For a nontrivial solution $y(x)$ of (31.1) consider the function $g = y^2 + \frac{1}{pq}(py')^2$; then $g' = -\left(\frac{y'}{q}\right)^2 (pq)'$. Now if $(pq)' \leq 0$ then $g' \geq 0$.

31.10. The self-adjoint form of (31.8) is $(xy')' + \left(\frac{x^2-a^2}{x}\right)y = 0$. Clearly, $p(x)q(x) = x^2 - a^2$ is increasing and positive in (a, ∞).

Lecture 32
Linear Boundary Value Problems

So far, we have concentrated only on initial value problems, in which for a given DE the supplementary conditions on the unknown function and its derivatives are prescribed at a fixed value x_0 of the independent variable x. However, as we have indicated in Lecture 1 there are a variety of other possible conditions that are important in applications. In many practical problems the additional requirements are given in the form of boundary conditions: the unknown function and some of its derivatives are fixed at more than one value of the independent variable x. The DE together with the boundary conditions are referred to as a boundary value problem.

Consider the second-order linear DE (6.6) in the interval $J = [\alpha, \beta]$, where, as we did earlier, we assume that the functions $p_0(x)$, $p_1(x)$, $p_2(x)$ and $r(x)$ are continuous in J. Together with the DE (6.6) we shall consider the boundary conditions of the form

$$\begin{aligned}
\ell_1[y] &= a_0 y(\alpha) + a_1 y'(\alpha) + b_0 y(\beta) + b_1 y'(\beta) = A \\
\ell_2[y] &= c_0 y(\alpha) + c_1 y'(\alpha) + d_0 y(\beta) + d_1 y'(\beta) = B,
\end{aligned} \qquad (32.1)$$

where a_i, b_i, c_i, d_i, $i = 0, 1$ and A, B are given constants. Throughout, we shall assume that these are essentially two conditions, i.e., there does not exist a constant c such that $(a_0\ a_1\ b_0\ b_1) = c(c_0\ c_1\ d_0\ d_1)$. Boundary value problem (6.6), (32.1) is called a nonhomogeneous two point linear boundary value problem, whereas the homogeneous DE (6.1) together with the homogeneous boundary conditions

$$\ell_1[y] = 0, \quad \ell_2[y] = 0 \qquad (32.2)$$

will be called a homogeneous boundary value problem.

Boundary conditions (32.1) are quite general and, in particular, include the

(i)   first boundary conditions (Dirichlet conditions)

$$y(\alpha) = A, \quad y(\beta) = B, \qquad (32.3)$$

(ii)   second boundary conditions (mixed conditions)

$$y(\alpha) = A, \quad y'(\beta) = B, \qquad (32.4)$$

R.P. Agarwal and D. O'Regan, *An Introduction to Ordinary Differential Equations*,
doi: 10.1007/978-0-387-71276-5_32, © Springer Science + Business Media, LLC 2008

or

$$y'(\alpha) \; = \; A, \quad y(\beta) \; = \; B, \tag{32.5}$$

(iii) separated boundary conditions (third boundary conditions)

$$\begin{aligned} a_0 y(\alpha) + a_1 y'(\alpha) \; &= \; A \\ d_0 y(\beta) + d_1 y'(\beta) \; &= \; B, \end{aligned} \tag{32.6}$$

where both $a_0^2 + a_1^2$ and $d_0^2 + d_1^2$ are different from zero, and
(iv) periodic boundary conditions

$$y(\alpha) \; = \; y(\beta), \quad y'(\alpha) \; = \; y'(\beta). \tag{32.7}$$

Boundary value problem (6.6), (32.1) is called *regular* if both α and β are finite, and the function $p_0(x) \neq 0$ for all $x \in J$. If $\alpha = -\infty$ and/or $\beta = \infty$ and/or $p_0(x) = 0$ for at least one point x in J, then the problem (6.6), (32.1) is said to be *singular*. We shall consider only regular boundary value problems.

By a solution of the boundary value problem (6.6), (32.1) we mean a solution of the DE (6.6) satisfying the boundary conditions (32.1).

The existence and uniqueness theory for the boundary value problems is more difficult than that of initial value problems. In fact, in the case of boundary value problems a slight change in the boundary conditions can lead to significant changes in the behavior of the solutions. For example, the initial value problem $y'' + y = 0$, $y(0) = c_1$, $y'(0) = c_2$ has a unique solution $y(x) = c_1 \cos x + c_2 \sin x$ for any set of values c_1, c_2. However, the boundary value problem $y'' + y = 0$, $y(0) = 0$, $y(\pi) = \epsilon (\neq 0)$ has no solution; the problem $y'' + y = 0$, $y(0) = 0$, $y(\beta) = \epsilon$, $0 < \beta < \pi$ has a unique solution $y(x) = \epsilon \sin x / \sin \beta$, while the problem $y'' + y = 0$, $y(0) = 0$, $y(\pi) = 0$ has an infinite number of solutions $y(x) = c \sin x$, where c is an arbitrary constant. Similarly, since for the DE $(1 + x^2) y'' - 2xy' + 2y = 0$ the general solution is $y(x) = c_1(x^2 - 1) + c_2 x$, there exists a unique solution satisfying the boundary conditions $y'(\alpha) = A$, $y'(\beta) = B$; an infinite number of solutions satisfying $y(-1) = 0 = y(1)$; and no solution satisfying $y(-1) = 0$, $y(1) = 1$.

Obviously, for the homogeneous problem (6.1), (32.2) the trivial solution always exists. However, from the above examples it follows that besides having the trivial solution, homogeneous boundary value problems may have nontrivial solutions also. Our first result provides necessary and sufficient condition so that the problem (6.1), (32.2) has only the trivial solution.

Theorem 32.1. Let $y_1(x)$ and $y_2(x)$ be any two linearly independent solutions of the DE (6.1). Then the homogeneous boundary value problem

(6.1), (32.2) has only the trivial solution if and only if

$$\Delta = \begin{vmatrix} \ell_1[y_1] & \ell_1[y_2] \\ \ell_2[y_1] & \ell_2[y_2] \end{vmatrix} \neq 0. \tag{32.8}$$

Proof. Any solution of the DE (6.1) can be written as

$$y(x) = c_1 y_1(x) + c_2 y_2(x).$$

This is a solution of the problem (6.1), (32.2) if and only if

$$\begin{aligned} \ell_1[c_1 y_1 + c_2 y_2] &= c_1 \ell_1[y_1] + c_2 \ell_1[y_2] = 0 \\ \ell_2[c_1 y_1 + c_2 y_2] &= c_1 \ell_2[y_1] + c_2 \ell_2[y_2] = 0. \end{aligned} \tag{32.9}$$

However, from Theorem 13.2, system (32.9) has only the trivial solution if and only if $\Delta \neq 0$. ∎

Clearly, Theorem 32.1 is independent of the choice of the solutions $y_1(x)$ and $y_2(x)$. Thus, for convenience we can always take $y_1(x)$ and $y_2(x)$ to be the solutions of (6.1) satisfying the initial conditions

$$y_1(\alpha) = 1, \quad y_1'(\alpha) = 0 \tag{32.10}$$

and

$$y_2(\alpha) = 0, \quad y_2'(\alpha) = 1. \tag{32.11}$$

Corollary 32.2. The homogeneous boundary value problem (6.1), (32.2) has an infinite number of nontrivial solutions if and only if $\Delta = 0$.

Example 32.1. Consider the boundary value problem

$$xy'' - y' - 4x^3 y = 0 \tag{32.12}$$

$$\begin{aligned} \ell_1[y] &= y(1) = 0 \\ \ell_2[y] &= y(2) = 0. \end{aligned} \tag{32.13}$$

For the DE (32.12), $y_1(x) = \cosh(x^2 - 1)$ and $y_2(x) = (1/2)\sinh(x^2 - 1)$ are two linearly independent solutions. Further, since for the boundary conditions (32.13), we have

$$\Delta = \begin{vmatrix} 1 & 0 \\ \cosh 3 & (1/2)\sinh 3 \end{vmatrix} \neq 0$$

the problem (32.12), (32.13) has only the trivial solution.

Example 32.2. Consider once again the DE (32.12) together with the boundary conditions

$$\begin{aligned} \ell_1[y] &= y'(1) = 0 \\ \ell_2[y] &= y'(2) = 0. \end{aligned} \tag{32.14}$$

Since $y_1'(x) = 2x \sinh(x^2 - 1)$ and $y_2'(x) = x \cosh(x^2 - 1)$, for the boundary conditions (32.14), we find

$$\Delta = \begin{vmatrix} 0 & 1 \\ 4\sinh 3 & 2\cosh 3 \end{vmatrix} \neq 0.$$

Thus, the problem (32.12), (32.14) has only the trivial solution.

Example 32.3. Consider the boundary value problem

$$y'' + 2y' + 5y = 0 \tag{32.15}$$

$$\begin{aligned} \ell_1[y] &= y(0) = 0 \\ \ell_2[y] &= y(\pi/2) = 0. \end{aligned} \tag{32.16}$$

For the DE (32.15), $y_1(x) = e^{-x}\cos 2x$ and $y_2(x) = e^{-x}\sin 2x$ are two linearly independent solutions. Further, since for the boundary conditions (32.16),

$$\Delta = \begin{vmatrix} 1 & 0 \\ -e^{-\pi/2} & 0 \end{vmatrix} = 0$$

the problem (32.15), (32.16) besides having the trivial solution also has nontrivial solutions. Indeed it has an infinite number of solutions $y(x) = ce^{-x}\sin 2x$, where c is an arbitrary constant.

Theorem 32.3. The nonhomogeneous boundary value problem (6.6), (32.1) has a unique solution if and only if the homogeneous boundary value problem (6.1), (32.2) has only the trivial solution.

Proof. Let $y_1(x)$ and $y_2(x)$ be any two linearly independent solutions of the DE (6.1) and $z(x)$ be a particular solution of (6.6). Then the general solution of (6.6) can be written as

$$y(x) = c_1 y_1(x) + c_2 y_2(x) + z(x). \tag{32.17}$$

This is a solution of the problem (6.6), (32.1) if and only if

$$\begin{aligned} \ell_1[c_1 y_1 + c_2 y_2 + z] &= c_1\ell_1[y_1] + c_2\ell_1[y_2] + \ell_1[z] = A \\ \ell_2[c_1 y_1 + c_2 y_2 + z] &= c_1\ell_2[y_1] + c_2\ell_2[y_2] + \ell_2[z] = B. \end{aligned} \tag{32.18}$$

However, from Theorem 13.2, nonhomogeneous system (32.18) has a unique solution if and only if $\Delta \neq 0$, i.e., if and only if the homogeneous system (32.9) has only the trivial solution. From Theorem 32.1, $\Delta \neq 0$ is equivalent to the homogeneous boundary value problem (6.1), (32.2) having only the trivial solution. ■

Example 32.4. Consider the boundary value problem

$$xy'' - y' - 4x^3 y = 1 + 4x^4 \tag{32.19}$$

$$\ell_1[y] = y(1) = 0$$
$$\ell_2[y] = y(2) = 1. \tag{32.20}$$

Since the corresponding homogeneous problem (32.12), (32.13) has only the trivial solution, Theorem 32.3 implies that the problem (32.19), (32.20) has a unique solution. Further, to find this solution once again we choose the linearly independent solutions of (32.12) to be $y_1(x) = \cosh(x^2 - 1)$ and $y_2(x) = (1/2)\sinh(x^2 - 1)$, and note that $z(x) = -x$ is a particular solution of (32.19). Thus, the system (32.18) for the boundary conditions (32.20) reduces to

$$c_1 - 1 = 0$$
$$\cosh 3\, c_1 + (1/2)\sinh 3\, c_2 - 2 = 1.$$

This system can be solved easily, and we obtain $c_1 = 1$ and $c_2 = 2(3 - \cosh 3)/\sinh 3$. Now substituting these quantities in (32.17) we find the solution of (32.19), (32.20) as

$$y(x) = \cosh(x^2 - 1) + \frac{(3 - \cosh 3)}{\sinh 3}\sinh(x^2 - 1) - x.$$

Problems

32.1. Solve the following boundary value problems:

(i)
$$y'' - y = 0$$
$$y(0) = 0,\ y(1) = 1.$$

(ii)
$$y'' + 4y' + 7y = 0$$
$$y(0) = 0,\ y'(1) = 1.$$

(iii)
$$y'' - 6y' + 25y = 0$$
$$y'(0) = 1,\ y(\pi/4) = 0.$$

(iv)
$$x^2 y'' + 7xy' + 3y = 0$$
$$y(1) = 1,\ y(2) = 2.$$

(v)
$$y'' + y = 0$$
$$y(0) + y'(0) = 10$$
$$y(1) + 3y'(1) = 4.$$

(vi)
$$y'' + y = x^2$$
$$y(0) = 0,\ y(\pi/2) = 1.$$

(vii)
$$y'' + 2y' + y = x$$
$$y(0) = 0,\ y(2) = 3.$$

(viii)
$$y'' + y' + y = x$$
$$y(0) + 2y'(0) = 1$$
$$y(1) - y'(1) = 8.$$

32.2. Show that the following boundary value problem has no solution:

$$y'' + y = x, \quad y(0) + y'(0) = 0, \quad y(\pi/2) - y'(\pi/2) = \pi/2.$$

32.3. Solve the following periodic boundary value problems:

(i)
$$y'' + 2y' + 10y = 0$$
$$y(0) = y(\pi/6)$$
$$y'(0) = y'(\pi/6).$$

(ii)
$$y'' + \pi^2 y = 0$$
$$y(-1) = y(1)$$
$$y'(-1) = y'(1).$$

32.4. Show that the boundary value problem $y'' = r(x)$, (32.6) has a unique solution if and only if

$$\Delta = a_0 d_0 (\beta - \alpha) + a_0 d_1 - a_1 d_0 \neq 0.$$

32.5. Determine the values of the constants β, A, and B so that the boundary value problem $y'' + 2py' + qy = 0$, $y(0) = A$, $y(\beta) = B$ with $p^2 - q < 0$ has only one solution.

32.6. Show that the boundary value problem $y'' + p(x)y = q(x)$, (32.3) where $p(x) \leq 0$ in $[\alpha, \beta]$ has a unique solution.

32.7. Let $z(x)$ be the solution of the initial value problem (6.6), $z(\alpha) = A$, $z'(\alpha) = 0$ and $y_2(x)$ be the solution of the initial value problem (6.1), (32.11). Show that the boundary value problem (6.6), (32.3) has a unique solution $y(x)$ if and only if $y_2(\beta) \neq 0$ and it can be written as

$$y(x) = z(x) + \frac{(B - z(\beta))}{y_2(\beta)} y_2(x).$$

32.8. Let $y_1(x)$ and $y_2(x)$ be the solutions of the initial value problems (6.1), $y_1(\alpha) = a_1$, $y_1'(\alpha) = -a_0$ and (6.1), $y_2(\beta) = -d_1$, $y_2'(\beta) = d_0$, respectively. Show that the boundary value problem (6.6), (32.6) has a unique solution if and only if $W(y_1, y_2)(\alpha) \neq 0$.

32.9. Let $y_1(x)$ and $y_2(x)$ be the solutions of the boundary value problems (6.1), (32.1) and (6.6), (32.2), respectively. Show that $y(x) = y_1(x) + y_2(x)$ is a solution of the problem (6.6), (32.1).

32.10. For the homogeneous DE

$$\mathcal{L}_2[y] = (x^2 + 1)y'' - 2xy' + 2y = 0 \tag{32.21}$$

x and $(x^2 - 1)$ are two linearly independent solutions. Use this information to show that the boundary value problem

$$\mathcal{L}_2[y] = 6(x^2 + 1)^2, \quad y(0) = 1, \quad y(1) = 2 \tag{32.22}$$

has a unique solution, and find it.

Answers or Hints

32.1. (i) $\frac{\sinh x}{\sinh 1}$. (ii) $\frac{e^{2(1-x)} \sin \sqrt{3}x}{(\sqrt{3}\cos\sqrt{3} - 2\sin\sqrt{3})}$. (iii) $\frac{1}{4}e^{3x}\sin 4x$. (iv) $\frac{1}{(2\sqrt{6} - 2 - \sqrt{6})} \times$ $[(16 - 2^{-\sqrt{6}})x^{-3+\sqrt{6}} + (2^{\sqrt{6}} - 16)x^{-3-\sqrt{6}}]$. (v) $\frac{1}{2\sin 1 + \cos 1}[\{5(\sin 1 + 3\cos 1) - 2\}\cos x + \{5(3\sin 1 - \cos 1) + 2\}\sin x]$. (vi) $2\cos x + \left(3 - \frac{\pi^2}{4}\right)\sin x + x^2 - 2$.

(vii) $e^{-x}\left[2+\left(\frac{3}{2}e^2-1\right)x\right]+x-2$. (viii) $\dfrac{18}{3\cos\frac{\sqrt{3}}{2}+\sqrt{3}\sin\frac{\sqrt{3}}{2}}e^{(1-x)/2}\cos\frac{\sqrt{3}}{2}x$
$+x-1$.

32.2. Leads to an inconsistent system of equations.

32.3. (i) Trivial solution. (ii) $c_1\cos\pi x+c_2\sin\pi x$, where c_1 and c_2 are arbitrary constants.

32.4. For the DE $y''=0$ two linearly independent solutions are 1, x. Now apply Theorem 32.3.

32.5. $\beta\neq\dfrac{n\pi}{\sqrt{q-p^2}}$, $e^{-px}\left[A\cos\sqrt{q-p^2}\,x+\dfrac{Be^{p\beta}-A\cos\sqrt{q-p^2}\,\beta}{\sin\sqrt{q-p^2}\,\beta}\sin\sqrt{q-p^2}\,x\right]$.

32.6. Use Theorem 32.3 and Example 31.1.

32.7. The function $y(x)=z_1(x)+cy_1(x)$ is a solution of the DE (6.6).

32.8. Use Theorem 32.3.

32.9. Verify directly.

32.10. Use variation of parameters to find the particular solution $z(x)=x^4+3x^2$. The solution of (32.22) is x^4+2x^2-2x+1.

Lecture 33
Green's Functions

The function $H(x,t)$ defined in (6.10) is a solution of the homogeneous DE (6.1) and it helps in finding an explicit representation of a particular solution of the nonhomogeneous DE (6.6) (also see Problem 18.9 for higher-order DEs). In this lecture, we shall find an analog of this function called *Green's function* $G(x,t)$ for the homogeneous boundary value problem (6.1), (32.2) and show that the solution of the nonhomogeneous boundary value problem (6.6), (32.2) can be explicitly expressed in terms of $G(x,t)$. The solution of the problem (6.6), (32.1) then can be obtained easily as an application of Problem 32.9. For this, in what follows throughout we shall assume that the problem (6.1), (32.2) has only the trivial solution. Green's function $G(x,t)$ for the boundary value problem (6.1), (32.2) is defined in the square $[\alpha, \beta] \times [\alpha, \beta]$ and possesses the following fundamental properties:

(i) $G(x,t)$ is continuous in $[\alpha, \beta] \times [\alpha, \beta]$.

(ii) $\partial G(x,t)/\partial x$ is continuous in each of the triangles $\alpha \leq x \leq t \leq \beta$ and $\alpha \leq t \leq x \leq \beta$; moreover

$$\frac{\partial G}{\partial x}(t^+, t) \; - \; \frac{\partial G}{\partial x}(t^-, t) \; = \; \frac{1}{p_0(t)},$$

where

$$\frac{\partial G}{\partial x}(t^+, t) \; = \; \lim_{\substack{x \to t \\ x > t}} \frac{\partial G(x,t)}{\partial x} \quad \text{and} \quad \frac{\partial G}{\partial x}(t^-, t) \; = \; \lim_{\substack{x \to t \\ x < t}} \frac{\partial G(x,t)}{\partial x}.$$

(iii) For every $t \in [\alpha, \beta]$, $z(x) = G(x,t)$ is a solution of the DE (6.1) in each of the intervals $[\alpha, t)$ and $(t, \beta]$.

(iv) For every $t \in [\alpha, \beta]$, $z(x) = G(x,t)$ satisfies the boundary conditions (32.2).

These properties completely characterize Green's function $G(x,t)$. To show this, let $y_1(x)$ and $y_2(x)$ be two linearly independent solutions of the DE (6.1). From the property (iii) there exist four functions, say, $\lambda_1(t)$, $\lambda_2(t)$, $\mu_1(t)$, and $\mu_2(t)$ such that

$$G(x,t) \; = \; \begin{cases} y_1(x)\lambda_1(t) + y_2(x)\lambda_2(t), & \alpha \leq x \leq t \\ y_1(x)\mu_1(t) + y_2(x)\mu_2(t), & t \leq x \leq \beta. \end{cases} \tag{33.1}$$

R.P. Agarwal and D. O'Regan, *An Introduction to Ordinary Differential Equations,*
doi: 10.1007/978-0-387-71276-5_33, © Springer Science + Business Media, LLC 2008

Now using properties (i) and (ii), we obtain the following two equations:

$$y_1(t)\lambda_1(t) + y_2(t)\lambda_2(t) = y_1(t)\mu_1(t) + y_2(t)\mu_2(t) \qquad (33.2)$$

$$y_1'(t)\mu_1(t) + y_2'(t)\mu_2(t) - y_1'(t)\lambda_1(t) - y_2'(t)\lambda_2(t) = \frac{1}{p_0(t)}. \qquad (33.3)$$

Let $\nu_1(t) = \mu_1(t) - \lambda_1(t)$ and $\nu_2(t) = \mu_2(t) - \lambda_2(t)$, so that (33.2) and (33.3) can be written as

$$y_1(t)\nu_1(t) + y_2(t)\nu_2(t) = 0 \qquad (33.4)$$

$$y_1'(t)\nu_1(t) + y_2'(t)\nu_2(t) = \frac{1}{p_0(t)}. \qquad (33.5)$$

Since $y_1(x)$ and $y_2(x)$ are linearly independent the Wronskian $W(y_1, y_2)(t)$ $\neq 0$ for all $t \in [\alpha, \beta]$. Thus, the relations (33.4), (33.5) uniquely determine $\nu_1(t)$ and $\nu_2(t)$.

Now using the relations $\mu_1(t) = \lambda_1(t) + \nu_1(t)$ and $\mu_2(t) = \lambda_2(t) + \nu_2(t)$, Green's function can be written as

$$G(x,t) = \begin{cases} y_1(x)\lambda_1(t) + y_2(x)\lambda_2(t), & \alpha \le x \le t \\ y_1(x)\lambda_1(t) + y_2(x)\lambda_2(t) + y_1(x)\nu_1(t) + y_2(x)\nu_2(t), & t \le x \le \beta. \end{cases}$$
$$(33.6)$$

Finally, using the property (iv), we find

$$\begin{aligned} \ell_1[y_1]\lambda_1(t) + \ell_1[y_2]\lambda_2(t) &= -b_0(y_1(\beta)\nu_1(t) + y_2(\beta)\nu_2(t)) \\ &\quad -b_1(y_1'(\beta)\nu_1(t) + y_2'(\beta)\nu_2(t)) \\ \ell_2[y_1]\lambda_1(t) + \ell_2[y_2]\lambda_2(t) &= -d_0(y_1(\beta)\nu_1(t) + y_2(\beta)\nu_2(t)) \\ &\quad -d_1(y_1'(\beta)\nu_1(t) + y_2'(\beta)\nu_2(t)). \end{aligned} \qquad (33.7)$$

Since the problem (6.1), (32.2) has only the trivial solution, from Theorem 32.1 it follows that the system (33.7) uniquely determines $\lambda_1(t)$ and $\lambda_2(t)$.

From the above construction it is clear that no other function exists which has properties (i)–(iv), i.e., Green's function $G(x,t)$ of the boundary value problem (6.1), (32.2) is unique.

As claimed earlier, we shall now show that the unique solution $y(x)$ of the problem (6.6), (32.2) can be represented in terms of $G(x,t)$ as follows:

$$y(x) = \int_\alpha^\beta G(x,t)r(t)dt = \int_\alpha^x G(x,t)r(t)dt + \int_x^\beta G(x,t)r(t)dt. \qquad (33.8)$$

Since $G(x,t)$ is differentiable with respect to x in each of the intervals, we find

$$y'(x) = G(x,x)r(x) + \int_\alpha^x \frac{\partial G(x,t)}{\partial x}r(t)dt - G(x,x)r(x) + \int_x^\beta \frac{\partial G(x,t)}{\partial x}r(t)dt$$

$$= \int_\alpha^x \frac{\partial G(x,t)}{\partial x}r(t)dt + \int_x^\beta \frac{\partial G(x,t)}{\partial x}r(t)dt$$

$$= \int_\alpha^\beta \frac{\partial G(x,t)}{\partial x}r(t)dt.$$

(33.9)

Next since $\partial G(x,t)/\partial x$ is a continuous function of (x,t) in the triangles $\alpha \le t \le x \le \beta$ and $\alpha \le x \le t \le \beta$, for any point (s,s) on the diagonal of the square, i.e., $t = x$ it is necessary that

$$\frac{\partial G}{\partial x}(s,s^-) = \frac{\partial G}{\partial x}(s^+,s)$$

(33.10)

and

$$\frac{\partial G}{\partial x}(s,s^+) = \frac{\partial G}{\partial x}(s^-,s).$$

(33.11)

Now differentiating the relation (33.9), we obtain

$$y''(x) = \frac{\partial G(x,x^-)}{\partial x}r(x) + \int_\alpha^x \frac{\partial^2 G(x,t)}{\partial x^2}r(t)dt$$
$$- \frac{\partial G(x,x^+)}{\partial x}r(x) + \int_x^\beta \frac{\partial^2 G(x,t)}{\partial x^2}r(t)dt,$$

which in view of (33.10) and (33.11) is the same as

$$y''(x) = \left[\frac{\partial G(x^+,x)}{\partial x} - \frac{\partial G(x^-,x)}{\partial x}\right]r(x) + \int_\alpha^\beta \frac{\partial^2 G(x,t)}{\partial x^2}r(t)dt.$$

Using property (ii) this relation gives

$$y''(x) = \frac{r(x)}{p_0(x)} + \int_\alpha^\beta \frac{\partial^2 G(x,t)}{\partial x^2}r(t)dt.$$

(33.12)

Thus, from (33.8), (33.9), and (33.12), and the property (iii), we get

$$p_0(x)y''(x) + p_1(x)y'(x) + p_2(x)y(x)$$
$$= r(x) + \int_\alpha^\beta \left[p_0(x)\frac{\partial^2 G(x,t)}{\partial x^2} + p_1(x)\frac{\partial G(x,t)}{\partial x} + p_2(x)G(x,t)\right]r(t)dt$$
$$= r(x),$$

i.e., $y(x)$ as given in (33.8) is a solution of the DE (6.6).

Finally, since

$$y(\alpha) = \int_\alpha^\beta G(\alpha, t) r(t) dt, \qquad y(\beta) = \int_\alpha^\beta G(\beta, t) r(t) dt$$

$$y'(\alpha) = \int_\alpha^\beta \frac{\partial G(\alpha, t)}{\partial x} r(t) dt, \qquad y'(\beta) = \int_\alpha^\beta \frac{\partial G(\beta, t)}{\partial x} r(t) dt,$$

it is easy to see that

$$\ell_1[y] = \int_\alpha^\beta \ell_1[G(x, t)] r(t) dt = 0 \quad \text{and} \quad \ell_2[y] = \int_\alpha^\beta \ell_2[G(x, t)] r(t) dt = 0$$

and hence $y(x)$ as given in (33.8) satisfies the boundary conditions (32.2) as well.

We summarize these results in the following theorem.

Theorem 33.1. Let the homogeneous problem (6.1), (32.2) have only the trivial solution. Then the following hold:

(i)   there exists a unique Green's function $G(x, t)$ for the problem (6.1), (32.2),

(ii) the unique solution $y(x)$ of the nonhomogeneous problem (6.6), (32.2) can be represented by (33.8).

Example 33.1. We shall construct Green's function of the problem

$$y'' = 0 \tag{33.13}$$

$$\begin{aligned} a_0 y(\alpha) + a_1 y'(\alpha) &= 0 \\ d_0 y(\beta) + d_1 y'(\beta) &= 0. \end{aligned} \tag{33.14}$$

For the DE (33.13) two linearly independent solutions are $y_1(x) = 1$ and $y_2(x) = x$. Thus, the problem (33.13), (33.14) has only the trivial solution if and only if $\Delta = a_0 d_0(\beta - \alpha) + a_0 d_1 - a_1 d_0 \neq 0$ (see Problem 32.4). Further, equalities (33.4) and (33.5) reduce to

$$\nu_1(t) + t\nu_2(t) = 0 \quad \text{and} \quad \nu_2(t) = 1.$$

Thus, $\nu_1(t) = -t$ and $\nu_2(t) = 1$.

Next for (33.13), (33.14) the system (33.7) reduces to

$$\begin{aligned} a_0 \lambda_1(t) + (a_0\alpha + a_1)\lambda_2(t) &= 0 \\ d_0 \lambda_1(t) + (d_0\beta + d_1)\lambda_2(t) &= -d_0(-t + \beta) - d_1, \end{aligned}$$

which easily determines $\lambda_1(t)$ and $\lambda_2(t)$ as

$$\lambda_1(t) = \frac{1}{\Delta}(a_0\alpha + a_1)(d_0\beta - d_0 t + d_1) \quad \text{and} \quad \lambda_2(t) = \frac{1}{\Delta}a_0(d_0 t - d_0\beta - d_1).$$

Substituting these functions in (33.6), we get the required Green's function

$$G(x,t) = \frac{1}{\Delta} \begin{cases} (d_0\beta - d_0t + d_1)(a_0\alpha - a_0x + a_1), & \alpha \le x \le t \\ (d_0\beta - d_0x + d_1)(a_0\alpha - a_0t + a_1), & t \le x \le \beta, \end{cases} \tag{33.15}$$

which is symmetric, i.e., $G(x,t) = G(t,x)$.

Example 33.2. Consider the periodic boundary value problem

$$y'' + k^2 y = 0, \quad k > 0 \tag{33.16}$$

$$\begin{aligned} y(0) &= y(\omega) \\ y'(0) &= y'(\omega), \quad \omega > 0. \end{aligned} \tag{33.17}$$

For the DE (33.16) two linearly independent solutions are $y_1(x) = \cos kx$ and $y_2(x) = \sin kx$. Hence, in view of Theorem 32.1 the problem (33.16), (33.17) has only the trivial solution if and only if

$$\Delta = 4k \sin^2 \frac{k\omega}{2} \ne 0, \quad \text{i.e.,} \quad \omega \in \left(0, \frac{2\pi}{k}\right).$$

Further, equalities (33.4) and (33.5) reduce to

$$\cos kt \, \nu_1(t) + \sin kt \, \nu_2(t) = 0$$
$$-k \sin kt \, \nu_1(t) + k \cos kt \, \nu_2(t) = 1.$$

These relations easily give

$$\nu_1(t) = -\frac{1}{k} \sin kt \quad \text{and} \quad \nu_2(t) = \frac{1}{k} \cos kt.$$

Next for (33.16), (33.17) the system (33.7) reduces to

$$\begin{aligned} (1 - \cos k\omega)\lambda_1(t) - \sin k\omega \, \lambda_2(t) &= \frac{1}{k} \sin k(\omega - t) \\ \sin k\omega \, \lambda_1(t) + (1 - \cos k\omega)\lambda_2(t) &= \frac{1}{k} \cos k(\omega - t), \end{aligned}$$

which determines $\lambda_1(t)$ and $\lambda_2(t)$ as

$$\lambda_1(t) = \frac{1}{2k \sin \frac{k}{2}\omega} \cos k\left(t - \frac{\omega}{2}\right) \quad \text{and} \quad \lambda_2(t) = \frac{1}{2k \sin \frac{k}{2}\omega} \sin k\left(t - \frac{\omega}{2}\right).$$

Substituting these functions in (33.6), we get Green's function of the boundary value problem (33.16), (33.17) as

$$G(x,t) = \frac{1}{2k \sin \frac{k}{2}\omega} \begin{cases} \cos k\left(x - t + \frac{\omega}{2}\right), & 0 \le x \le t \\ \cos k\left(t - x + \frac{\omega}{2}\right), & t \le x \le \omega \end{cases} \tag{33.18}$$

which as expected is symmetric.

Example 33.3. We shall construct Green's function for the boundary value problem (6.1), (33.14) where the DE (6.1) is assumed to be self-adjoint.

Let $y_1(x)$ and $y_2(x)$ be as in Problem 32.8. Since the homogeneous problem (6.1), (33.14) has only the trivial solution, from the same problem it follows that $y_1(x)$ and $y_2(x)$ are linearly independent solutions of the DE (6.1). Thus, in view of (6.11) the general solution of (6.6) can be written as

$$y(x) \; = \; c_1 y_1(x) + c_2 y_2(x) + \int_\alpha^x \frac{[y_1(t)y_2(x) - y_2(t)y_1(x)]}{p_0(t)W(y_1,y_2)(t)} r(t)dt. \quad (33.19)$$

However, since (6.1) is self-adjoint, from (30.23) we have $p_0(x)W(y_1,y_2)(x) = C$, a nonzero constant. Hence, (33.19) is the same as

$$y(x) \; = \; c_1 y_1(x) + c_2 y_2(x) + \frac{1}{C}\int_\alpha^x [y_1(t)y_2(x) - y_2(t)y_1(x)]r(t)dt. \quad (33.20)$$

This solution also satisfies the boundary conditions (33.14) if and only if

$$a_0(c_1 a_1 + c_2 y_2(\alpha)) + a_1(c_1(-a_0) + c_2 y_2'(\alpha)) \; = \; (a_0 y_2(\alpha) + a_1 y_2'(\alpha))c_2 \; = \; 0, \quad (33.21)$$

$$d_0\left(c_1 y_1(\beta) + c_2(-d_1) + \frac{1}{C}\int_\alpha^\beta [y_1(t)(-d_1) - y_2(t)y_1(\beta)]r(t)dt\right)$$

$$+ \, d_1\left(c_1 y_1'(\beta) + c_2 d_0 + \frac{1}{C}\int_\alpha^\beta [y_1(t)d_0 - y_2(t)y_1'(\beta)]r(t)dt\right)$$

$$= \; (d_0 y_1(\beta) + d_1 y_1'(\beta))c_1 - \frac{1}{C}\int_\alpha^\beta y_2(t)[d_0 y_1(\beta) + d_1 y_1'(\beta)]r(t)dt$$

$$= \; (d_0 y_1(\beta) + d_1 y_1'(\beta))\left[c_1 - \frac{1}{C}\int_\alpha^\beta y_2(t)r(t)dt\right] \; = \; 0.$$
$$(33.22)$$

But from our assumptions $a_0 y_2(\alpha) + a_1 y_2'(\alpha)$ as well as $d_0 y_1(\beta) + d_1 y_1'(\beta)$ is different from zero. Hence, equations (33.21) and (33.22) immediately determine

$$c_2 \; = \; 0 \quad \text{and} \quad c_1 \; = \; \frac{1}{C}\int_\alpha^\beta y_2(t)r(t)dt.$$

Substituting these constants in (33.20), we find the solution of the problem (6.6), (33.14) as

$$
\begin{aligned}
y(x) &= \frac{1}{C} \int_\alpha^\beta y_2(t) y_1(x) r(t) dt + \frac{1}{C} \int_\alpha^x [y_1(t) y_2(x) - y_2(t) y_1(x)] r(t) dt \\
&= \frac{1}{C} \int_\alpha^x y_1(t) y_2(x) r(t) dt + \frac{1}{C} \int_x^\beta y_2(t) y_1(x) r(t) dt \\
&= \int_\alpha^\beta G(x,t) r(t) dt.
\end{aligned}
$$

Hence, the required Green's function is

$$
G(x,t) = \frac{1}{C} \begin{cases} y_2(t) y_1(x), & \alpha \le x \le t \\ y_1(t) y_2(x), & t \le x \le \beta, \end{cases}
$$

which is also symmetric.

Problems

33.1. Show that

$$
G(x,t) = \begin{cases} -\cos t \sin x, & 0 \le x \le t \\ -\sin t \cos x, & t \le x \le \pi/2 \end{cases}
$$

is the Green function of the problem $y'' + y = 0$, $y(0) = y(\pi/2) = 0$. Hence, solve the boundary value problem

$$
y'' + y = 1 + x, \quad y(0) = y(\pi/2) = 1.
$$

33.2. Show that

$$
G(x,t) = \frac{1}{\sinh 1} \begin{cases} \sinh(t-1)\sinh x, & 0 \le x \le t \\ \sinh t \sinh(x-1), & t \le x \le 1 \end{cases}
$$

is the Green function of the problem $y'' - y = 0$, $y(0) = y(1) = 0$. Hence, solve the boundary value problem

$$
y'' - y = 2\sin x, \quad y(0) = 0, \quad y(1) = 2.
$$

33.3. Construct Green's function for each of the boundary value problems given in Problem 32.1 parts (vi) and (vii) and then find their solutions.

33.4. Verify that Green's function of the problem (32.21), $y(0) = 0$, $y(1) = 0$ is

$$
G(x,t) = \begin{cases} \dfrac{t(x^2-1)}{(t^2+1)^2}, & 0 \le t \le x \\ \dfrac{x(t^2-1)}{(t^2+1)^2}, & x \le t \le 1. \end{cases}
$$

Hence, solve the boundary value problem (32.22).

33.5. Show that the solution of the boundary value problem

$$y'' - \frac{1}{x}y' = r(x), \quad y(0) = 0, \quad y(1) = 0$$

can be written as

$$y(x) = \int_0^1 G(x,t)r(t)dt,$$

where

$$G(x,t) = \begin{cases} -\dfrac{(1-t^2)x^2}{2t}, & x \le t \\ -\dfrac{t(1-x^2)}{2}, & x \ge t. \end{cases}$$

33.6. Show that the solution of the boundary value problem

$$y'' - y = r(x), \quad y(-\infty) = 0, \quad y(\infty) = 0$$

can be written as

$$y(x) = \frac{1}{2}\int_{-\infty}^{\infty} e^{-|x-t|}r(t)dt.$$

33.7. Consider the DE

$$y'' = f(x, y, y') \tag{33.23}$$

together with the boundary conditions (32.3). Show that $y(x)$ is a solution of this problem if and only if

$$y(x) = \frac{(\beta - x)}{(\beta - \alpha)}A + \frac{(x - \alpha)}{(\beta - \alpha)}B + \int_\alpha^\beta G(x,t)f(t, y(t), y'(t))dt, \tag{33.24}$$

where $G(x,t)$ is the Green function of the problem $y'' = 0$, $y(\alpha) = y(\beta) = 0$ and is given by

$$G(x,t) = \frac{1}{(\beta - \alpha)}\begin{cases} (\beta - t)(\alpha - x), & \alpha \le x \le t \\ (\beta - x)(\alpha - t), & t \le x \le \beta. \end{cases} \tag{33.25}$$

Also establish the following:

(i) $G(x,t) \le 0$ in $[\alpha, \beta] \times [\alpha, \beta]$.

(ii) $|G(x,t)| \le \dfrac{1}{4}(\beta - \alpha)$.

(iii) $\displaystyle\int_\alpha^\beta |G(x,t)|dt = \frac{1}{2}(\beta - x)(x - \alpha) \le \frac{1}{8}(\beta - \alpha)^2$.

(iv) $\displaystyle\int_\alpha^\beta |G(x,t)| \sin\frac{\pi(t-\alpha)}{(\beta-\alpha)} dt = \frac{(\beta-\alpha)^2}{\pi^2}\sin\frac{\pi(x-\alpha)}{(\beta-\alpha)}.$

(v) $\displaystyle\int_\alpha^\beta \left|\frac{\partial G(x,t)}{\partial x}\right| dt = \frac{(x-\alpha)^2+(\beta-x)^2}{2(\beta-\alpha)} \leq \frac{1}{2}(\beta-\alpha).$

33.8. Consider the boundary value problem (33.23), (32.4). Show that $y(x)$ is a solution of this problem if and only if

$$y(x) = A + (x-\alpha)B + \int_\alpha^\beta G(x,t)f(t,y(t),y'(t))dt, \qquad (33.26)$$

where $G(x,t)$ is the Green function of the problem $y'' = 0$, $y(\alpha) = y'(\beta) = 0$ and is given by

$$G(x,t) = \begin{cases} (\alpha - x), & \alpha \leq x \leq t \\ (\alpha - t), & t \leq x \leq \beta. \end{cases} \qquad (33.27)$$

Also establish the following:

(i) $G(x,t) \leq 0$ in $[\alpha, \beta] \times [\alpha, \beta]$.

(ii) $|G(x,t)| \leq (\beta - \alpha)$.

(iii) $\displaystyle\int_\alpha^\beta |G(x,t)|dt = \frac{1}{2}(x-\alpha)(2\beta-\alpha-x) \leq \frac{1}{2}(\beta-\alpha)^2.$

(iv) $\displaystyle\int_\alpha^\beta \left|\frac{\partial G(x,t)}{\partial x}\right| dt = (\beta - x) \leq (\beta - \alpha).$

33.9. Consider the DE

$$y'' - ky = f(x,y,y'), \quad k > 0 \qquad (33.28)$$

together with the boundary conditions (32.3). Show that $y(x)$ is a solution of this problem if and only if

$$y(x) = \frac{\sinh\sqrt{k}(\beta-x)}{\sinh\sqrt{k}(\beta-\alpha)}A + \frac{\sinh\sqrt{k}(x-\alpha)}{\sinh\sqrt{k}(\beta-\alpha)}B + \int_\alpha^\beta G(x,t)f(t,y(t),y'(t))dt,$$

where $G(x,t)$ is the Green function of the problem $y'' - ky = 0$, $y(\alpha) = y(\beta) = 0$ and is given by

$$G(x,t) = \frac{-1}{\sqrt{k}\sinh\sqrt{k}(\beta-\alpha)} \begin{cases} \sinh\sqrt{k}(x-\alpha)\sinh\sqrt{k}(\beta-t), & \alpha \leq x \leq t \\ \sinh\sqrt{k}(t-\alpha)\sinh\sqrt{k}(\beta-x), & t \leq x \leq \beta. \end{cases}$$
$$(33.29)$$

Also establish the following:

(i) $G(x,t) \leq 0$ in $[\alpha, \beta] \times [\alpha, \beta]$.

(ii) $\displaystyle\int_\alpha^\beta |G(x,t)|dt \;=\; \frac{1}{k}\left(1 - \frac{\cosh\sqrt{k}\left(\frac{\beta+\alpha}{2} - x\right)}{\cosh\sqrt{k}\left(\frac{\beta-\alpha}{2}\right)}\right)$

$\displaystyle\qquad\qquad\le \frac{1}{k}\left(1 - \frac{1}{\cosh\sqrt{k}\left(\frac{\beta-\alpha}{2}\right)}\right).$

Answers or Hints

33.1. $1 + x - \frac{\pi}{2}\sin x.$

33.2. $\frac{(2+\sin 1)}{\sinh 1}\sinh x - \sin x.$

33.3. The associated Green's functions are as follows:

For Problem 32.1(vi), $\;G(x,t) = \begin{cases} -\cos t\sin x, & 0 \le x \le t \\ -\sin t\cos x, & t \le x \le \pi/2 \end{cases}$

For Problem 32.1(vii), $\;G(x,t) = \begin{cases} -\frac{x}{2}(2-t)e^{-(x-t)}, & 0 \le x \le t \\ -\frac{t}{2}(2-x)e^{-(x-t)}, & t \le x \le 2. \end{cases}$

33.4. Verify directly. $x^4 + 2x^2 - 2x + 1.$

33.5. Verify directly.

33.6. Verify directly.

33.7. Verify (33.24) directly. For part (ii) note that $|G(x,t)| \le (\beta-x)(x-\alpha)/(\beta-\alpha).$

33.8. Verify directly.

33.9. Verify directly.

Lecture 34
Degenerate Linear Boundary Value Problems

From Corollary 32.2 we know that if $\Delta = 0$, then the homogeneous boundary value problem (6.1), (32.2) has an infinite number of solutions. However, the following examples suggest that the situation is entirely different for the nonhomogeneous problem (6.6), (32.1).

Example 34.1. Consider the nonhomogeneous DE

$$y'' + 2y' + 5y = 4e^{-x} \tag{34.1}$$

together with the boundary conditions (32.16). As in Example 32.3 we take $y_1(x) = e^{-x}\cos 2x$ and $y_2(x) = e^{-x}\sin 2x$ as two linearly independent solutions of the homogeneous DE (32.15). It is easy to verify that $z(x) = e^{-x}$ is a solution of (34.1). Thus, the general solution of (34.1) can be written as

$$y(x) = c_1 e^{-x}\cos 2x + c_2 e^{-x}\sin 2x + e^{-x}.$$

This solution satisfies the boundary conditions (32.16) if and only if

$$\begin{aligned} c_1 + 1 &= 0 \\ -c_1 e^{-\pi/2} + e^{-\pi/2} &= 0, \end{aligned} \tag{34.2}$$

which is impossible. Hence, the problem (34.1), (32.16) has no solution.

Example 34.2. Consider the nonhomogeneous DE

$$y'' + 2y' + 5y = 4e^{-x}\cos 2x \tag{34.3}$$

together with the boundary conditions (32.16). For the DE (34.3), $z(x) = xe^{-x}\sin 2x$ is a particular solution, and hence as in Example 34.1 its general solution is

$$y(x) = c_1 e^{-x}\cos 2x + c_2 e^{-x}\sin 2x + xe^{-x}\sin 2x.$$

This solution satisfies the boundary conditions (32.16) if and only if

$$\begin{aligned} c_1 &= 0 \\ -c_1 e^{-\pi/2} &= 0, \end{aligned} \tag{34.4}$$

R.P. Agarwal and D. O'Regan, *An Introduction to Ordinary Differential Equations*, doi: 10.1007/978-0-387-71276-5_34, © Springer Science + Business Media, LLC 2008

i.e., $c_1 = 0$. Thus, the problem (34.3), (32.16) has an infinite number of solutions

$$y(x) = ce^{-x}\sin 2x + xe^{-x}\sin 2x,$$

where c is an arbitrary constant.

In systems (34.2) and (34.4) the unknowns are c_1 and c_2 and the coefficient matrix is

$$\begin{bmatrix} 1 & 0 \\ -e^{-\pi/2} & 0 \end{bmatrix},$$

whose rank is 1. Thus, from Theorem 13.3 the conclusions in the above examples are not surprising. As a matter of fact using this theorem we can provide necessary and sufficient conditions for the existence of at least one solution of the nonhomogeneous problem (6.6), (32.1). For this, it is convenient to write this problem in system form as

$$\begin{aligned} u' &= A(x)u + b(x) \\ L_0 u(\alpha) + L_1 u(\beta) &= \ell, \end{aligned} \tag{34.5}$$

where

$$A(x) = \begin{bmatrix} 0 & 1 \\ -\dfrac{p_2(x)}{p_0(x)} & -\dfrac{p_1(x)}{p_0(x)} \end{bmatrix}, \qquad b(x) = \begin{bmatrix} 0 \\ \dfrac{r(x)}{p_0(x)} \end{bmatrix},$$

$$L_0 = \begin{bmatrix} a_0 & a_1 \\ c_0 & c_1 \end{bmatrix}, \qquad L_1 = \begin{bmatrix} b_0 & b_1 \\ d_0 & d_1 \end{bmatrix}, \qquad \text{and} \quad \ell = \begin{bmatrix} A \\ B \end{bmatrix}.$$

Theorem 34.1. Let $\Psi(x)$ be a fundamental matrix solution of the homogeneous system $u' = A(x)u$, and let the rank of the matrix $P = L_0\Psi(\alpha) + L_1\Psi(\beta)$ be $2-m$ ($1 \le m \le 2$). Then the boundary value problem (34.5) has a solution if and only if

$$Q\ell - QL_1\Psi(\beta)\int_{\alpha}^{\beta}\Psi^{-1}(t)b(t)dt = 0, \tag{34.6}$$

where Q is a $m \times 2$ matrix whose row vectors are linearly independent vectors q^i, $1 \le i \le m$ satisfying $q^i P = 0$.

Further, if (34.6) holds, then any solution of (34.5) can be given by

$$u(x) = \sum_{i=1}^{m} k_i u^i(x) + \Psi(x)S\ell + \int_{\alpha}^{\beta} G(x,t)b(t)dt, \tag{34.7}$$

where k_i, $1 \le i \le m$ are arbitrary constants, $u^i(x)$, $1 \le i \le m$ are m linearly independent solutions of the homogeneous system $u' = A(x)u$

satisfying the homogeneous boundary conditions $L_0 u(\alpha) + L_1 u(\beta) = 0$, S is a 2×2 matrix independent of $b(x)$ and ℓ such that $PSv = v$ for any column vector v satisfying $Qv = 0$, and $G(x, t)$ is the piecewise continuous matrix called the *generalized Green's matrix*

$$G(x, t) = \begin{cases} -\Psi(x) S L_1 \Psi(\beta) \Psi^{-1}(t), & \alpha \le x \le t \\ \Psi(x)[I - S L_1 \Psi(\beta)] \Psi^{-1}(t), & t \le x \le \beta. \end{cases} \tag{34.8}$$

Proof. From (18.14) any solution of the above nonhomogeneous system can be written as

$$u(x) = \Psi(x) c + \Psi(x) \int_\alpha^x \Psi^{-1}(t) b(t) dt, \tag{34.9}$$

where c is an arbitrary constant vector.

Thus, the problem (34.5) has a solution if and only if the system

$$(L_0 \Psi(\alpha) + L_1 \Psi(\beta)) c + L_1 \Psi(\beta) \int_\alpha^\beta \Psi^{-1}(t) b(t) dt = \ell,$$

i.e.,

$$Pc = \ell - L_1 \Psi(\beta) \int_\alpha^\beta \Psi^{-1}(t) b(t) dt \tag{34.10}$$

has a solution. However, since the rank of the matrix P is $2 - m$, from Theorem 13.3 the system (34.10) has a solution if and only if (34.6) holds. This proves the first conclusion of the theorem.

When (34.6) holds, by Theorem 13.3 the constant vector c satisfying (34.10) can be given by

$$c = \sum_{i=1}^m k_i c^i + S \left[\ell - L_1 \Psi(\beta) \int_\alpha^\beta \Psi^{-1}(t) b(t) dt \right], \tag{34.11}$$

where k_i, $1 \le i \le m$ are arbitrary constants, c^i, $1 \le i \le m$ are m linearly independent column vectors satisfying $Pc^i = 0$, and S is a 2×2 matrix independent of

$$\ell - L_1 \Psi(\beta) \int_\alpha^\beta \Psi^{-1}(t) b(t) dt$$

such that $PSv = v$ for any column vector v satisfying $Qv = 0$.

Substituting (34.11) into (34.9), we obtain a solution of the problem

(34.5) as

$$
\begin{aligned}
u(x) &= \sum_{i=1}^{m} k_i \Psi(x) c^i + \Psi(x) S\ell - \Psi(x) S L_1 \Psi(\beta) \int_{\alpha}^{\beta} \Psi^{-1}(t) b(t) dt \\
&\quad + \Psi(x) \int_{\alpha}^{x} \Psi^{-1}(t) b(t) dt \\
&= \sum_{i=1}^{m} k_i u^i(x) + \Psi(x) S\ell + \int_{\alpha}^{x} \Psi(x) [I - S L_1 \Psi(\beta)] \Psi^{-1}(t) b(t) dt \\
&\quad - \int_{x}^{\beta} \Psi(x) S L_1 \Psi(\beta) \Psi^{-1}(t) b(t) dt \\
&= \sum_{i=1}^{m} k_i u^i(x) + \Psi(x) S\ell + \int_{\alpha}^{\beta} G(x,t) b(t) dt,
\end{aligned}
$$

where $u^i(x) = \Psi(x) c^i$ are evidently linearly independent solutions of the homogeneous system $u' - A(x)u$, and moreover since $Pc^i = 0$ it follows that

$$
\begin{aligned}
L_0 u^i(\alpha) + L_1 u^i(\beta) &= L_0 \Psi(\alpha) c^i + L_1 \Psi(\beta) c^i \\
&= (L_0 \Psi(\alpha) + L_1 \Psi(\beta)) c^i = Pc^i = 0. \quad \blacksquare
\end{aligned}
$$

Example 34.3. Consider the boundary value problem

$$
y'' + y = r(x) \tag{34.12}
$$

$$
\begin{aligned}
y(0) - y(2\pi) &= 0 \\
y'(0) - y'(2\pi) &= -\pi,
\end{aligned} \tag{34.13}
$$

which in the system form is the same as (34.5) with

$$
A(x) = \begin{bmatrix} 0 & 1 \\ -1 & 0 \end{bmatrix}, \qquad b(x) = \begin{bmatrix} 0 \\ r(x) \end{bmatrix},
$$

$$
L_0 = \begin{bmatrix} 1 & 0 \\ 0 & 1 \end{bmatrix}, \qquad L_1 = \begin{bmatrix} -1 & 0 \\ 0 & -1 \end{bmatrix}, \qquad \text{and} \quad \ell = \begin{bmatrix} 0 \\ -\pi \end{bmatrix}.
$$

For this problem we choose

$$
\Psi(x) = \begin{bmatrix} \cos x & \sin x \\ -\sin x & \cos x \end{bmatrix},
$$

so that

$$
P = \begin{bmatrix} 0 & 0 \\ 0 & 0 \end{bmatrix}
$$

whose rank is 0, i.e., $m = 2$. Let the matrix

$$Q = \begin{bmatrix} 1 & 0 \\ 0 & 1 \end{bmatrix}$$

whose row vectors q^1 and q^2 are linearly independent and satisfy $q^1 P = q^2 P = 0$. Thus, the condition (34.6) reduces to

$$\begin{bmatrix} 1 & 0 \\ 0 & 1 \end{bmatrix} \begin{bmatrix} 0 \\ -\pi \end{bmatrix} - \begin{bmatrix} 1 & 0 \\ 0 & 1 \end{bmatrix} \begin{bmatrix} -1 & 0 \\ 0 & -1 \end{bmatrix} \begin{bmatrix} 1 & 0 \\ 0 & 1 \end{bmatrix}$$
$$\times \int_0^{2\pi} \begin{bmatrix} \cos t & -\sin t \\ \sin t & \cos t \end{bmatrix} \begin{bmatrix} 0 \\ r(t) \end{bmatrix} dt = 0,$$

which is the same as

$$\int_0^{2\pi} \sin t \, r(t) dt = 0$$

$$\pi - \int_0^{2\pi} \cos t \, r(t) dt = 0. \tag{34.14}$$

Further,

$$u^1(x) = \begin{bmatrix} \cos x \\ -\sin x \end{bmatrix} \quad \text{and} \quad u^2(x) = \begin{bmatrix} \sin x \\ \cos x \end{bmatrix}$$

are linearly independent solutions of $u' = A(x)u$, and satisfy the boundary conditions $L_0 u(0) + L_1 u(2\pi) = 0$. Also, we note that $Qv = 0$ implies that $v = [0 \ 0]^T$, and hence we can choose the matrix

$$S = \begin{bmatrix} 0 & 0 \\ 0 & 0 \end{bmatrix}.$$

Thus, if the conditions (34.14) are satisfied, then any solution of the above problem can be written as

$$u(x) = k_1 \begin{bmatrix} \cos x \\ -\sin x \end{bmatrix} + k_2 \begin{bmatrix} \sin x \\ \cos x \end{bmatrix} + \int_0^{2\pi} G(x,t)b(t)dt,$$

where the generalized Green's matrix $G(x,t)$ is

$$G(x,t) = \begin{cases} 0, & 0 \le x \le t \\ \begin{bmatrix} \cos(x-t) & \sin(x-t) \\ -\sin(x-t) & \cos(x-t) \end{bmatrix}, & t \le x \le 2\pi. \end{cases}$$

Hence, any solution of (34.12), (34.13) is given by

$$y(x) = k_1 \cos x + k_2 \sin x + \int_0^x \sin(x - t) r(t) dt. \qquad (34.15)$$

In particular, for the function $r(x) = \cos x$, conditions (34.14) are satisfied, and (34.15) reduces to

$$y(x) = k_1 \cos x + k_2 \sin x + \frac{1}{2} x \sin x.$$

Example 34.4. Consider the boundary value problem

$$y'' = r(x) \qquad (34.16)$$

$$y(0) = 0$$
$$y(1) - y'(1) = 0, \qquad (34.17)$$

which in system form is the same as (34.5) with

$$A(x) = \begin{bmatrix} 0 & 1 \\ 0 & 0 \end{bmatrix}, \qquad b(x) = \begin{bmatrix} 0 \\ r(x) \end{bmatrix},$$

$$L_0 = \begin{bmatrix} 1 & 0 \\ 0 & 0 \end{bmatrix}, \qquad L_1 = \begin{bmatrix} 0 & 0 \\ 1 & -1 \end{bmatrix}, \quad \text{and} \quad \ell = \begin{bmatrix} 0 \\ 0 \end{bmatrix}.$$

For this problem we take

$$\Psi(x) = \begin{bmatrix} 1 & x \\ 0 & 1 \end{bmatrix},$$

so that

$$P = \begin{bmatrix} 1 & 0 \\ 1 & 0 \end{bmatrix}$$

whose rank is 1; i.e., $m = 1$. Let the matrix $Q = q^1 = (1 \quad -1)$ which satisfies the condition $q^1 P = 0$. Thus, the condition (34.6) reduces to

$$-(1 \quad -1) \begin{bmatrix} 0 & 0 \\ 1 & -1 \end{bmatrix} \begin{bmatrix} 1 & 1 \\ 0 & 1 \end{bmatrix} \int_0^1 \begin{bmatrix} 1 & -t \\ 0 & 1 \end{bmatrix} \begin{bmatrix} 0 \\ r(t) \end{bmatrix} dt = 0,$$

which is the same as

$$\int_0^1 t \, r(t) dt = 0. \qquad (34.18)$$

Further, $u^1(x) = [x \ 1]^T$ is a solution of $u' = A(x)u$, and satisfies the boundary conditions $L_0 u(0) + L_1 u(1) = 0$. Also, we note that $Qv = 0$ implies that we can take $v = [1 \ 1]^T$ and then

$$S = \begin{bmatrix} 1 & 0 \\ 0 & 0 \end{bmatrix}$$

satisfies the condition $PSv = v$. Thus, if the condition (34.18) is satisfied then any solution of the above problem can be written as

$$u(x) = k_1 \begin{bmatrix} x \\ 1 \end{bmatrix} + \int_0^1 G(x,t)b(t)dt,$$

where the generalized Green's matrix is

$$G(x,t) = \begin{cases} 0, & 0 \le x \le t \\ \begin{bmatrix} 1 & x-t \\ 0 & 1 \end{bmatrix}, & t \le x \le 1. \end{cases}$$

Hence, any solution of (34.16), (34.17) is given by

$$y(x) = k_1 x + \int_0^x (x-t)r(t)dt. \tag{34.19}$$

In particular, for the function $r(x) = 2 - 3x$ the condition (34.18) is satisfied, and (34.19) simplifies to

$$y(x) = k_1 x + x^2 - \frac{1}{2}x^3.$$

Problems

34.1.   Find necessary and sufficient conditions so that the following boundary value problems have a solution:

(i) $y'' + y = r(x)$
      $y(0) = y(\pi) = 0.$

(ii) $y'' = r(x)$
      $y'(0) = y'(1) = 0.$

(iii) $y'' + y = r(x)$
      $y(0) = y(2\pi)$
      $y'(0) = y'(2\pi).$

(iv) $y'' = r(x)$
      $y(-1) = y(1)$
      $y'(-1) = y'(1).$

(v) $y'' + y = r(x)$
      $y(0) = 1, \ y(\pi) = 1/2.$

(vi) $y'' = r(x)$
      $y'(0) = 1, \ y'(1) = 2.$

34.2. Solve the following boundary value problems:

(i) $y'' + y = \cos x$
      $y(0) = y(\pi) = 0.$

(ii) $y'' = \cos \pi x$
      $y'(0) = y'(1) = 0.$

(iii) $y'' + (1/4)y = \sin x/2$
      $y'(0) = -1, \; y(\pi) = 0.$

(iv) $y'' = x^3$
      $y(-1) - y(1) + (1/10) = 0$
      $y'(-1) - y'(1) = 0.$

34.3. Let the DE (6.1) be self-adjoint and $y_0(x)$ be a nontrivial solution of the homogeneous problem (6.1), (33.14). Show that the nonhomogeneous boundary value problem (6.6), (33.14) has a solution if and only if

$$\int_\alpha^\beta y_0(x)r(x)dx \; = \; 0.$$

Answers or Hints

34.1. (i) $\int_0^\pi r(x)\sin x dx = 0.$ (ii) $\int_0^1 r(x)dx = 0.$ (iii) $\int_0^{2\pi} r(x)\sin x dx = 0,$ $\int_0^{2\pi} r(x)\cos x dx = 0.$ (iv) $\int_{-1}^1 r(x)dx = 0.$ (v) $\int_0^\pi r(x)\sin x dx = \frac{3}{2}.$ (vi) $\int_0^1 r(x)dx = 1.$

34.2. (i) $k_1 \sin x + \frac{1}{2}x\sin x.$ (ii) $k_1 - \frac{1}{\pi^2}\cos \pi x.$ (iii) $k_1 \cos \frac{1}{2}x - x\cos \frac{1}{2}x.$ (iv) $k_1 + \frac{1}{20}x^5.$

34.3. Let $y_0(x),\, y_1(x)$ be linearly independent solutions of (6.1). Now write the general solution of (6.6) in the form (33.20).

Lecture 35
Maximum Principles

Maximum principles which are known for ordinary as well as partial differential inequalities play a key role in proving existence–uniqueness results and in the construction of solutions of DEs. In this lecture, we shall discuss the known maximum principle for a function satisfying a second-order differential inequality and extend it to a general form which is extremely useful in studying second-order initial and boundary value problems.

Theorem 35.1. If $y \in C^{(2)}[\alpha, \beta]$, $y''(x) \geq 0$ in (α, β), and $y(x)$ attains its maximum at an interior point of $[\alpha, \beta]$, then $y(x)$ is identically constant in $[\alpha, \beta]$.

Proof.   First, suppose that $y''(x) > 0$ in (α, β); if $y(x)$ attains its maximum at an interior point, say, x_0 of $[\alpha, \beta]$, then $y'(x_0) = 0$ and $y''(x_0) \leq 0$, which is a contradiction to our assumption that $y''(x) > 0$. Thus, if $y''(x) > 0$ in (α, β), then the function $y(x)$ cannot attain its maximum at an interior point of $[\alpha, \beta]$. Now suppose that $y''(x) \geq 0$ in (α, β) and that $y(x)$ attains its maximum at an interior point of $[\alpha, \beta]$, say, x_1. If $y(x_1) = M$, then $y(x) \leq M$ in $[\alpha, \beta]$. Suppose that there exists a point $x_2 \in (\alpha, \beta)$ such that $y(x_2) < M$. If $x_2 > x_1$, then we set $z(x) = \exp(\gamma(x - x_1)) - 1$, where γ is a positive constant. For this function $z(x)$, it is immediate that

$$z(x) < 0, \quad x \in [\alpha, x_1), \quad z(x_1) = 0, \quad z(x) > 0, \quad x \in (x_1, \beta] \quad (35.1)$$

and

$$z''(x) = \gamma^2 \exp(\gamma(x - x_1)) > 0, \quad x \in [\alpha, \beta].$$

Now we define $w(x) = y(x) + \epsilon z(x)$, where $0 < \epsilon < (M - y(x_2))/z(x_2)$. Since $y(x_2) < M$ and $z(x_2) > 0$, such an ϵ always exists. From (35.1), it follows that $w(x) < y(x) \leq M$, $x \in (\alpha, x_1)$, $w(x_2) = y(x_2) + \epsilon z(x_2) < M$, and $w(x_1) = M$.

Since $w''(x) = y''(x) + \epsilon z''(x) > 0$ in (α, x_2), the function $w(x)$ cannot attain a maximum in the interior of $[\alpha, x_2]$. However, since $w(\alpha) < M$, $w(x_2) < M$ and $w(x_1) = M$ where $x_1 \in (\alpha, x_2)$, $w(x)$ must attain a maximum greater than or equal to M at an interior point of (α, x_2), which is a contradiction. Therefore, there does not exist a point $x_2 \in (\alpha, \beta)$, $x_2 > x_1$ such that $y(x_2) < M$.

R.P. Agarwal and D. O'Regan, *An Introduction to Ordinary Differential Equations*,   258
doi: 10.1007/978-0-387-71276-5_35, © Springer Science + Business Media, LLC 2008

If $x_2 < x_1$, we can set $z(x) = \exp(-\gamma(x - x_1)) - 1$, where γ is a positive constant and again by similar arguments we can show that such an x_2 cannot exist. Therefore, $y(x) = M$ in $[\alpha, \beta]$. ■

The above result holds if we reverse the inequality and replace "maximum" by "minimum."

We shall now consider a more general inequality $y'' + p(x)y' + q(x)y \geq 0$. However, for this inequality the following examples show that no matter whether $q(x)$ is negative or positive the preceding result need not hold. Hence, we can at most expect a restricted form of maximum principle.

Example 35.1. The function $y(x) = \sin x$ is a solution of $y'' + y = 0$. However, in the interval $[0, \pi]$, $y(x)$ attains its maximum at $x = \pi/2$ which is an interior point.

Example 35.2. For $y'' - y = 0$, $y(x) = -e^x - e^{-x}$ is a solution which attains its maximum value -2 at $x = 0$ in the interval $[-1, 1]$.

Theorem 35.2. Let $y(x)$ satisfy the differential inequality

$$y''(x) + p(x)y'(x) + q(x)y(x) \ \geq \ 0, \quad x \in (\alpha, \beta) \tag{35.2}$$

in which $p(x)$ and $q(x)$ (< 0) are bounded in every closed subinterval of (α, β). If $y(x)$ assumes a nonnegative maximum value M at an interior point of $[\alpha, \beta]$, then $y(x) \equiv M$.

Proof. If the inequality in (35.2) is strict and $y(x)$ assumes a nonnegative maximum M at an interior point x_0 of $[\alpha, \beta]$, then $y(x_0) = M$, $y'(x_0) = 0$ and $y''(x_0) \leq 0$. Since $p(x)$ and $q(x)$ are bounded in a closed subinterval containing x_0 and $q(x) \leq 0$, we have

$$y''(x_0) + p(x_0)y'(x_0) + q(x_0)y(x_0) \ \leq \ 0,$$

contrary to our assumption of strict inequality in (35.2). Hence, if the inequality (35.2) is strict, $y(x)$ cannot attain its nonnegative maximum at an interior point of $[\alpha, \beta]$.

Now if (35.2) holds and $y(x_1) = M$ for some $x_1 \in (\alpha, \beta)$, we suppose that there exists a point $x_2 \in (\alpha, \beta)$ such that $y(x_2) < M$. If $x_2 > x_1$, then once again we set $z(x) = \exp(\gamma(x - x_1)) - 1$, where γ is a positive constant yet to be determined. This function $z(x)$ satisfies (35.1), and since $q(x) \leq 0$ it follows that

$$
\begin{aligned}
z'' &+ p(x)z' + q(x)z \\
&= [\gamma^2 + p(x)\gamma + q(x)(1 - \exp(-\gamma(x - x_1)))]\exp(\gamma(x - x_1)) \\
&\geq [\gamma^2 + p(x)\gamma + q(x)]\exp(\gamma(x - x_1)).
\end{aligned}
$$

We choose γ such that $\gamma^2 + p(x)\gamma + q(x) > 0$ in (α, β). This is always possible since $p(x)$ and $q(x)$ are bounded in every closed subinterval of (α, β). With such a choice of γ, we see that

$$z'' + p(x)z' + q(x)z > 0.$$

The rest of the proof is word for word the same as that of Theorem 35.1, except for the function w instead of $w''(x) > 0$ we now have $w''(x) + p(x)w'(x) + q(x)w(x) > 0$.

If $q(x)$ is not identically zero in (α, β), then the only nonnegative constant M for which $y(x) = M$ satisfies (35.2) is $M = 0$. For this, we have $y(x) = M \geq 0$, $y'(x) = y''(x) = 0$, $x \in (\alpha, \beta)$ and therefore $y''(x) + p(x)y'(x) + q(x)y(x) = q(x)M \geq 0$, but $q(x) \leq 0$, and hence it is necessary that $M = 0$. ∎

Next we shall prove the following corollaries.

Corollary 35.3. Suppose that $y(x)$ is a nonconstant solution of the differential inequality (35.2) having one-sided derivatives at α and β, and $p(x)$ and $q(x)$ (≤ 0) are bounded in every closed subinterval of (α, β). If $y(x)$ has a nonnegative maximum at α and if the function $p(x) + (x - \alpha)q(x)$ is bounded from below at α, then $y'(\alpha) < 0$. If $y(x)$ has a nonnegative maximum at β and if $p(x) - (\beta - x)q(x)$ is bounded from above at β, then $y'(\beta) > 0$.

Proof. Suppose that $y(x)$ has a nonnegative maximum at α, say, $y(\alpha) = M \geq 0$, then $y(x) \leq M$, $x \in [\alpha, \beta]$, and since $y(x)$ is nonconstant, there exists a $x_0 \in (\alpha, \beta)$ such that $y(x_0) < M$.

We define $z(x) = \exp(\gamma(x - \alpha)) - 1$, where γ is a positive constant yet to be determined. Then since $q(x) \leq 0$ and $1 - \exp(-\gamma(x - \alpha)) \leq \gamma(x - \alpha)$ for $x \geq \alpha$ it follows that

$$\begin{aligned}
z'' &+ p(x)z' + q(x)z \\
&= [\gamma^2 + p(x)\gamma + q(x)(1 - \exp(-\gamma(x - \alpha)))] \exp(\gamma(x - \alpha)) \\
&\geq [\gamma^2 + \gamma(p(x) + q(x)(x - \alpha))] \exp(\gamma(x - \alpha)).
\end{aligned}$$

We choose γ such that $\gamma^2 + \gamma(p(x) + q(x)(x - \alpha)) > 0$ for $x \in [\alpha, x_0]$. This is always possible since $p(x)$ and $q(x)$ are bounded in every closed subinterval and $p(x) + (x - \alpha)q(x)$ is bounded from below at α. Then $z'' + p(x)z' + q(x)z > 0$.

Now we define $w(x) = y(x) + \epsilon z(x)$, where $0 < \epsilon < (M - y(x_0))/z(x_0)$. Then $w(\alpha) = y(\alpha) = M$ which implies that $w(x)$ has a maximum greater than or equal to M in $[\alpha, x_0]$. However, since $w'' + p(x)w' + q(x)w > 0$, the nonnegative maximum must occur at one of the endpoints of $[\alpha, x_0]$.

Finally, $w(x_0) = y(x_0) + \epsilon z(x_0) < M$ implies that the maximum occurs at α. Therefore, the one-sided derivative of $w(x)$ at α cannot be positive, i.e., $w'(\alpha) \leq 0$, and $w'(\alpha) = y'(\alpha) + \epsilon z'(\alpha) \leq 0$. However, since $z'(\alpha) = \gamma > 0$, we must have $y'(\alpha) < 0$.

If the nonnegative maximum M of $y(x)$ occurs at β, then by similar arguments we can show that $y'(\beta) > 0$. ∎

Corollary 35.4. Suppose that $y(x)$ is a solution of the differential inequality (35.2), which is continuous in $[\alpha, \beta]$ and $y(\alpha) \leq 0$, $y(\beta) \leq 0$, and $p(x)$ and $q(x)$ (≤ 0) are bounded in every closed subinterval of (α, β). Then $y(x) < 0$ in (α, β) unless $y(x) = 0$ in $[\alpha, \beta]$.

Proof. If $y(x)$ has a negative maximum, then $y(x) < 0$ in $[\alpha, \beta]$. Otherwise, by Theorem 35.2, the nonnegative maximum of $y(x)$ must occur at the endpoints. However, since $y(\alpha) \leq 0$, $y(\beta) \leq 0$ we must have $y(x) < 0$ in (α, β) unless $y(x) = 0$ in $[\alpha, \beta]$. ∎

The following two examples illustrate how maximum principles can be applied to obtain lower and upper bounds for the solutions of DEs which cannot be solved explicitly.

Example 35.3. Consider the boundary value problem

$$y'' - x^2 y = 0, \quad x \in (\alpha, \beta)$$
$$y(\alpha) = \gamma_1, \quad y(\beta) = \gamma_2 \tag{35.3}$$

for which a unique solution $y(x)$ always exists.

Suppose there exists a function $z(x)$ such that

$$z'' - x^2 z \leq 0, \quad z(\alpha) \geq \gamma_1, \quad z(\beta) \geq \gamma_2. \tag{35.4}$$

For such a function $z(x)$, we define $w(x) = y(x) - z(x)$. Clearly, $w(x)$ satisfies

$$w'' - x^2 w \geq 0, \quad w(\alpha) \leq 0, \quad w(\beta) \leq 0 \tag{35.5}$$

and hence Corollary 35.4 is applicable, and we find that $w(x) \leq 0$ in $[\alpha, \beta]$, i.e., $y(x) \leq z(x)$ in $[\alpha, \beta]$.

Now we shall construct such a function $z(x)$ as follows: we set $z_1(x) = A\{2 - \exp(-\gamma(x - \alpha))\}$ where A and γ are constants yet to be determined. Since

$$z_1'' - x^2 z_1 = A\{(-\gamma^2 + x^2) \exp(-\gamma(x - \alpha)) - 2x^2\},$$

we choose $A = \max\{\gamma_1, \gamma_2, 0\}$, and $\gamma = \max\{|\alpha|, |\beta|\} + 1$, so that $A \geq 0$, $\gamma > 0$, $-\gamma^2 + x^2 < 0$, $x \in [\alpha, \beta]$. Thus, with this choice of A and γ, it follows that $z_1'' - x^2 z_1 \leq 0$, $z_1(\alpha) = A \geq \gamma_1$, $z_1(\beta) \geq A \geq \gamma_2$. Hence, $z_1(x)$ satisfies (35.4) and we have $y(x) \leq z_1(x)$, $x \in [\alpha, \beta]$.

Similarly, to obtain a lower bound for $y(x)$, we let $z_2(x) = B\{2 - \exp(-\gamma(x - \alpha))\}$ where γ is chosen as before and $B = \min\{\gamma_1, \gamma_2, 0\}$. Then $B \leq 0$, and $z_2(x)$ satisfies $z_2'' - x^2 z_2 \geq 0$, $z_2(\alpha) = B \leq \gamma_1$, $z_2(\beta) \leq B \leq \gamma_2$. Hence, $z_2(x)$ satisfies (35.4) with the inequalities reversed, and the function $w(x) = z_2(x) - y(x)$ satisfies (35.5). Therefore, it follows that $z_2(x) \leq y(x)$, $x \in [\alpha, \beta]$.

In conclusion, we have

$$B\phi(x) \leq y(x) \leq A\phi(x),$$

where $A = \max\{\gamma_1, \gamma_2, 0\}$, $B = \min\{\gamma_1, \gamma_2, 0\}$, $\phi(x) = 2 - \exp(-\gamma(x - \alpha))$ and $\gamma = \max\{|\alpha|, |\beta|\} + 1$.

Example 35.4. Once again we consider the DE $y'' - x^2 y = 0$ in the interval $(0, 1)$, but with initial conditions $y(0) = 1$, $y'(0) = 0$.

To obtain an upper bound for $y(x)$, it suffices to find a function $z_1(x)$ satisfying

$$z_1'' - x^2 z_1 \geq 0, \quad x \in (0, 1), \quad z_1(0) \geq 1, \quad z_1'(0) \geq 0. \qquad (35.6)$$

For this, we define $v_1(x) = z_1(x) - y(x)$, and note that

$$v_1'' - x^2 v_1 \geq 0, \quad x \in (0, 1), \quad v_1(0) \geq 0, \quad v_1'(0) \geq 0.$$

Since $v_1(0) \geq 0$, the function $v_1(x)$ has a nonnegative maximum in every subinterval $[0, x_0]$ of $[0, 1]$. Thus, from Theorem 35.2 it follows that this maximum must occur either at 0 or x_0. Since $v_1'(0) \geq 0$, from Corollary 35.3, the maximum must occur at x_0 unless $v_1(x)$ is constant in $[0, x_1]$. Hence, for $x_0 \in (0, 1)$, $v_1(x_0) \geq v_1(0) \geq 0$, and by Corollary 35.3 we find that $v_1'(x_0) \geq 0$. Therefore, it follows that for each $x \in (0, 1)$, $v_1(x) = z_1(x) - y(x) \geq v_1(0) \geq 0$, and hence $y(x) \leq z_1(x)$.

To construct such a function $z_1(x)$, we set $z_1(x) = c_1 x^2 + 1$, where c_1 is a constant yet to be determined. Since

$$z_1'' - x^2 z_1 = 2c_1 - x^2(c_1 x^2 + 1) = c_1(2 - x^4) - x^2$$

we need to choose c_1 such that $c_1 \geq x^2/(2 - x^4)$, $x \in [0, 1]$. Since $x^2/(2 - x^4) \leq 1$ for all $x \in [0, 1]$, we can let $c_1 = 1$. Then $z_1(x) = x^2 + 1$ and it satisfies (35.6). Therefore, it follows that $y(x) \leq x^2 + 1$, $x \in [0, 1]$.

Similarly, to obtain a lower bound we need to find a function $z_2(x)$ satisfying

$$z_2'' - x^2 z_2 \leq 0, \quad x \in (0, 1), \quad z_2(0) \leq 1, \quad z_2'(0) \leq 0. \qquad (35.7)$$

To construct such a function $z_2(x)$, once again we set $z_2(x) = c_2 x^2 + 1$, where c_2 is a constant yet to be determined. Since

$$z_2'' - x^2 z_2 = c_2(2 - x^4) - x^2$$

we need to choose c_2 such that $c_2 \leq x^2/(2-x^4)$, $x \in [0,1]$. Therefore, we can choose $c_2 = 0$, to obtain $z_2(x) = 1$ which satisfies (35.7). Hence, it follows that $1 \leq y(x)$, $x \in [0,1]$.

In conclusion, we have

$$1 \leq y(x) \leq 1+x^2, \quad x \in [0,1].$$

Finally, we remark that in Examples 35.3 and 35.4 above we can use polynomials, rational functions, exponentials, etc., for the construction of the functions $z_1(x)$ and $z_2(x)$.

Problems

35.1. The function $y = \sin x$, $x \in (0, \pi)$ attains its positive maximum at $x = \pi/2$, and is a solution of the DE

$$y'' + (\tan x)y' = 0.$$

Does this contradict Theorem 35.2?

35.2. Consider the DE

$$y'' + \alpha e^{\beta y} = -x^2, \quad x \in (0,1),$$

where α and β are positive constants. Show that its solution cannot attain a minimum in $(0,1)$.

35.3. Consider the DE

$$y'' - \alpha \cos(y') = \beta x^4, \quad x \in (-1,1),$$

where α and β are positive constants. Show that its solution cannot attain a maximum in $(-1,1)$.

35.4. Consider the boundary value problem

$$y'' + x^2 y' = -x^4, \quad x \in (0,1)$$
$$y(0) = 0 = y(1).$$

Show that its solution cannot attain a minimum in $(0,1)$. Further, show that $y'(0) > 0$, $y'(1) < 0$.

35.5. Show that the boundary value problem

$$y'' + p(x)y' + q(x)y = r(x), \quad x \in (\alpha, \beta)$$
$$y(\alpha) = A, \quad y(\beta) = B$$

where $p(x)$ and $q(x)$ are as in Theorem 35.2, has at most one solution.

35.6. Show that the solution $y(x)$ of the boundary value problem

$$y'' - xy = 0, \quad x \in (0,1), \quad y(0) = 0, \quad y(1) = 1$$

satisfies the inequalities $(x + x^2)/2 \le y(x) \le x, \ x \in [0,1]$.

35.7. Show that the solution $y(x)$ of the initial value problem

$$y'' + \frac{1}{x}y' - y = 0, \quad x \in (0,1), \quad y(0) = 1, \quad y'(0) = 0$$

satisfies the inequalities $1 + x^2/4 \le y(x) \le 1 + x^2/3, \ x \in [0,1]$.

Answers or Hints

35.1. No.

35.2. Use contradiction.

35.3. Use contradiction.

35.4. Use contradiction.

35.5. Use Corollary 35.4.

35.6. See Example 35.3.

35.7. See Example 35.4.

Lecture 36
Sturm–Liouville Problems

In our previous lectures we have seen that homogeneous boundary value problem (6.1), (32.2) may have nontrivial solutions. If the coefficients of the DE and/or of the boundary conditions depend upon a parameter, then one of the pioneer problems of mathematical physics is to determine the value(s) of the parameter for which such nontrivial solutions exist. These special values of the parameter are called *eigenvalues* and the corresponding nontrivial solutions are called *eigenfunctions*. The boundary value problem which consists of the self-adjoint DE

$$(p(x)y')' + q(x)y + \lambda r(x)y = \mathcal{P}_2[y] + \lambda r(x)y = 0 \qquad (36.1)$$

and the boundary conditions (33.14) is called the *Sturm–Liouville problem*. In the DE (36.1), λ is a parameter, and the functions q, $r \in C(J)$, $p \in C^1(J)$, and $p(x) > 0$, $r(x) > 0$ in J.

The problem (36.1), (33.14) satisfying the above conditions is said to be a *regular* Sturm–Liouville problem. Solving such a problem means finding values of λ (eigenvalues) and the corresponding nontrivial solutions $\phi_\lambda(x)$ (eigenfunctions). The set of all eigenvalues of a regular problem is called its *spectrum*.

The computation of eigenvalues and eigenfunctions is illustrated in the following examples.

Example 36.1. Consider the boundary value problem

$$y'' + \lambda y = 0 \qquad (36.2)$$

$$y(0) = y(\pi) = 0. \qquad (36.3)$$

If $\lambda = 0$, then the general solution of (36.2) is $y(x) = c_1 + c_2 x$, and this solution satisfies the boundary conditions (36.3) if and only if $c_1 = c_2 = 0$, i.e., $y(x) \equiv 0$ is the only solution of (36.2), (36.3). Hence, $\lambda = 0$ is not an eigenvalue of the problem (36.2), (36.3).

If $\lambda \neq 0$, it is convenient to replace λ by μ^2, where μ is a new parameter not necessarily real. In this case the general solution of (36.2) is $y(x) = c_1 e^{i\mu x} + c_2 e^{-i\mu x}$, and this solution satisfies the boundary conditions (36.3) if and only if

$$c_1 + c_2 = 0$$
$$c_1 e^{i\mu\pi} + c_2 e^{-i\mu\pi} = 0. \qquad (36.4)$$

R.P. Agarwal and D. O'Regan, *An Introduction to Ordinary Differential Equations*,
doi: 10.1007/978-0-387-71276-5_36, © Springer Science + Business Media, LLC 2008

The system (36.4) has a nontrivial solution if and only if

$$e^{-i\mu\pi} - e^{i\mu\pi} = 0. \tag{36.5}$$

If $\mu = a + ib$, where a and b are real, condition (36.5) reduces to

$$e^{b\pi}(\cos a\pi - i\sin a\pi) - e^{-b\pi}(\cos a\pi + i\sin a\pi)$$
$$= (e^{b\pi} - e^{-b\pi})\cos a\pi - i(e^{b\pi} + e^{-b\pi})\sin a\pi$$
$$= 2\sinh b\pi \cos a\pi - 2i\cosh b\pi \sin a\pi = 0,$$

i.e.,

$$\sinh b\pi \cos a\pi = 0 \tag{36.6}$$

and

$$\cosh b\pi \sin a\pi = 0. \tag{36.7}$$

Since $\cosh b\pi > 0$ for all values of b, equation (36.7) requires that $a = n$, where n is an integer. Further, for this choice of a, $\cos a\pi \neq 0$, and equation (36.6) reduces to $\sinh b\pi = 0$, i.e., $b = 0$. However, if $b = 0$, then we cannot have $a = 0$, because then $\mu = 0$, and we have seen that it is not an eigenvalue. Hence, $\mu = n$, where n is a nonzero integer. Thus, the eigenvalues of (36.2), (36.3) are $\lambda_n = \mu^2 = n^2$, $n = 1, 2, \ldots$. Further, from (36.4) since $c_2 = -c_1$ for $\lambda_n = n^2$ the corresponding nontrivial solutions of the problem (36.2), (36.3) are

$$\phi_n(x) = c_1(e^{inx} - e^{-inx}) = 2ic_1 \sin nx,$$

or, simply $\phi_n(x) = \sin nx$.

Example 36.2. Consider again the DE (36.2), but with the boundary conditions

$$y(0) + y'(0) = 0, \quad y(1) = 0. \tag{36.8}$$

If $\lambda = 0$, then the general solution $y(x) = c_1 + c_2 x$ of (36.2) also satisfies the boundary conditions (36.8) if and only if $c_1 + c_2 = 0$, i.e., $c_2 = -c_1$. Hence, $\lambda = 0$ is an eigenvalue of (36.2), (36.8) and the corresponding eigenfunction is $\phi_1(x) = 1 - x$.

If $\lambda \neq 0$, then once again we replace λ by μ^2 and note that the general solution $y(x) = c_1 e^{i\mu x} + c_2 e^{-i\mu x}$ of (36.2) satisfies the boundary conditions (36.8) if and only if

$$(c_1 + c_2) + i\mu(c_1 - c_2) = 0$$
$$c_1 e^{i\mu} + c_2 e^{-i\mu} = 0. \tag{36.9}$$

The system (36.9) has a nontrivial solution if and only if

$$(1 + i\mu)e^{-i\mu} - (1 - i\mu)e^{i\mu} = 0,$$

which is equivalent to

$$\tan \mu = \mu. \tag{36.10}$$

To find the real roots of (36.10) we graph the curves $y = \mu$ and $y = \tan \mu$ and observe the values of μ where these curves intersect.

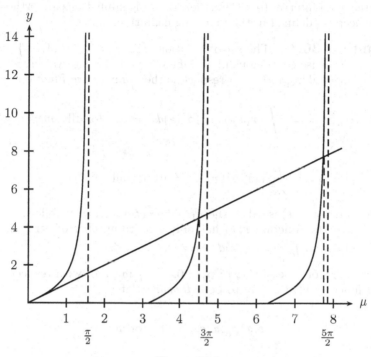

Figure 36.1

From Figure 36.1 it is clear that the equation (36.10) has an infinite number of positive roots μ_n, $n = 1, 2, \ldots$, which are approaching the odd multiples of $\pi/2$, i.e., $\mu_n \simeq (2n+1)\pi/2$. Further, since the equation (36.10) remains unchanged if μ is replaced by $-\mu$, we find that the only nonzero real roots of (36.10) are $\mu_n \simeq \pm(2n+1)\pi/2$, $n = 1, 2, \ldots$.

Thus, the problem (36.2), (36.8) also has an infinite number of eigenvalues, $\lambda_1 = 0$, $\lambda_{n+1} \simeq (2n+1)^2\pi^2/4$, $n = 1, 2, \ldots$. Further, from (36.9) since $c_2 = -c_1 e^{2i\mu}$ for these λ_n, $n > 1$ the corresponding nontrivial solutions of the problem (36.2), (36.8) are

$$y(x) = c_1 e^{i\sqrt{\lambda_n}\,x} - c_1 e^{-i\sqrt{\lambda_n}\,x} e^{2i\sqrt{\lambda_n}} = -2c_1 i e^{i\sqrt{\lambda_n}} \sin \sqrt{\lambda_n}(1-x).$$

Hence, the eigenfunctions of (36.2), (36.8) are

$$\phi_1(x) = 1 - x$$
$$\phi_n(x) = \sin \sqrt{\lambda_n}(1-x), \quad n = 2, 3, \ldots.$$

From Example 36.1 it is clear that the problem (36.2), (36.3) has an infinite number of real eigenvalues λ_n, which can be arranged as a monotonic increasing sequence $\lambda_1 < \lambda_2 < \cdots$ such that $\lambda_n \to \infty$ as $n \to \infty$. Also, corresponding to each eigenvalue λ_n of (36.2), (36.3) there exists a one-parameter family of eigenfunctions $\phi_n(x)$, which has exactly $(n-1)$ zeros in the open interval $(0, \pi)$. Further, these eigenfunctions are orthogonal. This concept is defined in the following definition.

Definition 36.1. The set of functions $\{\phi_n(x) : n = 0, 1, \ldots\}$ each of which is piecewise continuous in an infinite or a finite interval $[\alpha, \beta]$ is said to be *orthogonal* in $[\alpha, \beta]$ with respect to the nonnegative function $r(x)$ if

$$(\phi_m, \phi_n) = \int_\alpha^\beta r(x)\phi_m(x)\phi_n(x)dx = 0 \quad \text{for all} \quad m \neq n$$

and

$$\int_\alpha^\beta r(x)\phi_n^2(x)dx \neq 0 \quad \text{for all} \quad n.$$

The function $r(x)$ is called the *weight function*. In what follows we shall assume that the function $r(x)$ has only a finite number of zeros in $[\alpha, \beta]$ and the integrals $\int_\alpha^\beta r(x)\phi_n(x)dx$, $n = 0, 1, \ldots$ exist.

The orthogonal set $\{\phi_n(x) : n = 0, 1, \ldots\}$ in $[\alpha, \beta]$ with respect to the weight function $r(x)$ is said to be *orthonormal* if

$$\int_\alpha^\beta r(x)\phi_n^2(x)dx = 1 \quad \text{for all} \quad n.$$

Thus, orthonormal functions have the same properties as orthogonal functions, but, in addition, they have been normalized, i.e., each function $\phi_n(x)$ of the orthogonal set has been divided by the norm of that function, which is defined as

$$\|\phi_n\| = \left(\int_\alpha^\beta r(x)\phi_n^2(x)dx \right)^{1/2}.$$

Now since

$$\int_0^\pi \sin kx \sin \ell x \, dx = 0,$$

for all $k \neq \ell$, the set of eigenfunctions $\{\phi_n(x) = \sin nx, \ n = 1, 2, \ldots\}$ of (36.2), (36.3) is orthogonal in $[0, \pi]$ with the weight function $r(x) = 1$.

Clearly, the above properties of eigenvalues and eigenfunctions of (36.2), (36.3) are also valid for the problem (36.2), (36.8). In fact, these properties hold for the general regular Sturm–Liouville problem (36.1), (33.14). We shall state these properties as theorems and prove the results.

Theorem 36.1. The eigenvalues of the regular Sturm–Liouville problem (36.1), (33.14) are simple, i.e., if λ is an eigenvalue of (36.1), (33.14) and $\phi_1(x)$ and $\phi_2(x)$ are the corresponding eigenfunctions, then $\phi_1(x)$ and $\phi_2(x)$ are linearly dependent.

Proof. Since $\phi_1(x)$ and $\phi_2(x)$ both are solutions of (36.1), from (30.23) it follows that $p(x)W(\phi_1, \phi_2)(x) = c$ (constant). To find the value of c, we note that ϕ_1 and ϕ_2 satisfy the boundary conditions, and hence

$$a_0\phi_1(\alpha) + a_1\phi_1'(\alpha) = 0$$
$$a_0\phi_2(\alpha) + a_1\phi_2'(\alpha) = 0,$$

which implies $W(\phi_1, \phi_2)(\alpha) = 0$, and hence c is zero. Thus, $p(x)W(\phi_1, \phi_2)(x) \equiv 0$, i.e., $\phi_1(x)$ and $\phi_2(x)$ are linearly dependent. ∎

Theorem 36.2. Let λ_n, $n = 1, 2, \ldots$ be the eigenvalues of the regular Sturm–Liouville problem (36.1), (33.14) and $\phi_n(x)$, $n = 1, 2, \ldots$ be the corresponding eigenfunctions. Then the set $\{\phi_n(x) : n = 1, 2, \cdots\}$ is orthogonal in $[\alpha, \beta]$ with respect to the weight function $r(x)$.

Proof. Let λ_k and λ_ℓ, $(k \neq \ell)$ be eigenvalues, and $\phi_k(x)$ and $\phi_\ell(x)$ be the corresponding eigenfunctions of (36.1), (33.14). Since $\phi_k(x)$ and $\phi_\ell(x)$ are solutions of (36.1), we have

$$\mathcal{P}_2[\phi_k] + \lambda_k r(x)\phi_k(x) = 0$$

and

$$\mathcal{P}_2[\phi_\ell] + \lambda_\ell r(x)\phi_\ell(x) = 0.$$

Thus, from Green's identity (30.24) it follows that

$$(\lambda_\ell - \lambda_k) \int_\alpha^\beta r(x)\phi_k(x)\phi_\ell(x)dx = \int_\alpha^\beta (\phi_\ell \mathcal{P}_2[\phi_k] - \phi_k \mathcal{P}_2[\phi_\ell])dx$$

$$- p(x)[\phi_\ell(x)\phi_k'(x) - \phi_\ell'(x)\phi_k(x)]\Big|_\alpha^\beta.$$

$$(36.11)$$

Next since $\phi_k(x)$ and $\phi_\ell(x)$ satisfy the boundary conditions (33.14), i.e.,

$$a_0\phi_k(\alpha) + a_1\phi_k'(\alpha) = 0, \qquad d_0\phi_k(\beta) + d_1\phi_k'(\beta) = 0$$
$$a_0\phi_\ell(\alpha) + a_1\phi_\ell'(\alpha) = 0, \qquad d_0\phi_\ell(\beta) + d_1\phi_\ell'(\beta) = 0$$

it is necessary that

$$\phi_k(\alpha)\phi_\ell'(\alpha) - \phi_k'(\alpha)\phi_\ell(\alpha) = \phi_k(\beta)\phi_\ell'(\beta) - \phi_k'(\beta)\phi_\ell(\beta) = 0.$$

Hence, the identity (36.11) reduces to

$$(\lambda_\ell - \lambda_k) \int_\alpha^\beta r(x)\phi_k(x)\phi_\ell(x)dx = 0. \qquad (36.12)$$

However, since $\lambda_\ell \neq \lambda_k$, it follows that

$$\int_\alpha^\beta r(x)\phi_k(x)\phi_\ell(x)dx = 0. \quad \blacksquare$$

Corollary 36.3. Let λ_1 and λ_2 be two eigenvalues of the regular Sturm–Liouville problem (36.1), (33.14) and $\phi_1(x)$ and $\phi_2(x)$ be the corresponding eigenfunctions. Then $\phi_1(x)$ and $\phi_2(x)$ are linearly dependent only if $\lambda_1 = \lambda_2$.

Proof. The proof is a direct consequence of equality (36.12). $\blacksquare$

Theorem 36.4. For the regular Sturm–Liouville problem (36.1), (33.14) eigenvalues are real.

Proof. Let $\lambda = a + ib$ be a complex eigenvalue and $\phi(x) = \mu(x) + i\nu(x)$ be the corresponding eigenfunction of (36.1), (33.14). Then we have

$$(p(x)(\mu + i\nu)')' + q(x)(\mu + i\nu) + (a + ib)r(x)(\mu + i\nu) = 0$$

and hence

$$\mathcal{P}_2[\mu] + (a\mu(x) - b\nu(x))r(x) = 0$$
$$\mathcal{P}_2[\nu] + (b\mu(x) + a\nu(x))r(x) = 0$$
$$a_0\mu(\alpha) + a_1\mu'(\alpha) = 0, \quad d_0\mu(\beta) + d_1\mu'(\beta) = 0$$

and

$$a_0\nu(\alpha) + a_1\nu'(\alpha) = 0, \quad d_0\nu(\beta) + d_1\nu'(\beta) = 0.$$

Thus, as in Theorem 36.2, we find

$$\int_\alpha^\beta (\nu\mathcal{P}_2[\mu] - \mu\mathcal{P}_2[\nu])dx$$

$$= \int_\alpha^\beta [-(a\mu(x) - b\nu(x))\nu(x)r(x) + (b\mu(x) + a\nu(x))\mu(x)r(x)]dx$$

$$= b\int_\alpha^\beta (\nu^2(x) + \mu^2(x))r(x)dx$$

$$= p(x)(\nu\mu' - \nu'\mu)\Big|_\alpha^\beta = 0.$$

Hence, it is necessary that $b = 0$, i.e., λ is real. $\blacksquare$

Since (36.2), (36.8) is a regular Sturm–Liouville problem, from Theorem 36.4 it is immediate that the equation (36.10) has only real roots.

Lecture 37

Sturm–Liouville Problems (Contd.)

In Lecture 36 we have established several properties of the eigenvalues and eigenfunctions of the regular Sturm–Liouville problem (36.1), (33.14). In all these results the existence of eigenvalues is tacitly assumed. We now state and prove the following important result.

Theorem 37.1. For the regular Sturm–Liouville problem (36.1), (33.14) there exists an infinite number of eigenvalues λ_n, $n = 1, 2, \ldots$. These eigenvalues can be arranged as a monotonic increasing sequence $\lambda_1 < \lambda_2 < \cdots$ such that $\lambda_n \to \infty$ as $n \to \infty$. Further, eigenfunction $\phi_n(x)$ corresponding to the eigenvalue λ_n has exactly $(n-1)$ zeros in the open interval (α, β).

Proof. We shall establish this result first for the particular problem (36.1),

$$y(\alpha) = y(\beta) = 0. \tag{37.1}$$

For this, we observe the following:

(i)   If eigenvalues of (36.1), (37.1) exist, then these are all real numbers (cf. Theorem 36.4).

(ii)   For each fixed λ there exists a unique solution $y(x, \lambda)$ of the initial value problem (36.1),

$$y(\alpha, \lambda) = 0, \quad y'(\alpha, \lambda) = 1. \tag{37.2}$$

Further, $y(x, \lambda)$ as well as $y'(x, \lambda)$ vary continuously with λ (cf. Theorem 16.8).

(iii)   There exist constants p, P, q, Q, r, and R such that for all $x \in [\alpha, \beta]$, $0 < p \leq p(x) \leq P$, $q \leq q(x) \leq Q$, and $0 < r \leq r(x) \leq R$. Thus, for a fixed $\lambda > 0$ the solution $y(x, \lambda)$ of (36.1), (37.2) oscillates more rapidly than the solution $y_0(x, \lambda)$ of the problem

$$\begin{aligned} (Py_0')' + qy_0 + \lambda r y_0 &= 0 \\ y_0(\alpha, \lambda) = 0, \quad y_0'(\alpha, \lambda) &= 1 \end{aligned} \tag{37.3}$$

and less rapidly than the solution $y_1(x, \lambda)$ of the problem

$$\begin{aligned} (py_1')' + Qy_1 + \lambda R y_1 &= 0 \\ y_1(\alpha, \lambda) = 0, \quad y_1'(\alpha, \lambda) &= 1 \end{aligned} \tag{37.4}$$

R.P. Agarwal and D. O'Regan, *An Introduction to Ordinary Differential Equations*,
doi: 10.1007/978-0-387-71276-5_37, © Springer Science + Business Media, LLC 2008

(cf. Theorem 31.4). When λ is negative, r and R in (37.3) and (37.4) need to be interchanged.

The problems (37.3) and (37.4) are with constant coefficients (modified as indicated when λ is negative), and hence can be solved explicitly. If $\lambda > 0$ is so large that $0 < (q+\lambda r)/P = a^2$, then the solution of the problem (37.3) is $y_0(x) = (1/a)\sin a(x-\alpha)$, which vanishes at least once in the open interval (α, β) provided $a(\beta - \alpha) > \pi$, i.e., $(q+\lambda r)/P > \pi^2/(\beta-\alpha)^2$. Thus, for each

$$\lambda > \max\left\{0, \frac{1}{r}\left(\frac{P\pi^2}{(\beta-\alpha)^2} - q\right)\right\} = \lambda^0$$

the solution of the problem (36.1), (37.2) has at least one zero in (α, β). Similarly, if $\lambda < 0$ is so small that

$$-a^2 = \frac{Q+\lambda r}{P} < 0, \quad \text{i.e.,} \quad \lambda < \min\left\{0, -\frac{Q}{r}\right\} = \lambda^1,$$

then the solution of the modified problem (37.4) is $y(x) = (1/a)\sinh a(x - \alpha)$, which does not vanish again at $x = \beta$ or, indeed, anywhere else. Hence, for each $\lambda < \lambda^1$ the solution of the problem (36.1), (37.2) has no zero in the interval $(\alpha, \beta]$. Now since the solution $y(x, \lambda)$ of (36.1), (37.2) varies continuously with λ, if $y(x, \lambda)$ has a zero in (α, β), then its position also varies continuously with λ. Thus, if λ increases steadily from λ^1 (for which the solution of (36.1), (37.2) has no zero in $(\alpha, \beta]$) towards λ^0, then there will be a specific value of λ, say, λ_1, for which $y(x, \lambda)$ first vanishes at $x = \beta$. This proves that there exists a smallest eigenvalue λ_1 of the problem (36.1), (37.1) and $y(x, \lambda_1)$ the solution of (36.1), (37.2) is the corresponding eigenfunction. By allowing λ to increase from the value λ_1 one argues that there is a number $\lambda_2 > \lambda_1$ for which $y(x, \lambda_2)$, the solution of (36.1), (37.2) has precisely one zero in (α, β) and $y(\beta, \lambda_2) = 0$. As λ continues to increase there results a sequence of eigenvalues $\lambda_1 < \lambda_2 < \cdots$ and a corresponding sequence of eigenfunctions $y(x, \lambda_1)$, $y(x, \lambda_2), \ldots$. Further, $y(x, \lambda_n)$ will have precisely $(n-1)$ zeros in the open interval (α, β). This completes the proof of Theorem 37.1 for the problem (36.1), (37.1).

Next for the problem (36.1),

$$a_0 y(\alpha) + a_1 y'(\alpha) = 0, \quad y(\beta) = 0 \tag{37.5}$$

we note that the above proof holds if the solution $y(x, \lambda)$ of (36.1), (37.2) is replaced by the solution $z(x, \lambda)$ of the initial value problem (36.1),

$$z(\alpha, \lambda) = a_1, \quad z'(\alpha, \lambda) = -a_0. \tag{37.6}$$

Thus, the problem (36.1), (37.5) also has a sequence of eigenvalues $\lambda_1' < \lambda_2' < \cdots$ and a corresponding sequence of eigenfunctions $z(x, \lambda_1')$, $z(x, \lambda_2')$, $\cdots$ such that $z(x, \lambda_n')$ has precisely $(n-1)$ zeros in the open interval (α, β).

Finally, we shall consider the problem (36.1), (33.14). For the solution $z(x, \lambda)$ of (36.1), (37.6), Theorem 16.8 implies that $\partial z(x, \lambda)/\partial \lambda$ is the solution of the initial value problem

$$\mathcal{P}_2 \left[\frac{\partial}{\partial \lambda} z(x, \lambda) \right] + \lambda r(x) \frac{\partial}{\partial \lambda} z(x, \lambda) + r(x) z(x, \lambda) = 0$$

$$\frac{\partial}{\partial \lambda} z(\alpha, \lambda) = \frac{\partial}{\partial \lambda} z'(\alpha, \lambda) = 0.$$

Thus, from Green's identity (30.24) it follows that

$$\int_\alpha^\beta \left(\frac{\partial}{\partial \lambda} z(x, \lambda) \mathcal{P}_2[z(x, \lambda)] - z(x, \lambda) \mathcal{P}_2 \left[\frac{\partial}{\partial \lambda} z(x, \lambda) \right] \right) dx$$

$$= \int_\alpha^\beta r(x) z^2(x, \lambda) dx$$

$$= \left(p(x) \left(\frac{\partial}{\partial \lambda} z(x, \lambda) z'(x, \lambda) - \frac{\partial}{\partial \lambda} z'(x, \lambda) z(x, \lambda) \right) \right) \Big|_\alpha^\beta$$

$$= p(\beta) W \left(\frac{\partial}{\partial \lambda} z(\beta, \lambda), z(\beta, \lambda) \right);$$

i.e.,

$$W \left(\frac{\partial}{\partial \lambda} z(\beta, \lambda), z(\beta, \lambda) \right) > 0.$$

Now in the interval $(\lambda'_n, \lambda'_{n+1})$ we know that $z(\beta, \lambda) \neq 0$, thus for all $\lambda \in (\lambda'_n, \lambda'_{n+1})$ the function $\phi(\lambda) = z'(\beta, \lambda)/z(\beta, \lambda)$ is well defined. Further, since

$$\phi'(\lambda) = - \frac{W \left(\frac{\partial}{\partial \lambda} z(\beta, \lambda), z(\beta, \lambda) \right)}{z^2(\beta, \lambda)}$$

it follows that $\phi'(\lambda) < 0$, i.e., in the interval $(\lambda'_n, \lambda'_{n+1})$ the function $\phi(\lambda)$ monotonically decreases. Also, since $z(\beta, \lambda'_n) = z(\beta, \lambda'_{n+1}) = 0$, $z'(\beta, \lambda'_n) \neq 0$, and $z'(\beta, \lambda'_{n+1}) \neq 0$, it is necessary that $\phi(\lambda'_n) = \infty$, and $\phi(\lambda'_{n+1}) = -\infty$, i.e., $\phi(\lambda)$ monotonically decreases from $+\infty$ to $-\infty$. Therefore, there exists a unique $\lambda''_n \in (\lambda'_n, \lambda'_{n+1})$ such that

$$\frac{z'(\beta, \lambda''_n)}{z(\beta, \lambda''_n)} = - \frac{d_0}{d_1}.$$

Hence, for the problem (36.1), (33.14) there exists a sequence of eigenvalues $\lambda''_1 < \lambda''_2 < \cdots$ such that $\lambda''_n \in (\lambda'_n, \lambda'_{n+1})$, and $z(x, \lambda''_n)$, $n = 1, 2, \ldots$ are the corresponding eigenfunctions. Obviously, $z(x, \lambda''_n)$ has exactly $(n-1)$ zeros in (α, β). ∎

Now we shall give some examples of singular Sturm–Liouville problems which show that the properties of eigenvalues and eigenfunctions for regular problems do not always hold.

Example 37.1. For the singular Sturm–Liouville problem (36.2),

$$y(0) = 0, \quad |y(x)| \leq M < \infty \quad \text{for all} \quad x \in (0, \infty) \qquad (37.7)$$

each $\lambda \in (0, \infty)$ is an eigenvalue and $\sin \sqrt{\lambda} x$ is the corresponding eigenfunction. Thus, in comparison with the regular problems where the spectrum is always discrete, the singular problems may have continuous spectrum.

Example 37.2. Consider the singular Sturm–Liouville problem (36.2),

$$y(-\pi) = y(\pi), \quad y'(-\pi) = y'(\pi). \qquad (37.8)$$

This problem has eigenvalues $\lambda_1 = 0$, $\lambda_{n+1} = n^2$, $n = 1, 2, \ldots$. The eigenvalue $\lambda_1 = 0$ is simple and 1 is its corresponding eigenfunction. The eigenvalue $\lambda_{n+1} = n^2$, $n \geq 1$ is not simple and two independent eigenfunctions are $\sin nx$ and $\cos nx$. Thus, in contrast with regular problems where the eigenvalues are simple, there may be multiple eigenvalues for singular problems.

Finally, we remark that the properties of the eigenvalues and eigenfunctions of regular Sturm–Liouville problems can be extended under appropriate assumptions to singular problems also in which the function $p(x)$ is zero at α or β, or both, but remains positive in (α, β). This case includes, in particular, the following examples.

Example 37.3. Consider the singular Sturm–Liouville problem

$$(1 - x^2)y'' - 2xy' + \lambda y = ((1 - x^2)y')' + \lambda y = 0 \qquad (37.9)$$

$$\lim_{x \to -1} y(x) < \infty, \quad \lim_{x \to 1} y(x) < \infty. \qquad (37.10)$$

The eigenvalues of this problem are $\lambda_n = n(n - 1)$, $n = 1, 2, \ldots$ and the corresponding eigenfunctions are the *Legendre polynomials* $P_{n-1}(x)$ which in terms of *Rodrigues' formula* are defined by

$$P_n(x) = \frac{1}{2^n \, (n)!} \frac{d^n}{dx^n} (x^2 - 1)^n, \quad n = 0, 1, \ldots. \qquad (37.11)$$

Example 37.4. Consider the singular Sturm–Liouville problem (37.9),

$$y'(0) = 0, \quad \lim_{x \to 1} y(x) < \infty. \qquad (37.12)$$

The eigenvalues of this problem are $\lambda_n = (2n - 2)(2n - 1)$, $n = 1, 2, \ldots$ and the corresponding eigenfunctions are the even Legendre polynomials $P_{2n-2}(x)$.

Example 37.5. Consider the singular Sturm–Liouville problem (37.9),

$$y(0) = 0, \quad \lim_{x \to 1} y(x) < \infty. \qquad (37.13)$$

The eigenvalues of this problem are $\lambda_n = (2n-1)(2n)$, $n = 1, 2, \ldots$ and the corresponding eigenfunctions are the odd Legendre polynomials $P_{2n-1}(x)$.

Example 37.6. Consider the singular Sturm–Liouville problem

$$y'' - 2xy' + \lambda y = 0 = \left(e^{-x^2} y'\right)' + \lambda e^{-x^2} y \qquad (37.14)$$

$$\lim_{x \to -\infty} \frac{y(x)}{|x|^k} < \infty, \quad \lim_{x \to \infty} \frac{y(x)}{x^k} < \infty \quad \text{for some positive integer } k.$$

$$(37.15)$$

The eigenvalues of this problem are $\lambda_n = 2(n-1)$, $n = 1, 2, \ldots$ and the corresponding eigenfunctions are the *Hermite polynomials* $H_{n-1}(x)$ which in terms of *Rodrigues' formula* are defined by

$$H_n(x) = (-1)^n e^{x^2} \frac{d^n}{dx^n} e^{-x^2}, \quad n = 0, 1, \ldots. \qquad (37.16)$$

Example 37.7. Consider the singular Sturm–Liouville problem

$$xy'' + (1-x)y' + \lambda y = 0 = \left(xe^{-x} y'\right)' + \lambda e^{-x} y \qquad (37.17)$$

$$\lim_{x \to 0} |y(x)| < \infty, \quad \lim_{x \to \infty} \frac{y(x)}{x^k} < \infty \quad \text{for some positive integer } k.$$

$$(37.18)$$

The eigenvalues of this problem are $\lambda_n = n-1$, $n = 1, 2, \ldots$ and the corresponding eigenfunctions are the *Laguerre polynomials* $L_{n-1}(x)$ which in terms of *Rodrigues' formula* are defined by

$$L_n(x) = \frac{e^x}{n!} \frac{d^n}{dx^n} \left(x^n e^{-x}\right). \qquad (37.19)$$

Problems

37.1. Show that the set $\{1, \cos nx, \ n = 1, 2, \ldots\}$ is orthogonal on $[0, \pi]$ with $r(x) = 1$.

37.2. Show that the set $\left\{\sqrt{\frac{2}{\pi}} \sin nx, \ n = 1, 2, \ldots\right\}$ is orthonormal on $[0, \pi]$ with $r(x) = 1$.

37.3. Show that the set $\left\{\frac{1}{\sqrt{2\pi}}, \frac{1}{\sqrt{\pi}} \cos nx, \frac{1}{\sqrt{\pi}} \sin nx, \ n = 1, 2, \ldots\right\}$ is orthonormal on $[-\pi, \pi]$ with $r(x) = 1$.

37.4. Show that the substitution $x = \cos\theta$ transforms the *Cheby-shev DE*

$$(1 - x^2)y'' - xy' + n^2 y = 0 \qquad (37.20)$$

into an equation with constant coefficients. Hence, find its linearly independent solutions $\cos(n\cos^{-1}x)$ and $\sin(n\cos^{-1}x)$. Further, deduce that

(i) $T_n(x) = \cos(n\cos^{-1}x) = \dfrac{n}{2}\sum_{m=0}^{[n/2]}(-1)^m\dfrac{(n-m-1)!}{m!\,(n-2m)!}(2x)^{n-2m}, \quad n \geq 1;$

(ii) $\displaystyle\int_{-1}^{1}(1-x^2)^{-1/2}T_m(x)T_n(x)dx = \begin{cases} 0, & m \neq n \\ \pi/2, & m = n \neq 0 \\ \pi, & m = n = 0. \end{cases}$

37.5. Show that for the Legendre polynomials $P_n(x)$, $n = 0,1,\ldots$ defined in (37.11) the following hold:

$$\int_{-1}^{1}P_n(x)P_m(x)dx = \begin{cases} 0 & \text{if } m \neq n \\ \dfrac{2}{2n+1} & \text{if } m = n. \end{cases}$$

37.6. Find the eigenvalues and eigenfunctions of the problem (36.2) with the following boundary conditions:

(i) $y(0) = 0, \ y'(\beta) = 0.$
(ii) $y'(0) = 0, \ y(\beta) = 0.$
(iii) $y'(0) = 0, \ y'(\beta) = 0.$
(iv) $y(0) = 0, \ y(\beta) + y'(\beta) = 0.$
(v) $y(0) - y'(0) = 0, \ y'(\beta) = 0.$
(vi) $y(0) - y'(0) = 0, \ y(\beta) + y'(\beta) = 0.$

37.7. Find the eigenvalues and eigenfunctions of each of the following Sturm–Liouville problems:

(i) $y'' + \lambda y = 0, \quad y(0) = y(\pi/2) = 0.$
(ii) $y'' + (1+\lambda)y = 0, \quad y(0) = y(\pi) = 0.$
(iii) $y'' + 2y' + (1-\lambda)y = 0, \quad y(0) = y(1) = 0.$
(iv) $(x^2y')' + \lambda x^{-2}y = 0, \quad y(1) = y(2) = 0.$
(v) $x^2y'' + xy' + (\lambda x^2 - (1/4))y = 0, \quad y(\pi/2) = y(3\pi/2) = 0.$
(vi) $((x^2+1)y')' + \lambda(x^2+1)^{-1}y = 0, \quad y(0) = y(1) = 0.$

37.8. Consider the boundary value problem

$$\begin{aligned} x^2y'' + xy' + \lambda y &= 0, \quad 1 < x < e \\ y(1) &= 0, \quad y(e) = 0. \end{aligned} \tag{37.21}$$

(i)   Show that (37.21) is equivalent to the Sturm–Liouville problem

$$\begin{aligned} (xy')' + \frac{\lambda}{x}y &= 0, \quad 1 < x < e \\ y(1) &= 0, \quad y(e) = 0. \end{aligned} \tag{37.22}$$

(ii)   Verify that for (37.22) the eigenvalues are $\lambda_n = n^2\pi^2$, $n = 1, 2, \ldots$
and the corresponding eigenfunctions are $\phi_n(x) = \sin(n\pi \ln x)$.

(iii)   Show that

$$\int_1^e \frac{1}{x}\phi_m(x)\phi_n(x)dx = \begin{cases} 0, & m \neq n \\ 1/2, & m = n. \end{cases}$$

37.9.   Verify that for the Sturm–Liouville problem

$$(xy')' + \frac{\lambda}{x}y = 0, \quad 1 < x < e^{2\pi}$$

$$y'(1) = 0, \quad y'(e^{2\pi}) = 0$$

the eigenvalues are $\lambda_n = n^2/4$, $n = 0, 1, \ldots$ and the corresponding eigenfunctions are $\phi_n(x) = \cos\left(\frac{n \ln x}{2}\right)$. Show that

$$\int_1^{e^{2\pi}} \frac{1}{x}\phi_m(x)\phi_n(x)dx = 0, \quad m \neq n.$$

37.10.   Consider Mathieu's DE (see Example 20.1)

$$y'' + (\lambda + 16d\cos 2x)y = 0, \quad 0 \le x \le \pi$$

together with the periodic boundary conditions

$$y(0) = y(\pi), \quad y'(0) = y'(\pi).$$

Show that the eigenfunctions of this problem are orthogonal.

37.11.   Consider the DE

$$x^4 y'' + k^2 y = 0. \tag{37.23}$$

(i)   Verify that the general solution of (37.23) is

$$y(x) = x\left(A\cos\frac{k}{x} + B\sin\frac{k}{x}\right).$$

(ii)   Find the eigenvalues and eigenfunctions of the Sturm–Liouville problem (37.23), $y(\alpha) = y(\beta) = 0$, $0 < \alpha < \beta$.

37.12.   Show that the problem

$$y'' - 4\lambda y' + 4\lambda^2 y = 0, \quad y(0) = 0, \quad y(1) + y'(1) = 0$$

has only one eigenvalue, and find the corresponding eigenfunction.

37.13. Show that for the singular Sturm–Liouville problem (36.1), (32.7) with $p(\alpha) = p(\beta)$ eigenfunctions corresponding to different eigenvalues are orthogonal in $[\alpha, \beta]$ with respect to the weight function $r(x)$.

37.14. Solve the following singular Sturm–Liouville problems

(i) $y'' + \lambda y = 0$, $y'(0) = 0$, $|y(x)| < \infty$ for all $x \in (0, \infty)$

(ii) $y'' + \lambda y = 0$, $|y(x)| < \infty$ for all $x \in (-\infty, \infty)$.

Answers or Hints

37.1. Verify directly.

37.2. Verify directly.

37.3. Verify directly.

37.4. Equation (37.20) reduces to $\frac{d^2y}{d\theta^2} + n^2 y = 0$. (i) Use induction and the identity
$$\cos((n+1)\cos^{-1}x) + \cos((n-1)\cos^{-1}x) = 2\cos(n\cos^{-1}x)\cos(\cos^{-1}x),$$
i.e., $T_{n+1}(x) = 2xT_n(x) - T_{n-1}(x)$. (ii) Use $x = \cos\theta$.

37.5. From (37.11) it follows that
$$2^n\,n!\int_{-1}^1 P_m(x)P_n(x)dx = \int_{-1}^1 P_m(x)\frac{d^n}{dx^n}(x^2-1)^n dx$$
$$= -\int_{-1}^1 \frac{d}{dx}P_m(x)\frac{d^{n-1}}{dx^{n-1}}(x^2-1)^n dx.$$

37.6. (i) $\left(\frac{2n-1}{2\beta}\right)^2\pi^2$, $\sin\left(\frac{2n-1}{2\beta}\right)\pi x$. (ii) $\left(\frac{2n-1}{2\beta}\right)^2\pi^2$, $\cos\left(\frac{2n-1}{2\beta}\right)\pi x$
(iii) $\left(\frac{n-1}{\beta}\right)^2\pi^2$, $\cos\left(\frac{n-1}{\beta}\right)\pi x$. (iv) λ_n^2, where $\lambda = \lambda_n$ is a solution of $\tan\lambda\beta + \lambda = 0$, $\sin\lambda_n x$. (v) λ_n^2, where $\lambda = \lambda_n$ is a solution of $\cot\lambda\beta = \lambda$, $\sin\lambda_n x + \lambda_n\cos\lambda_n x$. (vi) λ_n^2, where $\lambda = \lambda_n$ is a solution of $\tan\lambda\beta = 2\lambda/(\lambda^2-1)$, $\sin\lambda_n x + \lambda_n\cos\lambda_n x$.

37.7. (i) $4n^2$, $\sin 2nx$. (ii) $n^2 - 1$, $\sin nx$. (iii) $-n^2\pi^2$, $e^{-x}\sin n\pi x$. (iv) $4n^2\pi^2$, $\sin 2n\pi\left(1 - \frac{1}{x}\right)$. (v) n^2, $\frac{1}{\sqrt{x}}\sin n\left(x - \frac{\pi}{2}\right)$. (vi) $16n^2$, $\sin(4n\tan^{-1}x)$.

37.8. Verify directly.

37.9. Verify directly.

37.10. Follow the proof of Theorem 36.2.

37.11. (i) Verify directly. (ii) $k_n = \frac{n\pi\alpha\beta}{\beta-\alpha}$, $x\sin\left[\frac{n\pi\beta(x-\alpha)}{x(\beta-\alpha)}\right]$.

37.12. -1, xe^{-2x}.

37.13. Use (36.11).

37.14. (i) $\lambda \geq 0$, $\phi(x) = \cos\sqrt{\lambda}x$. (ii) $\lambda \geq 0$, $\phi(x) = c_1\cos\sqrt{\lambda}x + c_2\sin\sqrt{\lambda}x$.

Lecture 38
Eigenfunction Expansions

The basis $\{e^1, \ldots, e^n\}$ (e^k is the unit vector) of $\mathbb{R}^n$ has an important characteristic, namely, for every $u \in \mathbb{R}^n$ there is a unique choice of constants $\alpha_1, \ldots, \alpha_n$ for which

$$u = \sum_{i=1}^{n} \alpha_i e^i.$$

Further, from the orthonormality of the vectors e^i, $1 \leq i \leq n$ we can determine α_i, $1 \leq i \leq n$ as follows:

$$<u, e^j> = \left\langle \sum_{i=1}^{n} \alpha_i e^i, e^j \right\rangle = \sum_{i=1}^{n} \alpha_i <e^i, e^j> = \alpha_j, \quad 1 \leq j \leq n.$$

Thus, the vector u has a unique representation

$$u = \sum_{i=1}^{n} <u, e^i> e^i.$$

A natural generalization of this result which is widely applicable and has led to a vast amount of advanced mathematics can be stated as follows: Let $\{\phi_n(x), \; n = 0, 1, 2, \ldots\}$ be an orthogonal set of functions in the interval $[\alpha, \beta]$ with respect to the weight function $r(x)$. Then an arbitrary function $f(x)$ can be expressed as an infinite series involving orthogonal functions $\phi_n(x), \; n = 0, 1, 2, \ldots$ as

$$f(x) = \sum_{n=0}^{\infty} c_n \phi_n(x). \tag{38.1}$$

It is natural to ask the meaning of equality in (38.1), i.e., the type of convergence, if any, of the infinite series on the right so that we will have some idea as to how well this represents $f(x)$. We shall also determine the constant coefficients $c_n, \; n = 0, 1, 2, \ldots$ in (38.1).

Let us first proceed formally without considering the question of convergence. We multiply (38.1) by $r(x)\phi_m(x)$ and integrate from α to β, to obtain

$$\int_{\alpha}^{\beta} r(x)\phi_m(x)f(x)dx = \int_{\alpha}^{\beta} \sum_{n=0}^{\infty} c_n r(x)\phi_n(x)\phi_m(x)dx.$$

R.P. Agarwal and D. O'Regan, *An Introduction to Ordinary Differential Equations*,
doi: 10.1007/978-0-387-71276-5_38, © Springer Science + Business Media, LLC 2008

Now assuming that the operations of integration and summation on the right of the above equality can be interchanged, we find

$$\int_\alpha^\beta r(x)\phi_m(x)f(x)dx \;=\; \sum_{n=0}^{\infty} c_n \int_\alpha^\beta r(x)\phi_m(x)\phi_n(x)dx$$

$$=\; c_m \int_\alpha^\beta r(x)\phi_m^2(x)dx \;=\; c_m\|\phi_m\|^2.$$

Thus, under suitable convergence conditions, the constant coefficients c_n, $n = 0, 1, 2, \ldots$ are given by the formula

$$c_n \;=\; \int_\alpha^\beta r(x)\phi_n(x)f(x)dx \;\bigg/\; \|\phi_n\|^2. \tag{38.2}$$

However, if the set $\{\phi_n(x)\}$ is orthonormal, so that $\|\phi_n\| = 1$, then we have

$$c_n \;=\; \int_\alpha^\beta r(x)\phi_n(x)f(x)dx. \tag{38.3}$$

If the series $\sum_{n=0}^{\infty} c_n\phi_n(x)$ converges uniformly to $f(x)$ in $[\alpha, \beta]$, then the above formal procedure is justified, and then the coefficients c_n are given by (38.2).

The coefficients c_n obtained in (38.2) are called the *Fourier coefficients* of the function $f(x)$ with respect to the orthogonal set $\{\phi_n(x)\}$ and the series $\sum_{n=0}^{\infty} c_n\phi_n(x)$ with coefficients (38.2) is called the *Fourier series* of $f(x)$.

We shall write

$$f(x) \;\sim\; \sum_{n=0}^{\infty} c_n\phi_n(x)$$

which, in general, is just a correspondence, i.e., often $f(x) \neq \sum_{n=0}^{\infty} c_n\phi_n(x)$, unless otherwise proved.

Example 38.1. In Problem 37.5 we have seen that the set of Legendre polynomials $\{\phi_n(x) = P_n(x), \quad n = 0, 1, \ldots\}$ is orthogonal on $[-1, 1]$ with $r(x) = 1$. Also,

$$\|P_n\|^2 \;=\; \int_{-1}^{1} P_n^2(x)dx \;=\; \frac{2}{2n+1}.$$

Thus, from (38.2) for a given function $f(x)$ the coefficients in the *Fourier–Legendre series* $f(x) \sim \sum_{n=0}^{\infty} c_n P_n(x)$ are given by

$$c_n \;=\; \frac{2n+1}{2} \int_{-1}^{1} P_n(x)f(x)dx, \quad n \geq 0.$$

Example 38.2. The set of functions

$$\left\{1, \ \cos\frac{n\pi x}{L}, \ \sin\frac{n\pi x}{L}, \ L > 0, \ n \geq 1\right\}$$

is orthogonal with respect to the weight function $r(x) = 1$ in the interval $[-L, L]$. For the norms of these functions, we have

$$\int_{-L}^{L} \cos^2\frac{n\pi x}{L}dx \ = \ \begin{cases} 2L, & n = 0 \\ L, & n \geq 1 \end{cases}$$

$$\int_{-L}^{L} \sin^2\frac{n\pi x}{L}dx \ = \ L, \ n \geq 1.$$

The general *trigonometric–Fourier series* of a given function $f(x)$ is defined to be

$$f(x) \ \sim \ \frac{1}{2}a_0 + \sum_{n=1}^{\infty}\left(a_n\cos\frac{n\pi x}{L} + b_n\sin\frac{n\pi x}{L}\right), \qquad (38.4)$$

where

$$a_n \ = \ \frac{1}{L}\int_{-L}^{L} f(x)\cos\frac{n\pi x}{L}dx, \quad n \geq 0$$

$$b_n \ = \ \frac{1}{L}\int_{-L}^{L} f(x)\sin\frac{n\pi x}{L}dx, \quad n \geq 1. \qquad (38.5)$$

Now we shall examine the convergence of the Fourier series to the function $f(x)$. For this, to make the analysis widely applicable we assume that the functions $\phi_n(x)$, $n = 0, 1, \ldots$ and $f(x)$ are only piecewise continuous in $[\alpha, \beta]$. Let the sum of first $N + 1$ terms $\sum_{n=0}^{N} c_n\phi_n(x)$ be denoted by $S_N(x)$. We consider the difference $|S_N(x) - f(x)|$ for various values of N and x. If for an arbitrary $\epsilon > 0$ there is an integer $N(\epsilon) > 0$ such that $|S_N(x) - f(x)| < \epsilon$, then the Fourier series converges (uniformly) to $f(x)$ for all x in $[\alpha, \beta]$. On the other hand, if N depends on x and ϵ both, then the Fourier series converges pointwise to $f(x)$. However, for the moment both of these type of convergence are too demanding, and we will settle for something less. To this end, we need the following definition.

Definition 38.1. Let each of the functions $\psi_n(x)$, $n \geq 0$ and $\psi(x)$ be piecewise continuous in $[\alpha, \beta]$. We say that the sequence $\{\psi_n(x)\}$ *converges in the mean* to $\psi(x)$ (with respect to the weight function $r(x)$ in the interval $[\alpha, \beta]$) if

$$\lim_{n\to\infty} \|\psi_n - \psi\|^2 \ = \ \lim_{n\to\infty}\int_{\alpha}^{\beta} r(x)(\psi_n(x) - \psi(x))^2 dx \ = \ 0. \qquad (38.6)$$

Thus, the Fourier series converges in the mean to $f(x)$ provided

$$\lim_{N\to\infty}\int_{\alpha}^{\beta} r(x)(S_N(x) - f(x))^2 dx \ = \ 0. \qquad (38.7)$$

Before we prove the convergence of the Fourier series, let us consider the possibility of representing $f(x)$ by a series of the form $\sum_{n=0}^{\infty} d_n \phi_n(x)$, where the coefficients d_n are not necessarily the Fourier coefficients. Let

$$T_N(x; d_0, d_1, \ldots, d_N) = \sum_{n=0}^{N} d_n \phi_n(x)$$

and let e_N be the quantity $\|T_N - f\|$. Then from the orthogonality of the functions $\phi_n(x)$ it is clear that

$$
\begin{aligned}
e_N^2 &= \|T_N - f\|^2 = \int_\alpha^\beta r(x) \left(\sum_{n=0}^{N} d_n \phi_n(x) - f(x) \right)^2 dx \\
&= \sum_{n=0}^{N} d_n^2 \int_\alpha^\beta r(x) \phi_n^2(x) dx - 2 \sum_{n=0}^{N} d_n \int_\alpha^\beta r(x) \phi_n(x) f(x) dx \\
&\quad + \int_\alpha^\beta r(x) f^2(x) dx \\
&= \sum_{n=0}^{N} d_n^2 \|\phi_n\|^2 - 2 \sum_{n=0}^{N} d_n c_n \|\phi_n\|^2 + \|f\|^2 \\
&= \sum_{n=0}^{N} \|\phi_n\|^2 (d_n - c_n)^2 - \sum_{n=0}^{N} \|\phi_n\|^2 c_n^2 + \|f\|^2.
\end{aligned}
$$

$$(38.8)$$

Thus, the quantity e_N is least when $d_n = c_n$ for $n = 0, 1, \ldots, N$. Therefore, we have established the following theorem.

Theorem 38.1. For any given nonnegative integer N, the best approximation in the mean to a function $f(x)$ by an expression of the form $\sum_{n=0}^{N} d_n \phi_n(x)$ is obtained when the coefficients d_n are the Fourier coefficients of $f(x)$.

Now in (38.8) let $d_n = c_n$, $n = 0, 1, \ldots, N$ to obtain

$$\|S_N - f\|^2 = \|f\|^2 - \sum_{n=0}^{N} \|\phi_n\|^2 c_n^2. \tag{38.9}$$

Thus, it follows that

$$\|T_N - f\|^2 = \sum_{n=0}^{N} \|\phi_n\|^2 (d_n - c_n)^2 + \|S_N - f\|^2. \tag{38.10}$$

Hence, we find

$$0 \leq \|S_N - f\| \leq \|T_N - f\|. \tag{38.11}$$

If the series $\sum_{n=0}^{\infty} d_n \phi_n(x)$ converges in the mean to $f(x)$, i.e., if $\lim_{N\to\infty} \|T_N - f\| = 0$, then from (38.11) it is clear that the Fourier series converges in the mean to $f(x)$, i.e., $\lim_{N\to\infty} \|S_N - f\| = 0$. However, then (38.10) implies that

$$\lim_{N\to\infty} \sum_{n=0}^{N} \|\phi_n\|^2 (d_n - c_n)^2 = 0.$$

But this is possible only if $d_n = c_n$, $n = 0, 1, \ldots$. Thus, we have proved the following result.

Theorem 38.2. If a series of the form $\sum_{n=0}^{\infty} d_n \phi_n(x)$ converges in the mean to $f(x)$, then the coefficients d_n must be the Fourier coefficients of $f(x)$.

Now from the equality (38.9) we note that

$$0 \le \|S_{N+1} - f\| \le \|S_N - f\|.$$

Thus, the sequence $\{\|S_N - f\|, N = 0, 1, \ldots\}$ is nonincreasing and bounded below by zero, and therefore, it must converge. If it converges to zero, then the Fourier series of $f(x)$ converges in the mean to $f(x)$. Further, from (38.9) we have the inequality

$$\sum_{n=0}^{N} \|\phi_n\|^2 c_n^2 \le \|f\|^2.$$

Since the sequence $\{C_N, N = 0, 1, \ldots\}$ where $C_N = \sum_{n=0}^{N} \|\phi_n\|^2 c_n^2$ is nondecreasing and bounded above by $\|f\|^2$, it must converge. Therefore, we have

$$\sum_{n=0}^{\infty} \|\phi_n\|^2 c_n^2 \le \|f\|^2. \tag{38.12}$$

Hence, from (38.9) we see that the Fourier series of $f(x)$ converges in the mean to $f(x)$ if and only if

$$\|f\|^2 = \sum_{n=0}^{\infty} \|\phi_n\|^2 c_n^2. \tag{38.13}$$

For the particular case when $\phi_n(x)$, $n = 0, 1, 2, \ldots$ are orthonormal, (38.12) reduces to *Bessel's inequality*

$$\sum_{n=0}^{\infty} c_n^2 \le \|f\|^2 \tag{38.14}$$

and (38.13) becomes the well-known *Parseval's equality*

$$\|f\|^2 = \sum_{n=0}^{\infty} c_n^2. \tag{38.15}$$

We summarize the above considerations in the following theorem.

Theorem 38.3. Let $\{\phi_n(x), \quad n = 0, 1, \ldots\}$ be an orthonormal set, and let c_n be the Fourier coefficients of $f(x)$ given in (38.3). Then the following hold:

(i)   The series $\sum_{n=0}^{\infty} c_n^2$ converges, and therefore

$$\lim_{n \to \infty} c_n = \lim_{n \to \infty} \int_{\alpha}^{\beta} r(x)\phi_n(x)f(x)dx = 0.$$

(ii)   The Bessel inequality (38.14) holds.

(iii)   The Fourier series of $f(x)$ converges in the mean to $f(x)$ if and only if Parseval's equality (38.15) holds.

Now let $C_p[\alpha, \beta]$ be the space of all piecewise continuous functions in $[\alpha, \beta]$. The orthogonal set $\{\phi_n(x), \quad n = 0, 1, \ldots\}$ is said to be *complete* in $C_p[\alpha, \beta]$ if for every function $f(x)$ of $C_p[\alpha, \beta]$ its Fourier series converges in the mean to $f(x)$. Clearly, if $\{\phi_n(x), \quad n = 0, 1, \ldots\}$ is orthonormal then it is complete if and only if Parseval's equality holds for every function in $C_p[\alpha, \beta]$. The following property of an orthogonal set is fundamental.

Theorem 38.4. If an orthogonal set $\{\phi_n(x), \quad n = 0, 1, \ldots\}$ is complete in $C_p[\alpha, \beta]$, then any function of $C_p[\alpha, \beta]$ that is orthogonal to every $\phi_n(x)$ must be zero except possibly at a finite number of points in $[\alpha, \beta]$.

Proof. Without loss of generality, let the set $\{\phi_n(x), \quad n = 0, 1, \ldots\}$ be orthonormal. If $f(x)$ is orthogonal to every $\phi_n(x)$, then from (38.3) all Fourier coefficients c_n of $f(x)$ are zero. But, then from the Parseval equality (38.15) the function $f(x)$ must be zero except possibly at a finite number of points in $[\alpha, \beta]$. ∎

The importance of this result lies in the fact that if we delete even one member from an orthogonal set, then the remaining functions cannot form a complete set. For example, the set $\{\cos nx, \quad n = 1, 2, \ldots\}$ is not complete in $[0, \pi]$ with respect to the weight function $r(x) = 1$.

Unfortunately, there is no single procedure for establishing the completeness of a given orthogonal set. However, the following results are known.

Theorem 38.5. The orthogonal set $\{\phi_n(x), \quad n = 0, 1, \ldots\}$ in the interval $[\alpha, \beta]$ with respect to the weight function $r(x)$ is complete in $C_p[\alpha, \beta]$ if $\phi_n(x)$ is a polynomial of degree n.

As a consequence of this result, it is clear that the Fourier–Legendre series of a piecewise continuous function $f(x)$ in $[-1, 1]$ converges in the mean to $f(x)$.

Theorem 38.6.   The set of all eigenfunctions $\{\phi_n(x),\ n = 1, 2, \ldots\}$ of the regular Sturm–Liouville problem (36.1), (33.14) is complete in the space $C_p[\alpha, \beta]$.

Theorem 38.6 can be extended to encompass the periodic eigenvalue problem (36.1), (32.7). In such a case, if necessary, two linearly independent mutually orthogonal eigenfunctions corresponding to one eigenvalue are chosen. Thus, from the problem (36.2), $y(-L) = y(L)$, $y'(-L) = y'(L)$ (cf. Example 37.2) it is clear that the set $\{1,\ \cos(n\pi x/L),\ \sin(n\pi x/L),\ L > 0,\ n \geq 1\}$ considered in Example 38.2 is complete in $C_p[-L, L]$, and therefore the trigonometric–Fourier series of any function $f(x)$ in $C_p[-L, L]$ converges in the mean to $f(x)$.

Lecture 39

Eigenfunction Expansions (Contd.)

The analytical discussions of uniform and pointwise convergence of the Fourier series of the function $f(x)$ to $f(x)$ are too difficult to be included here. Therefore, we state the following result without its proof.

Theorem 39.1. Let $\{\phi_n(x), \ n = 1, 2, \ldots\}$ be the set of all eigenfunctions of the regular Sturm–Liouville problem (36.1), (33.14). Then the following hold:

(i) The Fourier series of $f(x)$ converges to $[f(x+)+f(x-)]/2$ at each point in the open interval (α, β) provided $f(x)$ and $f'(x)$ are piecewise continuous in $[\alpha, \beta]$.

(ii) The Fourier series of $f(x)$ converges uniformly and absolutely to $f(x)$ in $[\alpha, \beta]$ provided $f(x)$ is continuous having a piecewise continuous derivative $f'(x)$ in $[\alpha, \beta]$, and is such that $f(\alpha) = 0$ if $\phi_n(\alpha) = 0$ and $f(\beta) = 0$ if $\phi_n(\beta) = 0$.

Example 39.1. To obtain the Fourier series of the function $f(x) = 1$ in the interval $[0, \pi]$ in terms of the eigenfunctions $\phi_n(x) = \sin nx, \ n = 1, 2, \ldots$ of the eigenvalue problem (36.2), (36.3) we recall that

$$\|\phi_n\|^2 = \int_0^\pi \sin^2 nx\, dx = \frac{\pi}{2}.$$

Thus, it follows that

$$c_n = \frac{1}{\|\phi_n\|^2} \int_0^\pi f(x)\, \sin nx\, dx = \frac{2}{\pi} \int_0^\pi \sin nx\, dx = \frac{2}{n\pi}\left(1 - (-1)^n\right).$$

Hence, we have

$$1 = \frac{4}{\pi} \sum_{n=1}^{\infty} \frac{1}{(2n-1)} \sin(2n-1)x. \tag{39.1}$$

From Theorem 39.1 it is clear that equality in (39.1) holds at each point of the open interval $(0, \pi)$.

Example 39.2. We shall obtain the Fourier series of the function $f(x) = x - x^2$, $x \in [0, 1]$ in terms of the eigenfunctions $\phi_1(x) = 1 - x$,

R.P. Agarwal and D. O'Regan, *An Introduction to Ordinary Differential Equations*, doi: 10.1007/978-0-387-71276-5_39, © Springer Science + Business Media, LLC 2008

$\phi_n(x) = \sin \sqrt{\lambda_n}(1 - x)$, $n = 2, 3, \ldots$ of the eigenvalue problem (36.2), (36.8). For this, we note that

$$\|\phi_1\|^2 = \int_0^1 (1-x)^2 dx = \frac{1}{3},$$

$$\|\phi_n\|^2 = \int_0^1 \sin^2 \sqrt{\lambda_n}(1-x) dx = \frac{1}{2} \int_0^1 (1 - \cos 2\sqrt{\lambda_n}(1-x)) dx$$

$$= \frac{1}{2} \left[x + \frac{1}{2\sqrt{\lambda_n}} \sin 2\sqrt{\lambda_n}(1-x) \right]\Big|_0^1 = \frac{1}{2} \left[1 - \frac{1}{2\sqrt{\lambda_n}} \sin 2\sqrt{\lambda_n} \right]$$

$$= \frac{1}{2} \left[1 - \frac{1}{2\sqrt{\lambda_n}} 2 \sin \sqrt{\lambda_n} \cos \sqrt{\lambda_n} \right] = \frac{1}{2}[1 - \cos^2 \sqrt{\lambda_n}]$$

$$= \frac{1}{2} \sin^2 \sqrt{\lambda_n}, \quad n \geq 2,$$

where we have used the fact that $\tan \sqrt{\lambda_n} = \sqrt{\lambda_n}$.

Thus, it follows that

$$c_1 = 3 \int_0^1 (1-x)(x - x^2) dx = \frac{1}{4}$$

and for $n \geq 2$,

$$c_n = \frac{2}{\sin^2 \sqrt{\lambda_n}} \int_0^1 (x - x^2) \sin \sqrt{\lambda_n}(1 - x) dx$$

$$= \frac{2}{\sin^2 \sqrt{\lambda_n}} \left[(x - x^2) \frac{\cos \sqrt{\lambda_n}(1-x)}{\sqrt{\lambda_n}} \Big|_0^1 - \int_0^1 (1 - 2x) \frac{\cos \sqrt{\lambda_n}(1-x)}{\sqrt{\lambda_n}} dx \right]$$

$$= \frac{-2}{\sqrt{\lambda_n} \sin^2 \sqrt{\lambda_n}} \left[(1 - 2x) \frac{\sin \sqrt{\lambda_n}(1-x)}{-\sqrt{\lambda_n}} \Big|_0^1 - \int_0^1 -2 \frac{\sin \sqrt{\lambda_n}(1-x)}{-\sqrt{\lambda_n}} dx \right]$$

$$= \frac{-2}{\sqrt{\lambda_n} \sin^2 \sqrt{\lambda_n}} \left[\frac{\sin \sqrt{\lambda_n}}{\sqrt{\lambda_n}} - \frac{2}{\sqrt{\lambda_n}} \frac{\cos \sqrt{\lambda_n}(1 - x)}{\sqrt{\lambda_n}} \Big|_0^1 \right]$$

$$= \frac{-2}{\lambda_n^{3/2} \sin^2 \sqrt{\lambda_n}} \left[\sqrt{\lambda_n} \sin \sqrt{\lambda_n} - 2 + 2 \cos \sqrt{\lambda_n} \right]$$

$$= \frac{-2}{\lambda_n^{3/2} \sin^2 \sqrt{\lambda_n}} \left[\lambda_n \cos \sqrt{\lambda_n} - 2 + 2 \cos \sqrt{\lambda_n} \right]$$

$$= \frac{2}{\lambda_n^{3/2} \sin^2 \sqrt{\lambda_n}} \left[2 - (2 + \lambda_n) \cos \sqrt{\lambda_n} \right].$$

Hence, we have

$$x - x^2 = \frac{1}{4}(1-x) + \sum_{n=2}^{\infty} \frac{2}{\lambda_n^{3/2} \sin^2 \sqrt{\lambda_n}} (2 - (2 + \lambda_n) \cos \sqrt{\lambda_n}) \sin \sqrt{\lambda_n}(1-x).$$

$$(39.2)$$

From Theorem 39.1 we find that equality in (39.2) holds uniformly in $[0, 1]$.

The convergence of Fourier–Legendre and trigonometric–Fourier series cannot be concluded from Theorem 39.1. For these, we have the following results.

Theorem 39.2. Let $f(x)$ and $f'(x)$ be piecewise continuous in the interval $[-1, 1]$. Then the Fourier–Legendre series of $f(x)$ converges to $[f(x+) + f(x-)]/2$ at each point in the open interval $(-1, 1)$, and at $x = -1$ the series converges to $f(-1+)$ and at $x = 1$ it converges to $f(1-)$.

Theorem 39.3. Let $f(x)$ and $f'(x)$ be piecewise continuous in the interval $[-L, L]$ $(L > 0)$. Then the trigonometric–Fourier series of $f(x)$ converges to $[f(x+) + f(x-)]/2$ at each point in the open interval $(-L, L)$ and at $x = \pm L$ the series converges to $[f(-L+) + f(L-)]/2$.

Example 20.3. Consider the function

$$f(x) = \begin{cases} 0, & x \in [-\pi, 0) \\ 1, & x \in [0, \pi]. \end{cases}$$

Clearly, $f(x)$ is piecewise continuous in $[-\pi, \pi]$, with a single jump discontinuity at 0. From (38.5), we obtain $a_0 = 1$, and for $n \geq 1$,

$$a_n = \frac{1}{\pi} \int_0^\pi \cos nx\, dx = 0, \quad b_n = \frac{1}{\pi} \int_0^\pi \sin nx\, dx = \frac{2}{n\pi}(1 - (-1)^n).$$

Thus, we have

$$f(x) = \frac{1}{2} + \frac{2}{\pi} \sum_{n=1}^\infty \frac{1}{(2n-1)} \sin(2n-1)x. \tag{39.3}$$

From Theorem 39.3, equality (39.3) holds at each point in the open intervals $(-\pi, 0)$ and $(0, \pi)$, whereas at $x = 0$ the right-hand side is $1/2$, which is the same as $[f(0+) + f(0-)]/2$. Also, at $x = \pm\pi$ the right-hand side is again $1/2$, which is the same as $[f(-\pi+) + f(\pi-)]/2$.

Now we shall consider the nonhomogeneous self-adjoint DE

$$(p(x)y')' + q(x)y + \mu r(x)y = \mathcal{P}_2[y] + \mu r(x)y = f(x) \tag{39.4}$$

together with the homogeneous boundary conditions (33.14). In (39.4) the functions $p(x)$, $q(x)$ and $r(x)$ are assumed to satisfy the same restrictions as in (36.1), μ is a given constant and $f(x)$ is a given function in $[\alpha, \beta]$. For the nonhomogeneous boundary value problem (39.4), (33.14) we shall assume that the solution $y(x)$ can be expanded in terms of eigenfunctions $\phi_n(x)$, $n = 1, 2, \ldots$ of the corresponding homogeneous Sturm–Liouville problem (36.1), (33.14), i.e., $y(x) = \sum_{n=1}^{\infty} c_n \phi_n(x)$. To compute the coefficients c_n in this expansion first we note that the infinite series $\sum_{n=1}^{\infty} c_n \phi_n(x)$ does satisfy the boundary conditions (33.14) since each $\phi_n(x)$ does so. Next consider the DE (39.4) that $y(x)$ must satisfy. For this, we have

$$\mathcal{P}_2 \left[\sum_{n=1}^{\infty} c_n \phi_n(x) \right] + \mu r(x) \sum_{n=1}^{\infty} c_n \phi_n(x) = f(x).$$

Thus, if we can interchange the operations of summation and differentiation, then

$$\sum_{n=1}^{\infty} c_n \mathcal{P}_2[\phi_n(x)] + \mu r(x) \sum_{n=1}^{\infty} c_n \phi_n(x) = f(x).$$

Since $\mathcal{P}_2[\phi_n(x)] = -\lambda_n r(x) \phi_n(x)$, this relation is the same as

$$\sum_{n=1}^{\infty} (\mu - \lambda_n) c_n \phi_n(x) = \frac{f(x)}{r(x)}. \tag{39.5}$$

Now we assume that the function $f(x)/r(x)$ satisfies the conditions of Theorem 39.1, so that it can be written as

$$\frac{f(x)}{r(x)} = \sum_{n=1}^{\infty} d_n \phi_n(x),$$

where from (38.2) the coefficients d_n are given by

$$d_n = \frac{1}{\|\phi_n\|^2} \int_{\alpha}^{\beta} r(x) \phi_n(x) \frac{f(x)}{r(x)} dx = \frac{1}{\|\phi_n\|^2} \int_{\alpha}^{\beta} \phi_n(x) f(x) dx. \tag{39.6}$$

With this assumption (39.5) takes the form

$$\sum_{n=1}^{\infty} [(\mu - \lambda_n) c_n - d_n] \phi_n(x) = 0.$$

Since this equation holds for each x in $[\alpha, \beta]$, it is necessary that

$$(\mu - \lambda_n) c_n - d_n = 0, \quad n = 1, 2, \ldots. \tag{39.7}$$

Thus, if μ is not equal to any eigenvalue of the corresponding homogeneous Sturm–Liouville problem (36.1), (33.14), i.e., $\mu \neq \lambda_n$, $n = 1, 2, \ldots$, then

$$c_n = \frac{d_n}{\mu - \lambda_n}, \quad n = 1, 2, \ldots. \tag{39.8}$$

Hence, the solution $y(x)$ of the nonhomogeneous problem (39.4), (33.14) can be written as

$$y(x) = \sum_{n=1}^{\infty} \frac{d_n}{\mu - \lambda_n} \phi_n(x). \tag{39.9}$$

Of course, the convergence of (39.9) is yet to be established.

If $\mu = \lambda_m$, then for $n = m$ equation (39.7) is of the form $0 \cdot c_m - d_m = 0$. Thus, if $d_m \neq 0$ then it is impossible to solve (39.7) for c_m, and hence the nonhomogeneous problem (39.4), (33.14) has no solution. Further, if $d_m = 0$ then (39.7) is satisfied for any arbitrary value of c_m, and hence the nonhomogeneous problem (39.4), (33.14) has an infinite number of solutions. From (39.6), $d_m = 0$ if and only if

$$\int_\alpha^\beta \phi_m(x)f(x)dx = 0,$$

i.e., $f(x)$ in (39.4) is orthogonal to the eigenfunction $\phi_m(x)$.

This formal discussion for the problem (39.4), (33.14) is summarized in the following theorem.

Theorem 39.4. Let $f(x)$ be continuous in the interval $[\alpha, \beta]$. Then the nonhomogeneous boundary value problem (39.4), (33.14) has a unique solution provided μ is different from all eigenvalues of the corresponding homogeneous Sturm–Liouville problem (36.1), (33.14). This solution $y(x)$ is given by (39.9), and the series converges for each x in $[\alpha, \beta]$. If μ is equal to an eigenvalue λ_m of the corresponding homogeneous Sturm–Liouville problem (36.1), (33.14), then the nonhomogeneous problem (39.4), (33.14) has no solution unless $f(x)$ is orthogonal to $\phi_m(x)$, i.e., unless

$$\int_\alpha^\beta \phi_m(x)f(x)dx = 0.$$

Further, in this case the solution is not unique.

Alternatively, this result can be stated as follows.

Theorem 39.5 (Fredholm's Alternative). For a given constant μ and a continuous function $f(x)$ in $[\alpha, \beta]$ the nonhomogeneous problem (39.4), (33.14) has a unique solution, or else the corresponding homogeneous problem (36.1), (33.14) has a nontrivial solution.

Example 39.4. Consider the nonhomogeneous boundary value problem

$$\begin{aligned} y'' + \pi^2 y &= x - x^2 \\ y(0) + y'(0) &= 0 = y(1). \end{aligned} \tag{39.10}$$

This problem can be solved directly to obtain the unique solution

$$y(x) = \frac{2}{\pi^4}\cos \pi x - \frac{1}{\pi^3}\left(1 + \frac{4}{\pi^2}\right)\sin \pi x + \frac{1}{\pi^2}\left(x - x^2 + \frac{2}{\pi^2}\right). \tag{39.11}$$

From Example 36.2 we know that π^2 is not an eigenvalue of the Sturm–Liouville problem (36.2), (36.8). Thus, from Theorem 39.4 the nonhomogeneous problem (39.10) has a unique solution. To find this solution in terms of the eigenvalues λ_n and eigenfunctions $\phi_n(x)$ of (36.2), (36.8) we note that the function $f(x) = x - x^2$ has been expanded in Example 39.2, and hence from (39.2) we have

$$d_1 = \frac{1}{4}, \quad d_n = \frac{2}{\lambda_n^{3/2} \sin^2 \sqrt{\lambda_n}}(2 - (2 + \lambda_n)\cos \sqrt{\lambda_n}), \quad n \geq 2.$$

Thus, from (39.9) we find that the solution $y(x)$ of (39.10) has the expansion

$$\begin{aligned} y(x) &= \frac{1}{4\pi^2}(1 - x) + \sum_{n=2}^{\infty} \frac{2}{(\pi^2 - \lambda_n)\lambda_n^{3/2} \sin^2 \sqrt{\lambda_n}} \\ &\quad \times (2 - (2 + \lambda_n)\cos \sqrt{\lambda_n}) \sin \sqrt{\lambda_n}(1 - x). \end{aligned} \tag{39.12}$$

Problems

39.1. For a given function $f(x)$ find the Fourier coefficients that correspond to the set of Chebyshev polynomials $T_n(x)$ defined in Problem 37.4.

39.2. Expand a given piecewise continuous function $f(x)$, $x \in [0, \pi]$

(i) in a *Fourier–cosine series*

$$f(x) \sim \frac{a_0}{2} + \sum_{n=1}^{\infty} a_n \cos nx \quad \text{where} \quad a_n = \frac{2}{\pi} \int_0^{\pi} f(t) \cos ntdt, \quad n \geq 0;$$

(ii) in a *Fourier–sine series*

$$f(x) \sim \sum_{n=1}^{\infty} b_n \sin nx \quad \text{where} \quad b_n = \frac{2}{\pi} \int_0^{\pi} f(t) \sin ntdt, \quad n \geq 1.$$

39.3. Show that the Fourier–Legendre series of the function $f(x) = \cos \pi x/2$ up to $P_4(x)$ in the interval $[-1, 1]$ is

$$\frac{2}{\pi}P_0(x) - \frac{10}{\pi^3}(12 - \pi^2)P_2(x) + \frac{18}{\pi^5}(\pi^4 - 180\pi^2 + 1680)P_4(x).$$

39.4. Find the trigonometric–Fourier series of each of the following functions:

(i) $f(x) = \begin{cases} 1, & -\pi < x < 0 \\ 2, & 0 < x < \pi. \end{cases}$ (ii) $f(x) = x - \pi, \quad -\pi < x < \pi.$

(iii) $f(x) = |x|$, $\quad -\pi < x < \pi$. $\qquad$ (iv) $f(x) = x^2$, $\quad -\pi < x < \pi$.

(v) $f(x) = \begin{cases} x, & -\pi < x < 0 \\ 2, & x = 0 \\ e^{-x}, & 0 < x < \pi. \end{cases}$ $\qquad$ (vi) $f(x) = x^4$, $\quad -\pi < x < \pi$.

39.5. (i) Let $\psi_n(x) = n\sqrt{x}e^{-nx^2/2}$, $x \in [0,1]$. Show that $\psi_n(x) \to 0$ as $n \to \infty$ for each x in $[0,1]$. Further, show that

$$e_n^2 = \int_0^1 (\psi_n(x) - 0)^2 dx = \frac{n}{2}(1 - e^{-n}),$$

and hence $e_n \to \infty$ as $n \to \infty$. Thus, pointwise convergence does not imply convergence in the mean.

(ii) Let $\psi_n(x) = x^n$, $x \in [0,1]$, and $f(x) = 0$ in $[0,1]$. Show that

$$e_n^2 = \int_0^1 (\psi_n(x) - f(x))^2 dx = \frac{1}{2n+1},$$

and hence $\psi_n(x)$ converges in the mean to $f(x)$. Further, show that $\psi_n(x)$ does not converge to $f(x)$ pointwise in $[0,1]$. Thus, mean convergence does not imply pointwise convergence.

39.6. Show that the sequence $\{x/(x+n)\}$ converges pointwise on $[0, \infty)$ and uniformly on $[0, a]$, $a > 0$.

39.7. Let $f(x) = \begin{cases} 0, & x \in [-1, 0) \\ 1, & x \in [0, 1]. \end{cases}$ Show that

$$\int_{-1}^1 \left(f(x) - \frac{1}{2} - \frac{3}{4}x \right)^2 dx \leq \int_{-1}^1 (f(x) - c_0 - c_1 x - c_2 x^2)^2 dx$$

for any set of constants c_0, c_1 and c_2.

39.8. Show that the following cannot be the Fourier series representation for any piecewise continuous function:

(i) $\displaystyle\sum_{n=1}^{\infty} n^{1/n}\phi_n(x)$. $\quad$ (ii) $\displaystyle\sum_{n=1}^{\infty} \frac{1}{\sqrt{n}}\phi_n(x)$.

39.9. Find Parseval's equality for the function $f(x) = 1$, $x \in [0, c]$ with respect to the orthonormal set $\left\{ \sqrt{\frac{2}{c}} \sin \frac{n\pi x}{c}, \ n = 1, 2, \ldots \right\}$.

39.10. Let $f(x)$ and $g(x)$ be piecewise continuous in the interval $[\alpha, \beta]$ and have the same Fourier coefficients with respect to a complete orthonormal set. Show that $f(x) = g(x)$ at each point of $[\alpha, \beta]$ where both functions are continuous.

39.11. Let $f(x)$ be a twice continuously differentiable, periodic function with a period 2π. Show the following:

(i)   The trigonometric–Fourier coefficients a_n and b_n of $f(x)$ satisfy

$$|a_n| \leq \frac{M}{n^2} \quad \text{and} \quad |b_n| \leq \frac{M}{n^2}, \quad n = 1, 2, \ldots,$$

where $M = \frac{1}{\pi} \int_{-\pi}^{\pi} |f''(x)| dx$.

(ii)   The trigonometric–Fourier series of $f(x)$ converges uniformly to $f(x)$ on $[-\pi, \pi]$.

39.12. Solve the following nonhomogeneous boundary value problems by means of an eigenfunction expansion:

(i) $y'' + 3y = e^x$, $y(0) = 0 = y(1)$.

(ii) $y'' + 2y = -x$, $y'(0) = 0 = y(1) + y'(1)$.

Answers or Hints

39.1. $c_n = \frac{2d_n}{\pi} \int_{-1}^{1} \frac{f(x)T_n(x)}{\sqrt{1-x^2}} dx$, where $d_0 = 1/2$ and $d_n = 1$ for $n \geq 1$.

39.2. (i) Use Problem 37.1. (ii) Use Problem 37.2.

39.3. Use (37.11) and $f(x) = \cos \pi x/2$ in Example 38.1.

39.4. (i) $\frac{3}{2} + \frac{2}{\pi} \left(\sin x + \frac{1}{3} \sin 3x + \cdots \right)$. (ii) $-\pi + \sum_{n=1}^{\infty} \frac{2(-1)^{n+1}}{n} \sin nx$.
(iii) $\frac{\pi}{2} + \sum_{n=1}^{\infty} \frac{2}{\pi n^2}((-1)^n - 1) \cos nx$. (iv) $\frac{\pi^2}{3} + 4 \sum_{n=1}^{\infty} \frac{(-1)^n}{n^2} \cos nx$.
(v) $-\left(\frac{e^{-\pi}-1}{2\pi} + \frac{\pi}{4} \right) + \frac{1}{\pi} \sum_{n=1}^{\infty} \left[\left(\frac{1+(-1)^{n+1}}{n^2} + \frac{1+(-1)^{n+1}e^{-\pi}}{1+n^2} \right) \cos nx \right.$
$\left. + \left(\frac{n}{1+n^2} (1 + (-1)^{n+1} e^{-\pi}) + \frac{\pi(-1)^{n+1}}{n} \right) \sin nx \right]$.
(vi) $\frac{\pi^4}{5} + 8 \sum_{n=1}^{\infty} \left(\frac{\pi^2}{n^2} - \frac{6}{n^4} \right) (-1)^n \cos nx$.

39.5. Verify directly.

39.6. Use definition.

39.7. For the given function Fourier–Legendre coefficients are $c_0 = 1/2$, $c_1 = 3/4$, $c_2 = 0$.

39.8. Use Theorem 38.3(i).

39.9. $\|f\|^2 = c$, $c_n^2 = \begin{cases} 0, & n \text{ even} \\ 8c/(n^2\pi^2), & n \text{ odd}. \end{cases}$ Thus, $c = \frac{8c}{\pi^2} \sum_{n=1}^{\infty} \frac{1}{(2n-1)^2}$.

39.10. Let $h(x) = f(x) - g(x)$ and $\{\phi_n(x)\}$ be a complete orthonormal set on the interval $[\alpha, \beta]$ with respect to the weight function $r(x)$. Note that for the function $h(x)$ Fourier coefficients $c_n = 0$, $n \geq 0$. Now apply Theorem 38.4.

39.11. (i) $a_n = \frac{1}{\pi} \int_{-\pi}^{\pi} f(x) \cos nx dx = -\frac{1}{n^2\pi} \int_{-\pi}^{\pi} f''(x) \cos nx dx.$ (ii) $\left| \frac{1}{2} a_0 \right.$
$\left. + \sum_{n=1}^{\infty} (a_n \cos nx + b_n \sin nx) \right| \le \frac{1}{2} |a_0| + \sum_{n=1}^{\infty} (|a_n| + |b_n|).$

39.12. (i) $2 \sum_{n=1}^{\infty} \frac{n\pi (1 + e(-1)^{n+1}) \sin n\pi x}{(1+n^2\pi^2)(3-n^2\pi^2)}.$ (ii) $2 \sum_{n=1}^{\infty} \frac{(2\cos\sqrt{\lambda_n} - 1)\cos\sqrt{\lambda_n} x}{\lambda_n(\lambda_n - 2)(1 + \sin^2\sqrt{\lambda_n})},$
where $\cot\sqrt{\lambda_n} = \sqrt{\lambda_n}.$

Lecture 40
Nonlinear Boundary
Value Problems

Theorem 32.3 provides necessary and sufficient conditions for the existence of a unique solution to the linear boundary value problem (6.6), (32.1). Unfortunately, this result depends on the explicit knowledge of two linearly independent solutions $y_1(x)$ and $y_2(x)$ to the homogeneous DE (6.1), which may not always be available. The purpose of this and the following lecture is to provide easily verifiable sets of sufficient conditions so that the second-order nonlinear DE

$$y'' = f(x, y) \qquad (40.1)$$

together with the boundary conditions (32.3) has at least and/or at most one solution.

We begin with the following examples which indicate possible difficulties that may arise in nonlinear problems.

Example 40.1. The boundary value problem

$$y'' = be^{ay}, \quad y(0) = y(1) = 0 \qquad (40.2)$$

arises in applications involving the diffusion of heat generated by positive temperature-dependent sources. For instance, if $a = 1$, it arises in the analysis of Joule losses in electrically conducting solids, with b representing the square of the constant current and e^y the temperature-dependent resistance, or in frictional heating with b representing the square of the constant shear stress and e^y the temperature dependent fluidity.

If $ab = 0$, the problem (40.2) has a unique solution:

(i) If $b = 0$, then $y(x) \equiv 0$.

(ii) If $a = 0$, then $y(x) = (b/2)x(x-1)$.

If $ab < 0$, the problem (40.2) has as many solutions as the number of roots of the equation $c = \sqrt{2|ab|} \cosh c/4$, and also for each such c_i, the solution is

$$y_i(x) = -\frac{2}{a}\left\{\ln\left(\cosh\left(\frac{1}{2}c_i\left(x-\frac{1}{2}\right)\right)\right) - \ln\left(\cosh\left(\frac{1}{4}c_i\right)\right)\right\}.$$

From the equation $c = \sqrt{2|ab|} \cosh c/4$, it follows that if

R.P. Agarwal and D. O'Regan, *An Introduction to Ordinary Differential Equations*,
doi: 10.1007/978-0-387-71276-5_40, © Springer Science + Business Media, LLC 2008

$$\sqrt{\frac{|ab|}{8}} \min_{c \geq 0} \frac{\cosh \frac{c}{4}}{\frac{c}{4}} \quad \begin{array}{ll} < 1, & (40.2) \quad \text{has two solutions} \\ = 1, & (40.2) \quad \text{has one solution} \\ > 1, & (40.2) \quad \text{has no solution.} \end{array}$$

If $ab > 0$, the problem (40.2) has a unique solution

$$y_1(x) = \frac{2}{a} \ln \left(c_1 / \cos \left(\frac{1}{2} c_1 \left(x - \frac{1}{2} \right) \right) \right) - \frac{1}{a} \ln(2ab),$$

where $c_1/4 \in (-\pi/2, \pi/2)$ is the root of the equation

$$\frac{c}{4} = \sqrt{\frac{ab}{8}} \cos \frac{c}{4}.$$

Example 40.2. Consider the nonlinear boundary value problem

$$y'' + |y| = 0, \quad y(0) = 0, \quad y(\beta) = B, \qquad (40.3)$$

where β and B are parameters.

It is clear that a solution $y(x)$ of $y'' + |y| = 0$ is a solution of $y'' - y = 0$ if $y(x) \leq 0$, and of $y'' + y = 0$ if $y(x) \geq 0$.

Since the function $f(x, y) = |y|$ satisfies the uniform Lipschitz condition (7.3), for each m the initial value problem $y'' + |y| = 0$, $y(0) = 0$, $y'(0) = m$ has a unique solution $y(x, m)$. Further, it is easy to obtain

$$y(x, m) = \begin{cases} 0 & \text{for all } x \in [0, \beta], \text{ if } m = 0 \\ m \sinh x & \text{for all } x \in [0, \beta], \text{ if } m < 0 \\ m \sin x & \text{for all } x \in [0, \pi], \text{ if } m > 0 \\ -m \sinh(x - \pi) & \text{for all } x \in [\pi, \beta], \text{ if } m > 0. \end{cases}$$

Thus, the boundary value problem (40.3) has a unique solution $y(x)$ if $\beta < \pi$, and it is given by

$$y(x) = \begin{cases} 0, & \text{if } B = 0 \\ B(\sinh \beta)^{-1} \sinh x & \text{if } B < 0 \\ B(\sin \beta)^{-1} \sin x & \text{if } B > 0. \end{cases}$$

If $\beta \geq \pi$, and $B > 0$ then (40.3) has no solution, whereas it has an infinite number of solutions $y(x) = c \sin x$ if $\beta = \pi$, and $B = 0$, where c is an arbitrary constant. If $\beta > \pi$, and $B = 0$ then $y(x) \equiv 0$ is the only solution of (40.3). Finally, if $\beta > \pi$, and $B < 0$ then (40.3) has two solutions $y_1(x)$ and $y_2(x)$ which are given by

$$y_1(x) = B(\sinh \beta)^{-1} \sinh x$$

and

$$y_2(x) = \begin{cases} -B(\sinh(\beta - \pi))^{-1}\sin x, & x \in [0, \pi] \\ B(\sinh(\beta - \pi))^{-1}\sinh(x - \pi), & x \in [\pi, \beta]. \end{cases}$$

The following result provides sufficient conditions on the function $f(x, y)$ so that the boundary value problem (40.1), (32.3) has at least one solution.

Theorem 40.1. Suppose that the continuous function $f(x, y)$ satisfies a uniform Lipschitz condition (7.3) in $[\alpha, \beta] \times \mathbb{R}$, and in addition, is bounded for all values of its arguments, i.e., $|f(x, y)| \leq M$. Then the boundary value problem (40.1), (32.3) has at least one solution.

Proof. From Theorem 15.3, for each m the initial value problem (40.1), $y(\alpha) = A$, $y'(\alpha) = m$ has a unique solution $y(x, m)$ in $[\alpha, \beta]$. Now since

$$\begin{aligned} y'(x, m) &= y'(\alpha, m) + \int_\alpha^x y''(t, m)dt \\ &= m + \int_\alpha^x f(t, y(t, m))dt \\ &\geq m - \int_\alpha^x M dt = m - M(x - \alpha) \end{aligned}$$

we find that

$$\begin{aligned} y(x, m) &= y(\alpha, m) + \int_\alpha^x y'(t, m)dt \\ &\geq A + \int_\alpha^x (m - M(t - \alpha))dt \\ &= A + m(x - \alpha) - \frac{M}{2}(x - \alpha)^2. \end{aligned}$$

Thus, in particular

$$y(\beta, m) \geq A + m(\beta - \alpha) - \frac{M}{2}(\beta - \alpha)^2. \tag{40.4}$$

Clearly, for $m = m_1$ sufficiently large and positive, (40.4) implies that $y(\beta, m_1) > B$. In the same way we obtain

$$y(\beta, m) \leq A + m(\beta - \alpha) + \frac{M}{2}(\beta - \alpha)^2$$

and hence, for $m = m_2$ sufficiently large and negative, $y(\beta, m_2) < B$. From Theorem 16.6 we know that $y(\beta, m)$ is a continuous function of m, so there is at least one m_3 such that $m_2 < m_3 < m_1$ and $y(\beta, m_3) = B$. Thus, the solution of the initial value problem (40.1), $y(\alpha) = A$, $y'(\alpha) = m_3$ is also a solution of the boundary value problem (40.1), (32.3). ■

Example 40.3. Since the function $f(x,y) = x\sin y$ satisfies the conditions of Theorem 40.1 in $[0, 2\pi] \times \mathbb{R}$, the boundary value problem

$$y'' = x\sin y, \quad y(0) = y(2\pi) = 0 \qquad (40.5)$$

has at least one solution. Indeed $y(x) \equiv 0$ is a solution of (40.5).

The following result gives sufficient conditions so that the problem (40.1), (32.3) has at most one solution.

Theorem 40.2. Suppose that the function $f(x,y)$ is continuous, and nondecreasing in y for all $(x,y) \in [\alpha, \beta] \times \mathbb{R}$. Then the boundary value problem (40.1), (32.3) has at most one solution.

Proof. Let $y_1(x)$ and $y_2(x)$ be two solutions of (40.1), (32.3). Then it follows that

$$y_1''(x) - y_2''(x) = f(x, y_1(x)) - f(x, y_2(x)),$$

which is the same as

$$(y_1(x) - y_2(x))(y_1''(x) - y_2''(x)) = (y_1(x) - y_2(x))(f(x, y_1(x)) - f(x, y_2(x))). \qquad (40.6)$$

Since $f(x,y)$ is nondecreasing in y, the right side of (40.6) is nonnegative. Thus, we have

$$\int_\alpha^\beta (y_1(x) - y_2(x))(y_1''(x) - y_2''(x))dx \ge 0,$$

i.e.,

$$(y_1(x) - y_2(x))(y_1'(x) - y_2'(x)) \Big|_\alpha^\beta - \int_\alpha^\beta (y_1'(x) - y_2'(x))^2 dx \ge 0. \qquad (40.7)$$

In (40.7) the first term is zero, and hence it is necessary that

$$\int_\alpha^\beta (y_1'(x) - y_2'(x))^2 dx = 0. \qquad (40.8)$$

Equation (40.8) holds if and only if $y_1'(x) - y_2'(x) \equiv 0$, i.e., $y_1(x) - y_2(x) = c$ (constant). However, since $y_1(\alpha) - y_2(\alpha) = 0$ the constant $c = 0$. Hence, $y_1(x) \equiv y_2(x)$. ∎

Example 40.4. If $ab > 0$, then the function be^{ay} is nondecreasing in y, and hence for this case Theorem 40.2 implies that the boundary value problem (40.2) has at most one solution.

Since the boundary value problem $y'' = -y$, $y(0) = y(\pi) = 0$ has an infinite number of solutions, conclusion of Theorem 40.2 does not remain true when $f(x, y)$ is decreasing with respect to y. Thus, in Theorem 40.2 "nondecreasing" cannot be replaced by "decreasing."

In Theorem 40.1 the condition that $f(x, y)$ is bounded for all values of its arguments in $[\alpha, \beta] \times \mathbb{R}$ makes it too restrictive. This condition is not satisfied in both the Examples 40.1 and 40.2. In fact even a linear function in y, i.e., $f(x, y) = p(x)y + q(x)$ does not meet this requirement. Thus, we state without proof a result which is very useful in applications.

Theorem 40.3. Suppose that $K > 0$ is a given number, and the function $f(x, y)$ is continuous in the set $D = \{(x, y) : \alpha \le x \le \beta, |y| \le 2K\}$, and hence there exists a $M > 0$ such that $|f(x, y)| \le M$ for all $(x, y) \in D$. Further, we assume that

$$\frac{1}{8}(\beta - \alpha)^2 M \le K, \quad \text{and} \quad \max\{|A|, |B|\} \le K. \quad (40.9)$$

Then the boundary value problem (40.1), (32.3) has at least one solution $y(x)$ such that $|y(x)| \le 2K$ for all $x \in [\alpha, \beta]$.

Corollary 40.4. Suppose that the function $f(x, y)$ is continuous and bounded, i.e., $|f(x, y)| \le M$ for all $(x, y) \in [\alpha, \beta] \times \mathbb{R}$. Then the boundary value problem (40.1), (32.3) has at least one solution.

Thus, we see that in Theorem 40.1 the hypothesis that $f(x, y)$ is uniform Lipschitz, is superfluous. We also note that for the given length of the interval $(\beta - \alpha)$, the inequality $(1/8)(\beta - \alpha)^2 M \le K$ in (40.9) restricts the upper bound M on the function $|f(x, y)|$ in D. Alternatively, for fixed M it provides an upper bound on the length of the interval, i.e.,

$$(\beta - \alpha) \le \sqrt{\frac{8K}{M}}. \quad (40.10)$$

Because of (40.10) Theorem 40.3 is called a local existence theorem, which corresponds to the local existence result for the initial value problems. Further, Corollary 40.4 is a global existence result.

Example 40.5. For the problem (40.2) the conditions of Theorem 40.3 are satisfied provided $(1/8)|b|e^{2|a|K} \le K$. Thus, in particular, the problem $y'' = e^y$, $y(0) = y(1) = 0$ has at least one solution $y(x)$ if $(1/8)e^{2K} \le K$, i.e., $K \simeq 1.076646182$. Further, $|y(x)| \le 2.153292364$.

Example 40.6. For the problem (40.3) the conditions of Theorem 40.3 hold provided $(1/8)\beta^2(2K) \le K$, i.e., $\beta \le 2$ and $|B| \le K$. Thus, as a special case the problem $y'' + |y| = 0$, $y(0) = 0$, $y(2) = 1$ has at least one solution $y(x)$, satisfying $|y(x)| \le 2$.

Lecture 41
Nonlinear Boundary
Value Problems (Contd.)

Picard's method of successive approximations for the initial value problems discussed in Lecture 8 is equally useful for the boundary value problem (40.1), (32.3). For this, from Problem 33.7 we note that this problem is equivalent to the integral equation

$$y(x) = \ell(x) + \int_\alpha^\beta G(x,t)f(t,y(t))dt, \qquad (41.1)$$

where

$$\ell(x) = \frac{(\beta - x)}{(\beta - \alpha)}A + \frac{(x - \alpha)}{(\beta - \alpha)}B \qquad (41.2)$$

and the Green's function $G(x,t)$ is defined in (33.25).

The following result provides sufficient conditions on the function $f(x,y)$ so that the sequence $\{y_m(x)\}$ generated by the iterative scheme

$$\begin{aligned} y_0(x) &= \ell(x) \\ y_{m+1}(x) &= \ell(x) + \int_\alpha^\beta G(x,t)f(t,y_m(t))dt, \quad m = 0,1,2,\ldots \end{aligned} \qquad (41.3)$$

converges to the unique solution of the integral equation (41.1).

Theorem 41.1. Suppose that the function $f(x,y)$ is continuous and satisfies a uniform Lipschitz condition (7.3) in $[\alpha, \beta] \times \mathbb{R}$, and in addition

$$\theta = \frac{1}{8}L(\beta - \alpha)^2 < 1. \qquad (41.4)$$

Then the sequence $\{y_m(x)\}$ generated by the iterative scheme (41.3) converges to the unique solution $y(x)$ of the boundary value problem (40.1), (32.3). Further, for all $x \in [\alpha, \beta]$ the following error estimate holds:

$$|y(x) - y_m(x)| \leq \frac{\theta^m}{1 - \theta} \max_{\alpha \leq x \leq \beta} |y_1(x) - y_0(x)|, \quad m = 0,1,2,\ldots. \qquad (41.5)$$

Proof. From (41.3) it is clear that the successive approximations $y_m(x)$ exist as continuous functions in $[\alpha, \beta]$. We need to prove that

$$|y_{m+1}(x) - y_m(x)| \leq \theta^m \max_{\alpha \leq x \leq \beta} |y_1(x) - y_0(x)|. \qquad (41.6)$$

R.P. Agarwal and D. O'Regan, *An Introduction to Ordinary Differential Equations*, doi: 10.1007/978-0-387-71276-5_41, © Springer Science + Business Media, LLC 2008

When $m = 1$, (41.3) gives

$$
\begin{aligned}
|y_2(x) - y_1(x)| &\leq \int_\alpha^\beta |G(x,t)||f(t, y_1(t)) - f(t, y_0(t))|dt \\
&\leq L \int_\alpha^\beta |G(x,t)||y_1(t) - y_0(t)|dt \\
&\leq L \max_{\alpha \leq x \leq \beta} |y_1(x) - y_0(x)| \int_\alpha^\beta |G(x,t)|dt \\
&\leq \frac{1}{8} L(\beta - \alpha)^2 \max_{\alpha \leq x \leq \beta} |y_1(x) - y_0(x)|,
\end{aligned}
$$

where we have used the Lipschitz condition and Problem 33.7. Thus, (41.6) holds for $m = 1$. Now let (41.6) be true for $m = k \geq 1$; then from (41.3), we have

$$
\begin{aligned}
|y_{k+2}(x) - y_{k+1}(x)| &\leq \int_\alpha^\beta |G(x,t)||f(t, y_{k+1}(t)) - f(t, y_k(t))|dt \\
&\leq L \int_\alpha^\beta |G(x,t)||y_{k+1}(t) - y_k(t)|dt \\
&\leq L\theta^k \max_{\alpha \leq x \leq \beta} |y_1(x) - y_0(x)| \int_\alpha^\beta |G(x,t)|dt \\
&\leq \frac{1}{8}(\beta - \alpha)^2 \theta^k \max_{\alpha \leq x \leq \beta} |y_1(x) - y_0(x)| \\
&\leq \theta^{k+1} \max_{\alpha \leq x \leq \beta} |y_1(x) - y_0(x)|.
\end{aligned}
$$

Hence, the inequality (41.6) is true for all m.

Now for $n > m$ inequality (41.6) gives

$$
\begin{aligned}
|y_n(x) - y_m(x)| &\leq \sum_{k=m}^{n-1} |y_{k+1}(x) - y_k(x)| \\
&\leq \sum_{k=m}^{n-1} \theta^k \max_{\alpha \leq x \leq \beta} |y_1(x) - y_0(x)| \qquad (41.7) \\
&\leq \frac{\theta^m}{1-\theta} \max_{\alpha \leq x \leq \beta} |y_1(x) - y_0(x)|.
\end{aligned}
$$

Since $\theta < 1$, an immediate consequence of (41.7) is that the sequence $\{y_m(x)\}$ is uniformly Cauchy in $[\alpha, \beta]$, and hence converges uniformly to a function $y(x)$ in $[\alpha, \beta]$. Clearly, this limit function $y(x)$ is continuous. Letting $m \to \infty$ in (41.3), it follows that $y(x)$ is a solution of (41.1). Also, letting $n \to \infty$ in (41.7) results in (41.5).

To prove the uniqueness of the solution $y(x)$ of (41.1), let $z(x)$ be another solution of (41.1). Then once again from Problem 33.7 it follows that

$$
\begin{aligned}
|y(x) - z(x)| &\leq \int_\alpha^\beta |G(x,t)||f(t,y(t)) - f(t,z(t))|dt \\
&\leq L \int_\alpha^\beta |G(x,t)||y(t) - z(t)|dt \qquad\qquad (41.8) \\
&\leq \frac{1}{8} L(\beta - \alpha)^2 \max_{\alpha \leq x \leq \beta} |y(x) - z(x)|.
\end{aligned}
$$

However, since $\theta < 1$ inequality (41.8) implies that $\max_{\alpha \leq x \leq \beta} |y(x) - z(x)| = 0$, i.e., $y(x) = z(x)$ for all $x \in [\alpha, \beta]$. This completes the proof of the theorem. ∎

For a given function the Lipschitz constant L is known, so the condition (41.4) restricts the length of the interval $(\beta - \alpha)$. Similarly, for the given boundary conditions the length of the interval $(\beta - \alpha)$ is known, so the condition (41.4) restricts the Lipschitz constant L. The problem of finding the largest interval of existence of the unique solution $y(x)$ of (40.1), (32.3) is of interest. By modifying the above proof it can be shown that the inequality (41.4) can indeed be replaced by

$$
\frac{1}{\pi^2} L(\beta - \alpha)^2 < 1. \qquad\qquad (41.9)
$$

Obviously, from Example 40.2 inequality (41.9) is the best possible in the sense that $<$ cannot be replaced by $\leq$.

Example 41.1. Consider the boundary value problem

$$
y'' = \sin y, \quad y(0) = 0, \quad y(1) = 1. \qquad\qquad (41.10)
$$

For this problem $L = 1$, $\beta - \alpha = 1$ and hence $\theta = 1/8$, so from Theorem 41.1, (41.10) has a unique solution. Further, since

$$
\begin{aligned}
y_0(x) &= x \\
y_1(x) &= x + \int_0^1 G(x,t) \sin t\, dt = x + x \sin 1 - \sin x
\end{aligned}
$$

it follows that

$$
|y_1(x) - y_0(x)| = |x \sin 1 - \sin x| \leq 0.06
$$

and hence from (41.5), we have

$$
|y(x) - y_m(x)| \leq \frac{8}{7}\left(\frac{1}{8}\right)^m (0.06), \quad m = 0, 1, 2, \ldots.
$$

Since the function be^{ay} satisfies Lipschitz condition (7.3) only in compact subsets of $[\alpha, \beta] \times \mathbb{R}$, Theorem 41.1 cannot be applied to Example 40.1. To accommodate such a situation we need to modify Theorem 40.1. For this, we note that if $y(x)$ is a solution of (40.1), (32.3) then $w(x) = y(x) - \ell(x)$ is a solution of the problem

$$w'' = F(x, w) \tag{41.11}$$

$$w(\alpha) = w(\beta) = 0, \tag{41.12}$$

where $F(x, w) = f(x, w + \ell(x))$. Clearly, F satisfies the Lipschitz condition (7.3) with the same Lipschitz constant as that for f. For the boundary value problem (41.11), (41.12) we state the following result which generalizes Theorem 41.1.

Theorem 41.2. Suppose that the function $F(x, w)$ is continuous and satisfies a uniform Lipschitz condition (7.3) in $[\alpha, \beta] \times [-N, N]$, where $N > 0$ is a constant. Further, let inequality (41.4) hold and either

$$\frac{1}{8}(\beta - \alpha)^2 \max_{\alpha \leq x \leq \beta} |F(x, 0)| \leq N(1 - \theta), \tag{41.13}$$

or

$$\frac{1}{8}(\beta - \alpha)^2 \max_{\substack{\alpha \leq x \leq \beta \\ |w| \leq N}} |F(x, w)| \leq N. \tag{41.14}$$

Then the boundary value problem (41.11), (41.12) has a unique solution $w(x)$ such that $|w(x)| \leq N$ for all $x \in [\alpha, \beta]$. Further, the iterative scheme

$$w_0(x) = 0$$
$$w_{m+1}(x) = \int_\alpha^\beta G(x, t) F(t, w_m(t)) dt, \quad m = 0, 1, 2, \ldots \tag{41.15}$$

converges to $w(x)$, and

$$|w(x) - w_m(x)| \leq \frac{\theta^m}{1 - \theta} \max_{\alpha \leq x \leq \beta} |w_1(x)|, \quad m = 0, 1, 2, \ldots. \tag{41.16}$$

Example 41.2. The function $F(x, w) = f(x, y) = -e^y$ satisfies Lipschitz condition in $[\alpha, \beta] \times [-N, N]$ with the Lipschitz constant $L = e^N$. Thus, for the problem (40.2) with $b = -1$, $a = 1$ conditions of Theorem 41.2 reduce to

$$\frac{1}{8}e^N < 1, \quad \text{i.e.,} \quad 0 < N \leq 2.0794 \tag{41.17}$$

and

$$\frac{1}{8} \leq N\left(1 - \frac{1}{8}e^N\right), \quad \text{i.e.,} \quad 0.14615 \leq N \leq 2.0154 \tag{41.18}$$

or

$$\frac{1}{8}e^N \leq N, \quad \text{i.e.,} \quad 0.14443 \leq N \leq 3.26167. \tag{41.19}$$

Thus, from (41.17) and (41.19) the problem $y'' + e^y = 0$, $y(0) = y(1) = 0$ has a unique solution $y(x)$ in the region $[0,1] \times [-2.0794, 2.0794]$, and $|y(x)| \leq 0.14443$. Further, since from (41.15), $w_0(x) = y_0(x) = 0$, $w_1(x) = y_1(x) = (1/2)x(1-x)$, (41.16) reduces to

$$|y(x) - y_m(x)| \leq \frac{\theta^m}{1-\theta} \left| \frac{1}{2}x(1-x) \right| \leq \frac{1}{8}\frac{\theta^m}{1-\theta}$$

and hence for $N = 0.14443$, i.e., $\theta \simeq 0.14441$, the error estimate becomes

$$|y(x) - y_m(x)| \leq (0.1461)(0.14441)^m, \quad m = 0,1,2,\ldots.$$

Finally, in this lecture we shall prove the following result.

Theorem 41.3. Suppose that $F(x,w)$ and $\partial F(x,w)/\partial w$ are continuous and $0 \leq \partial F(x,w)/\partial w \leq L$ in $[\alpha, \beta] \times \mathbb{R}$. Then the boundary value problem (41.11), (41.12) has a unique solution $w(x)$. Further, for any k such that $k \geq L$ the iterative scheme

$$w_0(x) = 0$$
$$w_{m+1}(x) = \int_\alpha^\beta G(x,t)[-kw_m(t) + F(t, w_m(t))]dt, \quad m = 0,1,2,\ldots$$
$$\tag{41.20}$$

converges to $w(x)$, where the Green's function $G(x,t)$ is defined in (33.29).

Proof. As in Theorem 41.1, first we shall show that the sequence $\{w_m(x)\}$ generated by (41.20) is a Cauchy sequence. For this, we have

$$w_{m+1}(x) - w_m(x)$$
$$= \int_\alpha^\beta G(x,t)[-k(w_m(t) - w_{m-1}(t)) + (F(t, w_m(t)) - F(t, w_{m-1}(t)))]dt$$
$$= -\int_\alpha^\beta G(x,t)\left[k - \frac{\partial F}{\partial w}(t, w_m(t) - \theta(t)(w_m(t) - w_{m-1}(t)))\right]$$
$$\times [w_m(t) - w_{m-1}(t)]dt,$$

where the mean value theorem has been used and $0 \leq \theta(t) \leq 1$. Since $0 \leq \partial F(x,w)/\partial w \leq L$ and $k \geq L$, we find that $0 \leq k - \partial F/\partial w \leq k$. Thus, from Problem 33.9 it follows that

$$|w_{m+1}(x) - w_m(x)| \leq \left(\int_\alpha^\beta |G(x,t)| k \, dt \right) Z_m$$

$$\leq \left(1 - \frac{\cosh \sqrt{k} \left(\frac{\beta+\alpha}{2} - x \right)}{\cosh \sqrt{k} \left(\frac{\beta-\alpha}{2} \right)} \right) Z_m$$

$$\leq \mu \, Z_m,$$

where $Z_m = \max_{\alpha \leq x \leq \beta} |w_m(x) - w_{m-1}(x)|$, and

$$\mu = \left(1 - \frac{1}{\cosh \sqrt{k} \left(\frac{\beta-\alpha}{2} \right)} \right) < 1.$$

From this it is immediate that

$$|w_{m+1}(x) - w_m(x)| \leq \mu^m \max_{\alpha \leq x \leq \beta} |w_1(x) - w_0(x)|$$

and hence $\{w_m(x)\}$ is a Cauchy sequence. Thus, in (41.20) we may take the limit as $m \to \infty$ to obtain

$$w(x) = \int_\alpha^\beta G(x,t)[-kw(t) + F(t, w(t))] dt,$$

which is equivalent to the boundary value problem (41.11), (41.12). The uniqueness of this solution $w(x)$ can be proved as in Theorem 41.1. ∎

Example 41.3. The linear boundary value problem

$$y'' = p(x)y + q(x), \quad y(\alpha) = y(\beta) = 0$$

where $p, q \in C[\alpha, \beta]$ and $p(x) \geq 0$ for all $x \in [\alpha, \beta]$ has a unique solution.

Problems

41.1. Show that the following boundary value problems have at least one solution:

(i) $y'' = 1 + x^2 e^{-|y|}$, $y(0) = 1$, $y(1) = 7$.
(ii) $y'' = \sin x \cos y + e^x$, $y(0) = 0$, $y(1) = 1$.

41.2. Use Theorem 40.3 to obtain optimum value of $\beta > 0$ so that the following boundary value problems have at least one solution:

(i) $y'' = y \cos y + \sin x$, $y(0) = y(\beta) = 0$.

(ii) $y'' = y^2 \sin x + e^{-x} \cos x$, $y(0) = 1$, $y(\beta) = 2$.

41.3. Show that the following boundary value problems have at most one solution:

(i) $y'' = y^3 + x$, $y(0) = 0$, $y(1) = 1$.

(ii) $y'' = y + \cos y + x^2$, $y(0) = 1$, $y(1) = 5$.

41.4. Find first two Picard's iterates for the following boundary value problems:

(i) $y'' + |y| = 0$, $y(0) = 0$, $y(1) = 1$.

(ii) $y'' + e^{-y} = 0$, $y(0) = y(1) = 0$.

Further, give a bound on the error introduced by stopping the computations at the second iterate.

41.5. State and prove a result analogous to Theorem 41.1 for the boundary value problem (40.1), (32.4).

41.6. Prove the following result: Suppose that the function $f(x, y, y')$ is continuous and satisfies a uniform Lipschitz condition

$$|f(x, y, y') - f(x, z, z')| \leq L|y - z| + M|y' - z'|$$

in $[\alpha, \beta] \times \mathbb{R}^2$, and in addition

$$\mu = \frac{1}{8}L(\beta - \alpha)^2 + \frac{1}{2}M(\beta - \alpha) < 1.$$

Then the sequence $\{y_m(x)\}$ generated by the iterative scheme

$$y_0(x) = \ell(x)$$

$$y_{m+1}(x) = \ell(x) + \int_\alpha^\beta G(x, t)f(t, y_m(t), y'_m(t))dt, \quad m = 0, 1, 2, \ldots$$

where $\ell(x)$ and $G(x, t)$ are, respectively, defined in (41.2) and (33.25), converges to the unique solution $y(x)$ of the boundary value problem (33.23), (32.3). Further, the following error estimate holds:

$$\|y - y_m\| \leq \frac{\mu^m}{1 - \mu}\|y_1 - y_0\|, \quad m = 0, 1, 2, \ldots,$$

where $\|y\| = L \max_{\alpha \leq x \leq \beta} |y(x)| + M \max_{\alpha \leq x \leq \beta} |y'(x)|$.

41.7. State and prove a result analogous to Problem 41.6 for the boundary value problem (33.23), (32.4).

41.8. Solve the following boundary value problems:

(i) $y'' = -2yy'$, $y(0) = 1$, $y(1) = 1/2$.

(ii) $y'' = -(y')^2/y$, $y(0) = 1$, $y'(1) = 3/4$.

(iii) $y'' = 2y^3$, $y(1) = 1$, $y(2) = 1/2$

(iv) $y'' = -(y')^3$, $y(0) = \sqrt{2}$, $y(1) = 2$.

*41.9. Suppose $f(x,y)$ is continuous and has a continuous first-order derivative with respect to y for $0 \le x \le 1$, $y \in \mathbb{R}$, and the boundary value problem $y'' = f(x,y)$, $y(0) = A$, $y(1) = B$ has a solution $y(x)$, $0 \le x \le 1$. If $\partial f(x,y)/\partial y > 0$ for $x \in [0,1]$ and $y \in \mathbb{R}$, show that there is an $\epsilon > 0$ such that the boundary value problem $y'' = f(x,y)$, $y(0) = A$, $y(1) = B_1$ has a solution for $0 \le x \le 1$ and $|B - B_1| \le \epsilon$.

Answers or Hints

41.1. (i) Use Corollary 40.4. (ii) Use Corollary 40.4.

41.2. (i) 2. (ii) $4\sqrt{17}/17$.

41.3. (i) Use Theorem 40.2. (ii) Use Theorem 40.2.

41.4. (i) $y_1(x) = \frac{1}{6}(7x - x^3)$, $y_2(x) = \frac{1}{360}(417x - 70x^3 + 3x^5)$, error $= \sqrt{3}/1{,}512$. (ii) $y_1(x) = \frac{1}{2}(x - x^2)$, $y_2(x) = (1-x)\int_0^x t \exp\left(-\frac{1}{2}(t - t^2)\right) dt + x\int_0^1 (1 - t)\exp\left(-\frac{1}{2}(t - t^2)\right) dt$, error $=$ same as in Example 41.2.

41.5. The statement and the proof remain the same, except now $\ell(x) = A + (x - \alpha)B$, $\theta = \frac{1}{2}L(\beta - \alpha)^2 < 1$ and the Green's function $G(x,t)$ is as in (33.27).

41.6. Use Problem 33.7 and modify the proof of Theorem 41.1.

41.7. The statement and the proof remain the same, except now $\ell(x) = A + (x - \alpha)B$, $\mu = \frac{1}{2}L(\beta - \alpha)^2 + M(\beta - \alpha) < 1$ and the Green's function $G(x,t)$ is as in (33.27).

41.8. (i) $1/(1 + x)$. (ii) $\sqrt{3x + 1}$. (iii) $1/x$. (iv) $\sqrt{2x + 2}$.

Lecture 42
Topics for Further Studies

We begin this lecture with a brief description of a selection of topics related to ordinary differential equations which have motivated a vast amount of research work in the last 30 years. It is clear from our previous lectures that one of the main areas of research in differential equations is the existence, uniqueness, oscillation, and stability of solutions to nonlinear initial value problems [2, 3, 10, 12, 18, 21–25, 33, 36–38], and the existence and uniqueness of solutions to nonlinear boundary value problems [1, 5, 6, 9, 11, 31, 32, 34, 35]. When modeling a physical or biological system one must first decide what structure best fits the underlying properties of the system under investigation. In the past a continuous approach was usually adopted. For example, if one wishes to model a fluid flow, a continuous approach would be appropriate; and the evolution of the system can then be described by ordinary or partial differential equations. On the other hand, if data are only known at distinct times, a discrete approach may be more appropriate; and the evolution of the system in this case can be described by difference equations [4–9, 14]. However, the model variables may evolve in time in a way which involves both discrete and continuous elements. For example, suppose the life span of a species of insect is one time unit, and at the end of its life span the insect mates, lays eggs, and then dies. Suppose the eggs lie dormant for a further one time unit before hatching. The *time scale* on which the insect population evolves is therefore best represented by a set of continuous intervals separated by discrete gaps. As a result, recently *time scale calculus* (differentiation and integration) has been introduced. This has led to the study of dynamic equations of so-called *time scales* which unifies the theories of differential and difference equations and to cases *in between* [15]. Other types of differential equations which have made a significant impact in mathematics and are being continuously studied are functional differential equations [8, 13, 19], impulsive differential equations [27], differential equations in abstract spaces [16, 26], set and multivalued differential equations [17, 29], and fuzzy differential equations [28]. Now, instead of going into detail on any one of these topics (which is outside the scope of this book), we will describe a number of physical problems which have motivated some of the current research in the presented literature.

First, we describe an initial value problem. Consider a spherical cloud of gas and denote its total pressure at a distance r from the center by $p(r)$. The total pressure is due to the usual gas pressure and a contribution from

R.P. Agarwal and D. O'Regan, *An Introduction to Ordinary Differential Equations*,
doi: 10.1007/978-0-387-71276-5_42, © Springer Science + Business Media, LLC 2008

radiation,

$$p = \frac{1}{3}aT^4 + \frac{RT}{v},$$

where a, T, R and v are, respectively, the radiation constant, the absolute temperature, the gas constant, and the volume. Pressure and density $\rho = v^{-1}$ vary with r and $p = K\rho^\gamma$, where γ and K are constants. Let m be the mass within a sphere of radius r and G be the constant of gravitation. The equilibrium equations for the configuration are

$$\frac{dp}{dr} = -\frac{Gm\rho}{r^2} \quad \text{and} \quad \frac{dm}{dr} = 4\pi r^2 \rho.$$

Elimination of m yields

$$\frac{1}{r^2}\frac{d}{dr}\left(\frac{r^2}{\rho}\frac{dp}{dr}\right) + 4\pi G\rho = 0.$$

Now let $\gamma = 1 + \mu^{-1}$ and set $\rho = \lambda\,\phi^\mu$, so that

$$p = K\rho^{1+\mu^{-1}} = K\lambda^{1+\mu^{-1}}\phi^{\mu+1}.$$

Thus, we have

$$\frac{1}{r^2}\frac{d}{dr}\left(r^2\frac{d\phi}{dr}\right) + k^2\phi^\mu = 0,$$

where

$$k^2 = \frac{4\pi G\lambda^{1-\mu^{-1}}}{(\mu+1)K}.$$

Next let $x = kr$, to obtain

$$\frac{d^2\phi}{dx^2} + \frac{2}{x}\frac{d\phi}{dr} + \phi^\mu = 0.$$

If we let $\lambda = \rho_0$, the density at $r = 0$, then we may take $\phi = 1$ at $x = 0$. By symmetry the other condition is $d\phi/dx = 0$ when $x = 0$. A solution of the differential equation satisfying these initial conditions is called a Lane–Emden function of index $\mu = (\gamma - 1)^{-1}$.

The differential equation

$$y'' + \frac{2}{t}y' + g(y) = 0 \qquad\qquad (42.1)$$

was first studied by Emden when he examined the thermal behavior of spherical clouds of gas acting on gravitational equilibrium subject to the laws of thermodynamics. The usual interest is in the case $g(y) = y^n$, $n \geq 1$, which was treated by Chandrasekhar in his study of stellar structure. The natural initial conditions for (42.1) are

$$y(0) = 1, \quad y'(0) = 0. \tag{42.2}$$

It is easy to check that (42.1), (42.2) can be solved exactly if $n = 1$ with the solution

$$y(t) = \frac{\sin t}{t}$$

and if $n = 5$ with the solution

$$y(t) = \left(1 + \frac{1}{3}t^2\right)^{-1/2}.$$

It is also of interest to note that the Emden differential equation $y'' - t^a y^b = 0$ arises in various astrophysical problems, including the study of the density of stars. Of course, one is interested only in positive solutions in the above models.

Next we describe four boundary value problems, namely, (i) a problem in membrane theory, (ii) a problem in non-Newtonian flow, (iii) a problem in spherical caps, and (iv) a problem in the theory of colloids. These problems have motivated the study of singular differential equations with boundary conditions over finite and infinite intervals, and have led to new areas in the qualitative theory of differential equations [6, 9, 11, 31, 32].

Our first problem examines the deformation shape of a membrane cap which is subjected to a uniform vertical pressure P and either a radial displacement or a radial stress on the boundary. Assuming the cap is shallow (i.e., nearly flat), the strains are small, the pressure P is small, and the undeformed shape of the membrane is radially symmetric and described in cylindrical coordinates by $z = C(1 - r^\gamma)$ ($0 \leq r \leq 1$ and $\gamma > 1$) where the undeformed radius is $r = 1$ and $C > 0$ is the height at the center of the cap. Then for any radially symmetric deformed state, the scaled radial stress S_r satisfies the differential equation

$$r^2 S_r'' + 3r S_r' = \frac{\lambda^2 r^{2\gamma - 2}}{2} + \frac{\beta \nu r^2}{S_r} - \frac{r^2}{8 S_r^2},$$

the regularity condition

$$S_r(r) \quad \text{bounded as} \quad r \to 0^+,$$

and the boundary condition

$$b_0 S_r(1) + b_1 S_r'(1) = A,$$

where λ and β are positive constants depending on the pressure P, the thickness of the membrane and Young's modulus, $b_0 > 0$, $b_1 \geq 0$, and

$A > 0$. For the stress problem $b_0 = 1$, $b_1 = 0$, whereas for the displacement problem, $b_0 = 1 - \nu$, $b_1 = 1$, where ν $(0 \leq \nu < 0.5)$ is the Poisson ratio.

For the second problem, recall that the Cauchy stress $\mathbf{T}$ in an incompressible homogeneous fluid of third grade has the form

$$\mathbf{T} = -p\mathbf{I} + \mu\mathbf{A}_1 + \alpha_1\mathbf{A}_2 + \alpha_2\mathbf{A}_1^2 + \beta_1\mathbf{A}_3 + \beta_2[\mathbf{A}_1\mathbf{A}_2 + \mathbf{A}_2\mathbf{A}_1] + \beta_3(tr\mathbf{A}_1^2)\mathbf{A}_2, \tag{42.3}$$

where $-p\mathbf{I}$ is the spherical stress due to the constraint of incompressibility, μ, α_1, α_2, β_1, β_2, β_3 are material moduli, and $\mathbf{A}_1$, $\mathbf{A}_2$, $\mathbf{A}_3$ are the first three Rivlin–Ericksen tensors given by

$$\mathbf{A}_1 = \mathbf{L} + \mathbf{L}^T, \qquad \mathbf{A}_2 = \frac{d\mathbf{A}_1}{dt} + \mathbf{L}^T\mathbf{A}_1 + \mathbf{A}_1\mathbf{L}$$

and

$$\mathbf{A}_3 = \frac{d\mathbf{A}_2}{dt} + \mathbf{L}^T\mathbf{A}_2 + \mathbf{A}_2\mathbf{L};$$

here $\mathbf{L}$ represents the spatial gradient of velocity and d/dt the material time derivative. Now consider the flow of a third grade fluid, obeying (42.3), maintained at a cylinder (of radius R) by its angular velocity (Ω). The steady state equation for this fluid is

$$0 = \mu\left[\frac{d^2\tilde{v}}{d\tilde{r}^2} + \frac{1}{\tilde{r}}\frac{d\tilde{v}}{d\tilde{r}} - \frac{\tilde{v}}{\tilde{r}^2}\right] + \beta\left(\frac{d\tilde{v}}{d\tilde{r}} - \frac{\tilde{v}}{\tilde{r}}\right)^2\left[6\frac{d^2\tilde{v}}{d\tilde{r}^2} - \frac{2}{\tilde{r}}\frac{d\tilde{v}}{d\tilde{r}} + \frac{2\tilde{v}}{\tilde{r}^2}\right]$$

with the boundary conditions

$$\tilde{v} = R\Omega \quad \text{at} \quad \tilde{r} = R, \quad \text{and} \quad \tilde{v} \to 0 \quad \text{as} \quad \tilde{r} \to \infty;$$

here $\tilde{v}$ is the nonzero velocity in polar coordinates and μ and β are material constants. Making the change of variables

$$r = \frac{\tilde{r}}{R} \quad \text{and} \quad v = \frac{\tilde{v}}{R\Omega},$$

our problem is transformed to

$$\frac{d^2v}{dr^2} + \frac{1}{r}\frac{dv}{dr} - \frac{v}{r^2} + \epsilon\left(\frac{dv}{dr} - \frac{v}{r}\right)^2\left[6\frac{d^2v}{dr^2} - \frac{2}{r}\frac{dv}{dr} + \frac{2v}{r^2}\right] = 0$$

for $1 < r < \infty$, with the boundary conditions

$$v = 1 \quad \text{if} \quad r = 1, \quad v \to 0 \quad \text{as} \quad r \to \infty;$$

here $\epsilon = \Omega^2\beta/\mu$. As a result our non-Newtonian fluid problem reduces to a second-order boundary value problem on the infinite interval.

Our third problem concerns

$$\begin{cases} y'' + \left(\dfrac{t^2}{32y^2} - \dfrac{\lambda^2}{8} \right) = 0, & 0 < t < 1 \\ y(0) = 0, \quad 2y'(1) - (1+\nu)y(1) = 0, & 0 < \nu < 1 \quad \text{and} \quad \lambda > 0, \end{cases}$$

which models the large deflection membrane response of a spherical cap. Here $S_r = y/t$ is the radial stress at points on the membrane, $d(\rho S_r)/d\rho$ is the circumferential stress ($\rho = t^2$), λ is the load geometry and ν is the Poisson ratio.

For our fourth problem we note that in the theory of colloids it is possible to relate particle stability with the charge on the colloidal particle. We model the particle and its attendant electrical double layer using Poisson's equation for a flat plate. If Ψ is the potential, ρ the charge density, D the dielectric constant, and y the displacement, then we have

$$\frac{d^2\Psi}{dy^2} = -\frac{4\pi\rho}{D}.$$

We assume that the ions are point charged and their concentrations in the double layer satisfies the Boltzmann distribution

$$c_i = c_i^* \exp\left(\frac{-z_i e\Psi}{\kappa T} \right),$$

where c_i is the concentration of ions of type i, $c_i^* = \lim_{\Psi \to 0} c_i$, κ the Boltzmann constant, T the absolute temperature, e the electrical charge, and z the valency of the ion. In the neutral case, we have

$$\rho = c_+ z_+ e + c_- z_- e, \quad \text{or} \quad \rho = ze(c_+ - c_-),$$

where $z = z_+ - z_-$. Then using

$$c_+ = c\exp\left(\frac{-ze\Psi}{\kappa T} \right) \quad \text{and} \quad c_- = c\exp\left(\frac{ze\Psi}{\kappa T} \right),$$

it follows that

$$\frac{d^2\Psi}{dy^2} = \frac{8\pi cze}{D} \sinh\left(\frac{ze\Psi}{\kappa T} \right),$$

where the potential initially takes some positive value $\Psi(0) = \Psi_0$ and tends to zero as the distance from the plate increases, i.e., $\Psi(\infty) = 0$. Using the transformation

$$\phi(y) = \frac{ze\Psi(y)}{\kappa T} \quad \text{and} \quad x = \sqrt{\frac{4\pi cz^2 e^2}{\kappa TD}}\, y,$$

the problem becomes

$$\frac{d^2\phi}{dx^2} = 2 \sinh \phi, \quad 0 < x < \infty$$

$$\phi(0) = c_1, \quad \lim_{x \to \infty} \phi(x) = 0,$$

(42.4)

where $c_1 = ze\Psi_0/(\kappa T) > 0$. From a physical point of view, we wish the solution ϕ in (42.4) also to satisfy $\lim_{x \to \infty} \phi'(x) = 0$.

Finally, we remark that the references provided include an up-to-date account on many areas of research on the theory of ordinary differential equations. An inquisitive reader can easily select a book which piques his interest and pursue further research in the field.

References

[1] R. P. Agarwal, *Boundary Value Problems for Higher-Order Differential Equations*, World Scientific, Singapore, 1986.

[2] R. P. Agarwal and R. C. Gupta, *Essentials of Ordinary Differential Equations*, McGraw–Hill, Singapore, 1991.

[3] R. P. Agarwal and V. Lakshmikantham, *Uniqueness and Nonuniqueness Criteria for Ordinary Differential Equations*, World Scientific, Singapore, 1993.

[4] R. P. Agarwal and P. J. Y. Wong, *Advanced Topics in Difference Equations*, Kluwer Academic Publishers, Dordrecht, Netherlands, 1997.

[5] R. P. Agarwal, *Focal Boundary Value Problems for Differential and Difference Equations*, Kluwer Academic Publishers, Dordrecht, Netherlands, 1998.

[6] R. P. Agarwal, D. O'Regan, and P. J. Y. Wong, *Positive Solutions of Differential, Difference and Integral Equations*, Kluwer Academic Publishers, Dordrecht, Netherlands, 1999.

[7] R. P. Agarwal, *Difference Equations and Inequalities*, 2nd ed., Marcel Dekker, New York, 2000.

[8] R. P. Agarwal, S. R. Grace and D. O'Regan, *Oscillation Theory for Difference and Functional Differential Equations*, Kluwer Academic Publishers, Dordrecht, Netherlands, 2000.

[9] R. P. Agarwal and D. O'Regan, *Infinite Interval Problems for Differential, Difference and Integral Equations*, Kluwer Academic Publishers, Dordrecht, Netherlands, 2001.

[10] R. P. Agarwal, S. R. Grace and D. O'Regan, *Oscillation Theory for Second-Order Linear, Half-linear, Superlinear and Sublinear Dynamic Equations*, Kluwer Academic Publishers, Dordrecht, Netherlands, 2002.

[11] R. P. Agarwal and D. O'Regan, *Singular Differential and Integral Equations with Applications*, Kluwer Academic Publishers, Dordrecht, Netherlands, 2003.

[12] R. P. Agarwal, S. R. Grace and D. O'Regan, *Oscillation Theory for Second-Order Dynamic Equations*, Taylor and Francis, London, 2003.

[13] R. P. Agarwal, M. Bohner and W.-T. Li, *Nonoscillation and Oscillation: Theory for Functional Differential Equations*, Marcel Dekker, New York, 2004.

[14] R. P. Agarwal, M. Bohner, S. R. Grace, and D. O'Regan, *Discrete Oscillation Theory*, Hindawi Publishing Corporation, New York, 2005.

[15] M. Bohner and A. C. Peterson, *Dynamic Equations on Time Scales: An Introduction with Applications*, Birkhäuser Boston, Cambridge, MA, 2001.

[16] K. Deimling, *Ordinary Differential Equations in Banach Spaces*, Springer-Verlag, New York, 1977.

[17] K. Deimling, *Multivalued Differential Equations*, Walter de Grutyer, Berlin, 1992.

[18] J. K. Hale, *Ordinary Differential Equations*, Wiley–Interscience, New York, 1969.

[19] J. K. Hale, *Functional Differential Equations*, Springer-Verlag, New York, 1971.

[20] P. Hartman, A differential equation with nonunique solutions, *Amer. Math. Monthly*, **70** (1963), 255–259.

[21] P. Hartman, *Ordinary Differential Equations*, Wiley, New York, 1964.

[22] E. Hille, *Lectures on Ordinary Differential Equations*, Addison–Wesley, Reading, MA, 1969.

[23] M. W. Hirsch and S. Smale, *Differential Equations, Dynamical Systems, and Linear Algebra*, Academic Press, New York, 1974.

[24] J. Kurzweil, *Ordinary Differential Equations*, Elsevier, Amsterdam, 1986.

[25] V. Lakshmikantham and S. Leela, *Differential and Integral Inequalities*, Vol. 1, Academic Press, New York, 1969.

[26] V. Lakshmikantham and S. Leela, *Nonlinear Differential Equations in Abstract Spaces*, Pergamon Press, Oxford, 1981.

[27] V. Lakshmikantham, D. D. Bainov, and P. S. Simenov, *Theory of Impulsive Differential Equations*, World Scientific, Singapore, 1989.

[28] V. Lakshmikantham and R. N. Mohapatra, *Theory of Fuzzy Differential Equations and Inclusions*, Taylor and Francis, London, 2003.

[29] V. Lakshmikantham, T. Gnana Bhaskar, and J. Vasundra Devi, *Theory of Set Differential Equations in Metric Spaces*, Cambridge Scientific, Cottenham, Cambridge, 2006.

[30] M. A. Lavrentev, Sur une équation differentielle du premier ordre, *Math. Z.*, **23** (1925), 197–209.

[31] D. O'Regan, *Theory of Singular Boundary Value Problems*, World Scientific, Singapore, 1994.

[32] D. O'Regan, Existence Theory for Nonlinear Ordinary Differential Equations, Kluwer Academic Publishers, Dordrecht, Netherlands, 1997.

[33] L. C. Piccinini, G. Stampacchia, and G. Vidossich, *Ordinary Differential Equations in $\mathbb{R}^n$*, Springer-Verlag, New York, 1984.

[34] M. H. Protter and H. F. Weinberger, *Maximum Principles in Differential Equations*, Prentice–Hall, Englewood Cliffs, 1967.

[35] I. Stakgold, *Green's Functions and Boundary Value Problems*, Wiley, New York, 1979.

[36] T. Yoshizawa, *Stability Theory by Liapunov's Second Method*, Mathematical Society of Japan, Tokyo, 1966.

[37] K. Yosida, *Lectures on Differential and Integral Equations*, Wiley–Interscience, New York, 1960.

[38] W. Walter, *Differential and Integral Inequalities*, Springer-Verlag, Berlin, 1970.

Index

Universitext